T0215513

Dynamic Fracture Mechanics

editor

Arun Shukla
University of Rhode Island, USA

Dynamic Fracture Mechanics

World Scientific

NEW JERSEY • LONDON • SINGAPORE • BEIJING • SHANGHAI • HONG KONG • TAIPEI • CHENNAI

Published by

World Scientific Publishing Co. Pte. Ltd.

5 Toh Tuck Link, Singapore 596224

USA office: 27 Warren Street, Suite 401-402, Hackensack, NJ 07601

UK office: 57 Shelton Street, Covent Garden, London WC2H 9HE

British Library Cataloguing-in-Publication Data
A catalogue record for this book is available from the British Library.

First published 2006 (Hardcover)
Reprinted 2016 (in paperback edition)
ISBN 978-981-3203-26-6

DYNAMIC FRACTURE MECHANICS

ISBN-13 978-981-256-840-3
ISBN-13 981-256-840-9

Printed in Singapore

Editor's Preface

This book consists of 9 chapters encompassing both the fundamental aspects of dynamic fracture mechanics as well as the studies of fracture mechanics in novel engineering materials. The chapters have been written by leading authorities in various fields of fracture mechanics from all over the world. These chapters have been arranged such that the first five deal with the more basic aspects of fracture including dynamic crack initiation, crack propagation and crack arrest, and the next four chapters present the application of dynamic fracture to advanced engineering materials.

The first chapter deals with the atomistic simulation of dynamic fracture in brittle materials as well as instabilities and crack dynamics at interfaces. This chapter also includes a brief introduction of atomistic modeling techniques and a short review of important continuum mechanics concepts of fracture. The second chapter discusses the initiation of cracks under dynamic conditions. The techniques used for studying dynamic crack initiation as well as typical results for some materials are presented. The third chapter discusses the most commonly used experimental technique, namely, strain gages in the study of dynamic fracture. This chapter presents the details of the strain gage method for studying dynamic fracture in isotropic homogenous materials, orthotropic materials, interfacial fracture between isotropic-isotropic materials and interfacial fracture between isotropic-orthotropic materials. In the fourth chapter important optical techniques used for studying propagating cracks in transparent and opaque materials are discussed. In particular, the techniques of photoelasticity, coherent gradient sensing and Moire' interferometry are discussed and the results of dynamic fracture from using these methods are presented. The fifth chapter presents a detailed discussion of the arrest of running cracks. The sixth chapter reviews the dynamics of fast fracture in brittle amorphous materials. The dynamics of crack-front interactions with localized material inhomogeneities are described. The seventh chapter investigates dynamic fracture initiation toughness at elevated temperatures. The experimental set-up for measuring dynamic fracture toughness at high temperatures and results from a new generation of titanium aluminide alloy are presented in detail. In chapter eight the dynamic crack propagation in materials with varying properties, i.e., functionally graded materials is presented. An elastodynamic solution for a propagating crack inclined to the direction of property variation is presented. Crack tip stress, strain and displacement fields are obtained through an asymptotic analysis coupled with displacement potential approach. Also, a systematic theoretical analysis is provided to incorporate the effect of transient nature of growing crack-tip on the crack-tip stress, strain and displacement fields. The ninth chapter presents dynamic dynamic fracture in a nanocomposite material. The complete history of crack propagation including crack initiation, propagation, arrest and crack branching in a nanocomposite fabricated from titanium dioxide particles and polymer matrix is presented.

I consider it an honor and privilege to have had the opportunity to edit this book. I am very thankful to all the authors for their outstanding contributions to this special volume on dynamic fracture mechanics. I am also thankful to my graduate student Mr. Srinivasan Arjun Tekalur for helping me put together this book. My special thanks to the funding agencies National Science Foundation, Office of Naval Research and the Air Force Office of Scientific Research for funding my research on dynamic fracture mechanics over the years.

Arun Shukla
Simon Ostrach Professor
Editor

Contents

Chapter 1

Modeling Dynamic Fracture Using Large-Scale Atomistic Simulations

Markus J. Buehler

Massachusetts Institute of Technology,
Department of Civil and Environmental Engineering
77 Massachusetts Avenue Room 1-272, Cambridge, MA., 02139, USA
mbuehler@MIT.EDU

Huajian Gao

Max Planck Institute for Metals Research
Heisenbergstrasse 3, D-70569 Stuttgart, Germany

We review a series of large-scale molecular dynamics studies of dynamic fracture in brittle materials, aiming to clarify questions such as the limiting speed of cracks, crack tip instabilities and crack dynamics at interfaces. This chapter includes a brief introduction of atomistic modeling techniques and a short review of important continuum mechanics concepts of fracture. We find that hyperelasticity, the elasticity of large strains, can play a governing role in dynamic fracture. In particular, hyperelastic deformation near a crack tip provides explanations for a number of phenomena including the "mirror-mist-hackle" instability widely observed in experiments as well as supersonic crack propagation in elastically stiffening materials. We also find that crack propagation along interfaces between dissimilar materials can be dramatically different from that in homogeneous materials, exhibiting various discontinuous transition mechanisms (mother-daughter and mother-daughter-granddaughter) to different admissible velocity regimes.

1. Introduction

Why and how cracks spread in brittle materials is of essential interest to numerous scientific disciplines and technological applications [1-3]. Large-scale molecular dynamics (MD) simulation [4-13] is becoming an increasingly useful tool to investigate some of the most fundamental aspects of dynamic fracture [14-20]. Studying rapidly propagating cracks using atomistic methods is particularly attractive, because cracks propagate at speeds of kilometers per second, corresponding to time-and length scales of nanometers per picoseconds readily accessible within classical MD methods. This similarity in time and length scales partly explains the success of MD in describing the physics and mechanics of dynamic fracture.

1.1 *Brief review: MD modeling of fracture*

Atomistic simulations of fracture were carried out as early as 1976, in first studies by Ashurst and Hoover [21]. Some important features of dynamic fracture were described in that paper, although the simulation sizes were extremely small, comprising of only 64x16 atoms with crack lengths around ten atoms. A later classical paper by Abraham and coworkers published in 1994 stimulated much further research in this field [22]. Abraham and coworkers reported molecular-dynamics simulations of fracture in systems up to 500,000 atoms, which was a significant number at the time. In these atomistic calculations, a Lennard-Jones (LJ) potential [23] was used. The results in [22, 24] were quite striking because the molecular-dynamics simulations were shown to reproduce phenomena that were discovered in experiments only a few years earlier [25]. An important classical phenomenon in dynamic fracture was the so-called "mirror-mist-hackle" transition. It was known since 1930s that the crack face morphology changes as the crack speed increases. This phenomenon is also referred to as dynamic instability of cracks. Up to a speed of about one third of the Rayleigh-wave speed, the crack surface is atomically flat (mirror regime). For higher crack speeds the crack starts to roughen (mist regime) and eventually becomes very rough (hackle regime), accompanied by extensive crack branching and perhaps severe plastic deformation near the macroscopic crack tip. Such phenomena were observed at similar velocities in both experiments and modeling [25]. Since the molecular-dynamics simulations are performed in atomically perfect lattices, it was concluded that the dynamic instabilities are a universal property of cracks, which have been subject to numerous further studies in the following years (e.g. [26]).

The last few years have witnessed ultra large-scale atomistic modeling of dynamical fracture with system sizes exceeding one billion atoms [4, 5, 12, 13, 27, 28]. Many aspects of fracture have been investigated, including crack limiting speed [10, 29-31], dynamic fracture toughness [32] and dislocation emission processes at crack tips and during nanoindentation [33, 34]. Recent progresses also include systematic atomistic-continuum studies of fracture [29-31, 35-38], investigations of the role of hyperelasticity in dynamic fracture [30, 39] and the instability dynamics of cracks [22, 24, 39]. A variety of numerical models have been proposed, including concurrent multi-scale schemes that combine atomistic and continuum domains within a single model [40-48].

1.2 *Outline of this chapter*

In this chapter, we mainly focus on the work involved by the authors in using simplistic interatomic potentials to probe crack dynamics in model materials, with an aim to gain broad insights into fundamental, physical aspects of dynamic fracture.

A particular focus of our studies is on understanding the effect of hyperelasticity on dynamic fracture. Most existing theories of fracture assume a linear elastic stress–strain law. However, the relation between stress and strain in real solids is strongly nonlinear due to large deformations near a moving crack tip, a phenomenon referred to here as hyperelasticity. Our studies strongly suggest that hyperelasticity, in contrast to most of the classical linear theories of fracture, indeed has a major impact on crack dynamics.

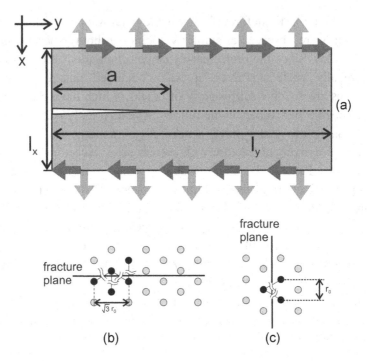

Fig. 1. Subplot (a): A schematic illustration of the simulation geometry used in our large-scale atomistic studies of fracture. The geometry is characterized by the slab width l_x and slab length l_y, and the initial crack length a. We consider different loading cases, including mode I, mode II (indicated in the figure) and mode III (not shown). Subplots (b) and (c) illustrate two different possible crack orientations. Configuration (b) is used for the studies discussed in Sections 3 and 5, whereas configuration (c) is used for the studies described in Section 4. The orientation shown in subplot (c) has lower fracture surface energy than the orientation shown in subplot (b).

The plan of this chapter is as follows. First, we review atomistic modeling techniques, in particular our approach of using simple model potentials to study dynamic fracture. We will then cover three topics: (i) confined crack dynamics along weak layers in homogeneous materials, focusing on the crack limiting speed (Section 3), (ii) instability dynamics of fracture, focusing on the critical speed for the onset of crack instability (Section 4), and (iii) dynamics of cracks at interfaces of dissimilar materials (Section 5). Whereas cracks are confined to propagate along a prescribed path in Section 3, they are completely unconstrained in

Section 4. Section 5 contains studies on both constrained and unconstrained crack propagation. We conclude with a discussion and outlook to future research in this area.

2. Large-scale atomistic modeling of dynamic fracture: A fundamental viewpoint

2.1 *Molecular dynamics simulations*

Our simulation tool is classical molecular-dynamics (MD) [23, 49]. For a more thorough review of MD and implementation on supercomputers, we refer the reader to other articles and books [19, 23, 50-52]. Here we only present a very brief review. MD predicts the motion of a large number of atoms governed by their mutual interatomic interaction, and it requires numerical integration of Newton's equations of motion usually via a Velocity Verlet algorithm [23] with time step Δt on the order of a few femtoseconds.

2.2 *Model potentials for brittle materials: A simplistic but powerful approach*

The most critical input parameter in MD is the choice of interatomic potentials [23, 49]. In the studies reported in this article, the objective is to develop an interatomic potential that yields generic elastic behaviors at small and large strains, which can then be linked empirically to the behavior of real materials and allows independent variation of parameters governing small-strain and large-strain properties. Interatomic potentials for a variety of different brittle materials exist, many of which are derived form first principles (see, for example [53-55]). However, it is difficult to identify generic relationships between potential parameters and macroscopic observables such as the crack limiting or instability speeds when using such complicated potentials. We deliberately avoid these complexities by adopting a simple pair potential based on a harmonic interatomic potential with spring constant

k_0. In this case, the interatomic potential between pairs of atoms is expressed as

$$\phi(r) = a_0 + \frac{1}{2} k_0 (r - r_0)^2, \qquad (1)$$

where r_0 denotes the equilibrium distance between atoms, for a 2D triangular lattice, as schematically shown in the inlay of Figure 1. This harmonic potential is a first-order approximation of the Lennard-Jones 12:6 potential [23], one of the simplest and most widely used pair potentials defined as

$$\phi(r) = 4\varepsilon \left(\left[\frac{\sigma}{r} \right]^{12} - \left[\frac{\sigma}{r} \right]^6 \right). \qquad (2)$$

We express all quantities in reduced units, so that lengths are scaled by the LJ-parameter σ which is assumed to be unity, and energies are scaled by the parameter ε, the depth of the minimum of the LJ potential. We note that corresponding to choosing $\varepsilon = 1$ in eq. (2), we choose $a_0 = -1$ in the harmonic approximation shown in eq. (1). Further, we note that $r_0 = \sqrt[6]{2} \approx 1.12246$ (see Figure 1). Note that the parameter σ is coupled to the lattice constant a as $\sigma = a / \sqrt{2} / \sqrt[6]{2}$. The reduced temperature is $k_B T / \varepsilon$ with k_B being the Boltzmann constant. The mass of each atoms in the models is assumed to be unity, relative to the reference mass m^*. The reference time unit is then given by $t^* = \sqrt{m^* \sigma^2 / \varepsilon}$. For example, when choosing electron volt as reference energy ($\varepsilon = 1\,\mathrm{eV}$), Angstrom as reference length ($\sigma = 1$ Å), and the atomic mass unit as reference mass ($m^* = 1$ amu), the reference time unit corresponds to $t^* \approx 1.01805\,\mathrm{E}\text{-}14$ seconds.

Although the choice of a simple harmonic interatomic potential can not lead to quantitative calculations of fracture properties in a particular material, it allows us to draw certain generic conclusions about fundamental, material-independent mechanisms that can help elucidating the physical foundations of brittle fracture. The harmonic potential leads to linear elastic material properties, and thus serves as the starting point when comparing simulation results to predictions by the classical linear theories of fracture.

Various flavors and modifications of the simplistic, harmonic model potentials as described in eq. (1) are used for the studies reviewed in this article. These modifications are discussed in detail in each section. The concept of using simplistic model potentials to understand the generic features of fracture was pioneered by Abraham and coworkers [10, 12, 22, 24, 31, 40, 56-58].

2.3 *Model geometry and simulation procedure*

We typically consider a crack in a two-dimensional geometry with slab size $l_x \times l_y$ and initial crack length a, as schematically shown in Fig. 1. The slab size is chosen large enough such that waves reflected from the boundary do not interfere with the propagating crack at early stages of the simulation. The slab size in the crack direction is chosen between 2 and 4 times larger than the size orthogonal to the initial crack.

The slab is initialized at zero temperature prior to simulation. Most of our studies are carried out in a two-dimensional, hexagonal lattice. The three-dimensional studies are performed in a FCC lattice, with periodic boundary conditions in the out-of-plane z direction.

The slab is slowly loaded with a constant strain rate of $\dot{\varepsilon}_x$ (corresponding to tensile, mode I loading), and/or $\dot{\varepsilon}_{xy}$ (corresponding to shear, mode II loading). We establish a linear velocity gradient prior to simulation to avoid shock wave generation from the boundaries. While the loading is applied, the stress σ_{ij} (corresponding to the specified loading case) steadily increases leading to a slowly increasing crack velocity upon fracture initiation. In the case of anti-plane shear loading (mode III), the load is applied in a similar way. It can be shown that the stress intensity factor remains constant in a strip geometry inside a region of crack lengths a [59]

$$\tfrac{1}{4}l_x < a < (l_y - \tfrac{1}{4}l_x). \tag{3}$$

Accurate determination of crack tip velocity is crucial as we need to be able to measure small changes in the propagation speed. The crack tip position a is determined by finding the surface atom with maximum y position in the interior of a search region inside the slab using a potential energy criterion. This quantity is averaged over small time intervals to eliminate very high frequency fluctuations. To obtain the steady state velocity of the crack, crack speed measurements are performed within a region of constant stress intensity factor.

3. Constrained cracks in homogeneous materials: How fast can cracks propagate?

In this section we focus on the limiting speed of cracks, discussing recent results on cracks faster than both shear and longitudinal wave speeds in an elastically stiffening material [10, 12, 30, 31].

The elasticity of a solid clearly depends on its state of deformation. Metals will weaken, or soften, and polymers may stiffen as the strain approaches the state of materials failure. It is only for infinitesimal deformation that the elastic moduli can be considered constant and the elasticity of the solid linear. However, many existing theories model fracture using linear elasticity. Certainly, this can be considered questionable since material is failing at the tip of a dynamic crack due to

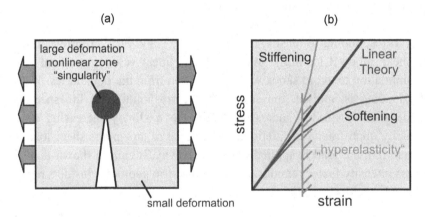

Fig. 2. The concept of hyperelasticity (a) and the region of large deformation near a moving crack (b). The linear elastic approximation is only valid for small deformation. Close to crack tips, material deformation is extremely large, leading to significant changes of local elasticity, referred to as "hyperelasticity". Our research results show that this local elasticity can, under certain conditions, completely govern the dynamics of fracture, in which case the assumption of linear elastic material behavior becomes insufficient to describe the physics of fracture [30, 60, 61].

the extreme deformation there. Here we show by large-scale atomistic simulations that hyperelasticity, the elasticity of large strains, can play a governing role in the dynamics of fracture. We introduce the concept of a characteristic length scale χ for the energy flux near the crack tip and demonstrate that the local hyperelastic wave speed governs the crack speed when the hyperelastic zone approaches this energy length scale χ [30].

3.1 *Introduction: The limiting speed of cracks*

We show by large-scale atomistic simulation that hyperelasticity, the elasticity of large strains, can play a governing role in the dynamics of brittle fracture [30, 62, 63]. This is in contrast to many existing theories of dynamic fracture where the linear elastic behavior of solids is assumed sufficient to predict materials failure [14]. Real solids have elastic properties that are significantly different for small and for large deformations. The concept of hyperelasticity, both for stiffening and softening material behavior, is reviewed in Figure 2, indicating the region close to a moving crack where hyperelastic material behavior is important.

A number of phenomena associated with rapidly propagating cracks are not thoroughly understood. Some experiments [25, 64] and computer simulations [22, 24] have shown a significantly reduced crack propagation speed in comparison with the predictions by the theory. In contrast, other experiments indicated that over 90% of the Rayleigh wave speed can be achieved [65]. Such discrepancies between theories, experiment and simulations cannot always be attributed to the fact that real solids have many different types of imperfections, such as grain boundaries and microcracks (either pre-existing or created during the crack propagation), as similar discrepancies also appear in molecular-dynamics simulations of cracks traveling in perfect atomic lattices.

Gao [62, 63] and Abraham [22, 24, 56] have independently proposed that hyperelastic effects at the crack tip may play an important role in the dynamics of fracture. Their suggestions have been used to help to explain phenomena related to crack branching and dynamic crack tip instability, as well as explaining the significantly lower maximum crack propagation

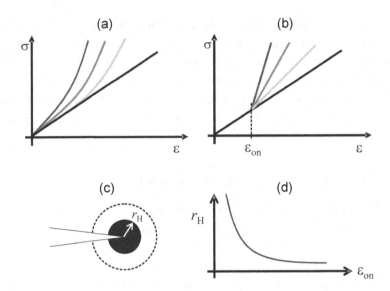

Fig. 3. Concept of increasingly strong hyperelastic effect (subplot (a)). The increasingly strong hyperelastic effect is modeled by using biharmonic potentials (subplot (b)). The bilinear or biharmonic model allows to tune the size of the hyperelastic region near a moving crack, as indicated in subplots (c) and (d) [30]. The local increase of elastic modulus and thus wave speeds can be tuned by changing the slope of the large-strain stress-strain curve ("local modulus" [30, 62, 63]).

speed observed in some experiments and many computer simulations. However, it is not generally accepted that hyperelasticity should play a significant role in dynamic fracture. One reason for this belief stems from the fact that the zone of large deformation in a loaded body with a crack is highly confined to the crack tip, so that the region where linear elastic theory does not hold is extremely small compared to the extensions of the specimen [14, 15]. This is demonstrated in Figure 2(a).

We have performed large-scale molecular-dynamics studies in conjunction with continuum mechanics concepts to demonstrate that hyperelasticity can be crucial for understanding dynamic fracture. Our study shows that local hyperelasticity around the crack tip can significantly influence the limiting speed of cracks by enhancing or reducing local energy flow. This is true even if the zone of hyperelasticity is small compared to the specimen dimensions. The hyperelastic theory completely changes the concept of the maximum crack velocity in the classical theories. For example, the classical theories [14, 15] clearly predict that mode I cracks are limited by the Rayleigh wave speed and mode II cracks are limited by the longitudinal wave speed. In mode III, theory predicts that the limiting speed of cracks is the shear wave speed [14, 15]. In contrast, super-Rayleigh mode I and supersonic mode II cracks are allowed by hyperelasticity and have been seen in computer simulations [13, 31]. We have also observed mode III cracks faster than the shear wave speed [66].

3.2 *Strategy of investigation*

Our strategy of study is to first establish harmonic reference systems with behaviors perfectly matching those predicted by the existing linear elastic theories of fracture. Using a biharmonic model potentials [30], we then introduce increasingly stronger nonlinearities and show that under certain conditions, the linear elastic theory breaks down as the material behaviour near crack tips deviates increasingly from the linear elastic approximation. This is further visualized in Figure 3.

In comparison with experiments, a major advantage of computer simulations is that they allow material behaviors to be fine tuned by introducing interatomic potentials that focus on specific aspects, one at a time, such as the large-strain elastic modulus, smoothing at bond breaking and cohesive stress. From this point of view, MD simulations can be regarded as computer experiments that are capable of testing theoretical concepts and identifying controlling factors in complex systems.

3.3 *Elastic properties: The link between the atomistic scale and continuum theory*

3.3.1 *The virial stress and strain*

Stresses are calculated according to the virial theorem [67, 68]. The atomistic stress is given by (note that Ω is the atomic volume)

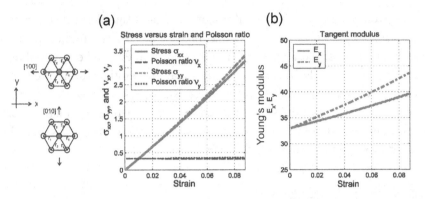

Fig. 4. Elastic properties of the two-dimensional model system with harmonic potential, including (a) stress versus strain and Poisson's ratio, and (b) tangent Young's modulus as a function of strain. The figure shows the results for two different crystal orientations (see left part of the figure) [70].

$$\sigma_{ij} = \frac{1}{\Omega}\left(\frac{1}{2}\left[\sum_{\alpha,\beta} -\frac{1}{r}\frac{\partial \phi}{\partial r} r_i r_j \bigg|_{r=r_{\alpha\beta}}\right]\right). \tag{4}$$

We also use the concept of an atomic strain to link the atomistic simulation results with continuum theory. For a 2D triangular lattice, we define the left Cauchy-Green strain tensor of an atom l [69]

$$b_{ij}^{l} = \left\{\frac{1}{3r_0^2}\sum_{k=1}^{N}\left[\left(x_i^l - x_i^k\right)\left(x_j^l - x_j^k\right)\right]\right\}, \tag{5}$$

with $N = 6$ nearest neighbors, and the variable x_i^l denoting the position vector of atom l. Unlike the virial stress, the virial strain is valid instantaneously in space and time. The maximum principal strain b_1 is obtained by diagonalization of the strain tensor b_{ij}. The maximum principal engineering strain is then given by $\varepsilon_1 = \sqrt{b_1} - 1$.

The definition of atomic stress and strain allows an immediate comparison of MD results with the prediction by continuum theory, such as changes in deformation field as a function of increasing crack speed [14]. This has been discussed extensively in the literature [29, 37, 51, 60, 61, 70, 71]. Some of these results will be reviewed later.

3.3.2 *Elastic properties associated with the harmonic potential: The reference systems*

The simulations considered here are all carried out in a two-dimensional triangular lattice using a harmonic interatomic potential introduced in eq. (1). For a harmonic two-dimensional elastic sheet of atoms, the Young's modulus is given by

$$E = \frac{2}{\sqrt{3}} k_0,$$ (6)

and the shear modulus is

$$\mu = \frac{\sqrt{3}}{4} k_0$$ (7)

(see e.g., ref. [38]). The Poisson's ratio for the two-dimensional lattice is $\nu \approx 0.33$. At large strain, this two-dimensional harmonic lattice shows a slight stiffening effect. The elastic properties for the harmonic potential are shown in Figure 4, indicating that the harmonic potential leads to linear-elastic properties as assumed in many classical fracture theories [14]. We note that when $k_0 = 36/\sqrt[3]{2} \approx 28.57$, $E \approx 33$ and $\mu \approx 12.4$. The shear wave speed is

$$c_s = \sqrt{\frac{\mu}{\rho}},$$ (8)

the longitudinal wave speed is

$$c_l = \sqrt{\frac{E}{\rho}},$$ (9)

and the Rayleigh-wave speed is $c_R \approx 0.9235 \cdot c_s$, where $\rho \approx 0.9165$ is the material density in the examples considered here.

3.3.3 *Elastic properties associated with the biharmonic model potential*

We adopt a biharmonic, interatomic potential composed of two spring constants $k_0 = 36/\sqrt[3]{2} \approx 28.57$ and $k_1 = 2k_0$ (all quantities given are in dimensionless units), as suggested in [30].

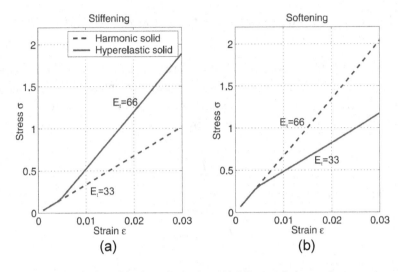

Fig. 5. Elastic properties of a triangular lattice with biharmonic interactions: stress versus strain in the x-direction (a) and in the y-direction (b). The stress state is uniaxial tension, with stress in the direction orthogonal to the loading relaxed to zero [30].

We consider two "model materials", one with elastic stiffening and the other with elastic softening behavior.

In the elastic stiffening system, the spring constant k_0 is associated with small perturbations from the equilibrium distance r_0, and the second spring constant k_1 is associated with large bond stretching for $r > r_{on}$. The role of k_0 and k_1 is reversed in the elastic softening system ($k_0 = 2k_1$, and $k_1 = 36/\sqrt[3]{2}$). Purely harmonic systems (corresponding to eq. (1)) are obtained if r_{on} is chosen to be larger than r_{break}. The interatomic potential is defined as

$$\phi(r) = \begin{cases} \frac{1}{2}k_0(r-r_0)^2 & \text{if } r < r_{on}, \\ a_2 + \frac{1}{2}k_1(r-r_1)^2 & \text{if } r \geq r_{on}. \end{cases} \tag{10}$$

The other parameters of the potential are determined to be

$$a_2 = \frac{1}{2}k_0(r_{on}-r_0)^2 - \frac{1}{2}k_1(r_{on}-r_1)^2, \tag{11}$$

and

$$r_1 = \frac{1}{2}(r_{on}+r_0) \tag{12}$$

from continuation conditions. The elastic properties associated with this potential are shown in Figure 5 for uniaxial stress in the x and y-directions. This potential allows to smoothly interpolate between harmonic potentials and strongly nonlinear potentials by changing the parameter r_{on} and/or the ratio of the spring constants k_1 / k_0.

3.3.4 *Fracture surface energy*

The fracture surface energy γ is an important quantity for nucleation and propagation of cracks. It is defined as the energy required to generate a unit distance of a pair of new surfaces (cracks can be regarded as sinks for energy, where elastic energy is converted into surface fracture energy). The Griffith criterion predicts that the crack tip begins to propagate when the crack tip energy release rate G reaches the fracture surface energy $G = 2\gamma$ [72]. Knowledge of the atomic lattice and the interatomic potential can also be used to define the fracture surface energy. This quantity depends on the crystallographic directions and is calculated to be

$$\gamma = -\frac{\Delta\phi}{d},\tag{13}$$

where $\Delta\phi$ the energy necessary to break atomic bonds as the crack advances a distance d. For the harmonic bond snapping potential as described above, the fracture surface energy is given by $\gamma \approx 0.0332$ for the direction of high surface energy (Figure 1(b)), and $\gamma \approx 0.0288$ for the other direction (Figure 1(c); about 15 % smaller than in the other direction).

3.4 *Harmonic reference systems*

After briefly defining the atomistic models in the previous sections, we now discuss simulation results using the harmonic potentials as the reference system.

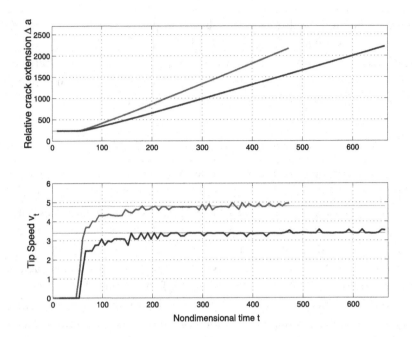

Fig. 6. Crack tip history (upper part) as well as the crack speed history (lower part) for a soft as well as a stiff harmonic material [29, 35].

Our results show that the harmonic systems behave as predicted by classical, linear elastic continuum theories of fracture [14]. For a mode I tensile crack, linear theory predicts that the energy release rate vanishes for all velocities in excess of the Rayleigh wave speed, implying that a mode I crack cannot move faster than the Rayleigh wave speed. This prediction is indeed confirmed in systems with the harmonic potential. The crack velocity approaches the Rayleigh wave speed independent of the slab size, provided that the applied strain is larger than 1.08 percent and the slab width is sufficiently large ($l_x > 1,000$). Results for a mode I crack and two different spring constants are shown in Figure 6.

In other studies, we find that the crack limiting speed of mode II cracks is the longitudinal wave speed. Mode III cracks are limited by the shear wave speed (details and results of those simulations are not described here). These results agree well with linear elastic fracture mechanics theories [14].

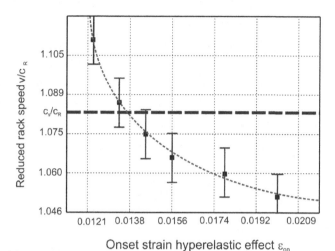

Fig. 7. Change of the crack speed as a function of ε_{on}. The smaller the ε_{on}, the larger the hyperelastic region, and the larger the crack speed [30]. These results indicate that the crack speed of mode I cracks is not limited by the Rayleigh-wave speed.

3.5 Biharmonic simulations: Elastic stiffening

Now we describe a series of computer experiments carried out with the biharmonic interatomic potential introduced in eq. (4) and (5). The results indicate that a local hyperelastic zone around the crack tip (as schematically shown in Figure 2(a)) can have significant effect on the velocity of the crack.

We consider hyperelastic effect of different strengths by using a biharmonic potential with different onset strains governed by the parameter r_{on}. The parameter r_{on} governs the onset strain of the hyperelastic effect $\varepsilon_{on} = (r_{on} - r_0)/r_0$. The simulations reveal crack propagation at super-Rayleigh velocities in steady-state with a local stiffening zone around the crack tip.

Fig. 7 plots the crack velocity as a function of the hyperelasticity onset strain ε_{on}. We observe that the earlier the hyperelastic effect is

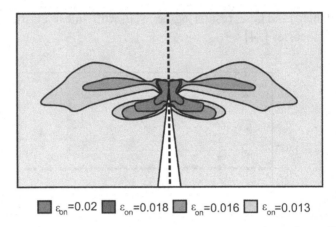

Fig. 8. Shape of the hyperelastic regions for different choices of ε_{on}. The smaller the ε_{on}, the larger the hyperelastic region. The hyperelastic region takes a complex butterfly-like shape [30].

turned on, the larger the limiting velocity. Measuring the hyperelastic area using the principal strain criterion, we find that the hyperelastic area grows as ε_{on} becomes smaller.

In Fig. 8, we depict the shape of the hyperelastic region near the crack tip for different choices of the parameter ε_{on}. The shape and size of the hyperelastic region is found to be independent of the slab width l_x. In all cases, the hyperelastic area remains confined to the crack tip and does not extend to the boundary of the simulation. The results presented in Fig. 6 show that the hyperelastic effect is sensitive to the potential parameter and the extension of the local hyperelastic zone.

Mode I cracks can travel at steady-state intersonic velocities if there exists a stiffening hyperelastic zone near the crack tip. For example, when the large-strain spring constant is chosen to be $k_1 = 4k_0$, with $r_{on} = 1.1375$ and $r_{break} = 1.1483$ (*i.e.* "stronger" stiffening and thus larger local wave speed than before), the mode I crack propagates about 20 percent faster than the Rayleigh speed of the soft material, and becomes intersonic, as shown by the Mach cones of shear wave front depicted in Fig. 9.

Fig. 9. Intersonic mode I crack. The plot shows a mode I crack in a strongly stiffening material ($k_1 = 4k_0$) propagating faster than the shear wave speed [30].

We have also simulated a shear-dominated mode II crack using the biharmonic stiffening potential. We define $r_{break} = 1.17$, and r_{on} is chosen slightly below r_{break} to keep the hyperelastic region small. The dynamic loading is stopped soon after the daughter crack is nucleated. The result is shown in Figure 10 which plots a sequence of snapshots of a moving mode II supersonic crack. The daughter crack nucleated from the mother crack propagates supersonically through the material, although the hyperelastic zone remains localized to the crack tip region. Supersonic mode II crack propagation has been observed previously by Abraham and co-workers [13] using an anharmonic stiffening potential. However, a clearly defined hyperelastic zone could not be specified in their simulations. Our result proves that a local hyperelastic stiffening effect at the crack tip causes supersonic crack propagation, in clear contrast to the linear continuum theory.

The observation of super-Rayleigh and intersonic mode I cracks, as well as supersonic mode II cracks, clearly contradicts the prediction by the classical linear elastic theories of fracture [14].

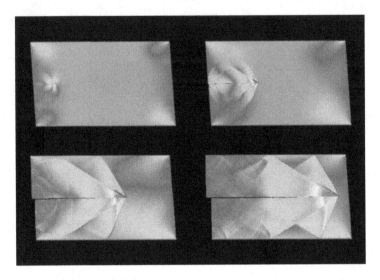

Fig. 10. Sequence of snapshots showing supersonic propagation of a crack under shear loading. A small localized hyperelastic region (not shown here; see reference [30]) at the crack tip leads to crack speeds faster than both shear and longitudinal wave speeds in the material [30].

3.6 *Characteristic energy length scale χ*

The problem of a super-Rayleigh mode I crack in an elastically stiffening material is somewhat analogous to Broberg's [73] problem of a mode I crack propagating in a stiff elastic strip embedded in a soft matrix. The geometry of this problem is shown in Fig. 11. Broberg [73] has shown that, when such a crack propagates supersonically with respect to the wave speeds of the surrounding matrix, the energy release rate can be expressed in the form

$$G = \frac{\sigma^2 h}{E} f(v, c_1, c_2) \qquad (14)$$

where σ is the applied stress, h is the half width of the stiff layer and f is a non-dimensional function of crack velocity and wave speeds in the strip and the surrounding matrix (c_1, c_2). The dynamic Griffith energy balance requires $G = 2\gamma$, indicating that crack propagation velocity is a function of the ratio h/χ where $\chi \sim \gamma E / \sigma^2$ can be defined as a

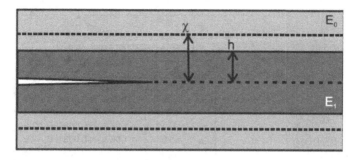

Fig. 11. Geometry of the Broberg problem [30, 73], consisting of a crack embedded in a stiff layer (high wave velocities, Young's modulus E_1) surrounded by soft medium (low wave velocities, Young's modulus E_0).

characteristic length scale for local energy flux. By dimensional analysis, the energy release rate of our hyperelastic stiffening material is expected to have similar features except that Broberg's strip width h should be replaced by a characteristic size of the hyperelastic region r_H. Therefore, we introduce the concept of a characteristic length

$$\chi = \beta \frac{\gamma E}{\sigma^2} \qquad (15)$$

for local energy flux near a crack tip. The coefficient β may depend on the ratio between hyperelastic and linear elastic properties. We have simulated the Broberg problem and found that the mode I crack speed reaches the local Rayleigh wave speed as soon as h/χ reaches unity. Numerous simulations verify that the scaling law in eq. (7) holds when γ, E and σ is changed independently. The results are shown in Figure 12. From the simulations, we estimate numerically $\beta \approx 87$ and the characteristic energy length scale $\chi \approx 750$.

The existence of a characteristic length for local energy flux near the crack tip has not been discussed in the literature and plays the central role in understanding the effect of hyperelasticity. Under a particular experimental or simulation condition, the relative importance of hyperelasticity is determined by the ratio r_H / χ. For small r_H / χ, the crack dynamics is dominated by the global linear elastic properties since much of the energy transport necessary to sustain crack motion occurs in

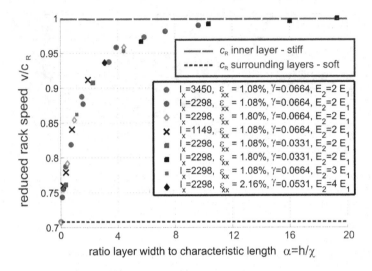

Fig. 12. Calculation results of the geometry defined in the Broberg problem [30]. The plot shows results of different calculations where the applied stress, elastic properties and fracture surface energy are independently varied. In accordance with the concept of the characteristic energy length scale, all points fall onto the same curve and the velocity depends only on the ratio h/χ.

the linear elastic region. However, when r_H/χ approaches unity, as is the case in some of our molecular dynamics simulations, the dynamics of the crack is dominated by local elastic properties because the energy transport required for crack motion occurs within the hyperelastic region.

The concept of energy characteristic length χ immediately provides an explanation how the classical barrier for transport of energy over large distances can be undone by rapid transport near the tip.

3.7 *Discussion and Conclusions*

We have shown that local hyperelasticity has a significant effect on the dynamics of brittle crack speeds and have discovered a characteristic length associated with energy transport near a crack tip. The assumption of linear elasticity fails if there is a hyperelastic zone in the vicinity of the crack tip comparable to the energy characteristic length. Therefore,

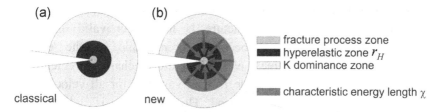

Fig. 13. Different length scales associated with dynamic fracture. Subplot (a) shows the classical picture, and subplot (b) shows the picture with the new concept of the characteristic energy length χ, describing from which region energy flows to a moving crack. The parameter χ is on the order of a few millimeters for 0.1% shear strain in PMMA [30]. Hyperelasticity governs dynamic fracture when the size of the hyperelastic region (r_H) is on the order of the region of energy transport (χ), or $r_H / \chi \gg 1$ [30].

we conclude that hyperelasticity is crucial for understanding and predicting the dynamics of brittle fracture. Our simulations prove that even if the hyperelastic zone extends only a small area around the crack tip, there may still be important hyperelastic effects on the limiting speed. If there is a local softening effect, we find that the limiting crack speed is lower than in the case of harmonic solid.

Our study has shown that hyperelasticity dominates the energy transport process when the zone of hyperelastic zone becomes comparable to the characteristic length

$$\chi \sim \frac{\gamma E}{\sigma_0^2}. \tag{16}$$

Under normal experimental conditions, the magnitude of stress may be one or two orders of magnitude smaller than that under MD simulations. In such cases, the characteristic length χ is relatively large and the effect of hyperelasticity on effective velocity of energy transport is relatively small. However, χ decreases with the square of the applied stress. At about one percent of elastic strain as in our simulations, this zone is already on the order of a few hundred atomic spacing and significant hyperelastic effects are observed. The concept of the characteristic length χ is summarized in Figure 13.

Our simulations indicate that the universal function $A(v/c_R)$ in the classical theory of dynamic fracture is no longer valid once the hyperelastic zone size r_H becomes comparable to the energy characteristic length χ. Linear elastic fracture mechanics predicts that the energy release rate of a mode I crack vanishes for all velocities in excess of the Rayleigh wave speed. However, this is only true if $r_H/\chi \ll 1$. A hyperelastic theory of dynamic fracture should incorporate this ratio into the universal function so that the function should be generalized as $A(v/c_R, r_H/\chi)$. The local hyperelastic zone changes not only the near-tip stress field within the hyperelastic region, but also induces a finite change in the integral of energy flux around the crack tip. A single set of global wave speeds is not capable of capturing all phenomena observed in dynamic fracture.

We believe that the length scale χ, heretofore missing in the existing theories of dynamic fracture, will prove to be helpful in forming a comprehensive picture of crack dynamics. In most engineering and geological applications, typical values of stress are much smaller than those in MD simulations. In such cases, the ratio r_H/χ is small and effective speed of energy transport is close to predictions by linear elastic theory. However, the effect of hyperelasticity will be important for highly stressed materials, such as thin films or nanostructured materials, as well as for materials under high speed impact.

4. Dynamical crack tip instabilities

Cracks moving at low speeds create atomically flat mirror-like surfaces, whereas cracks at higher speeds leave misty and hackly fracture surfaces. This change in fracture surface morphology is a universal phenomenon found in a wide range of different brittle materials. The underlying physical reason of this instability has been debated over an extensive period of time.

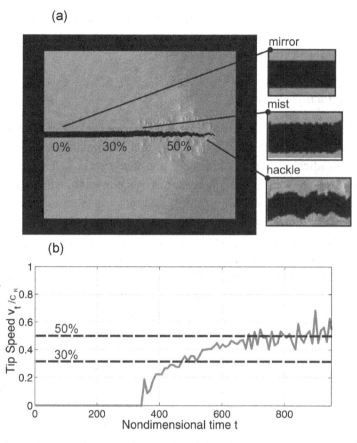

Fig. 14. Crack propagation in a Lennard-Jones (see eq. (2)) system as reported earlier [22, 24]. Subplot (a) shows the σ_{xx}-field and indicates the dynamical mirror-mist-hackle transition as the crack speed increases. The crack velocity history (normalized by the Rayleigh-wave speed) is shown in subplot (b).

Most existing theories of fracture assume a linear elastic stress-strain law. However, from a hyperelastic point of view, the relation of stress and strain in real solids is strongly nonlinear near a moving crack tip. Using massively parallel large-scale atomistic simulations, we show that hyperelasticity also plays an important role in dynamical crack tip instabilities. We find that the dynamical instability of cracks can be regarded as a competition between different instability mechanisms controlled by local energy flow and local stress field near the crack tip. We hope the simulation results can help explain controversial experimental and computational results.

4.1 *Introduction*

Here we focus on the instability dynamics of rapidly moving cracks. In several experimental [25, 64, 74, 75] and computational [20, 22, 26, 76, 77] studies, it was established that the crack face morphology changes as the crack speed increases, a phenomenon which has been referred to as the dynamical instability of cracks. Up to a critical speed, newly created crack surfaces are mirror flat (atomistically flat in MD simulations), whereas at higher speeds, the crack surfaces start to roughen (mist regime) and eventually becomes very rough (hackle regime). This is found to be a universal behavior that appears in various brittle materials, including ceramics, glasses and polymers. This dynamical crack instability has also been observed in computer simulation [20, 22, 26, 76, 77]. The result of a large-scale MD simulation illustrating the mirror-mist-hackle transition is shown in Fig. 14.

Despite extensive studies in the past, previous results have led to numerous discrepancies that remain largely unresolved up to date. For example, linear elasticity analyses first carried out by Yoffe [78] predict that the instability speed of cracks is about 73% of the Rayleigh-wave speed [14, 15], as the circumferential hoop stress exhibits a maximum at an inclined cleavage plane for crack speeds beyond this critical crack speed. This is in sharp contrast to observations in several experiments and computer simulations. Experiments have shown that the critical instability speed can be much lower in many materials. In 1992, Fineberg *et al.* [25, 64] observed an instability speed at about one third of the Rayleigh wave speed, which significantly deviates from Yoffe's theory [78]. Similar observations were made in a number of other experimental studies in different classes of materials (e.g. crystalline silicon, polymers such as PMMA). The mirror-mist-hackle transition at about one third of the Rayleigh-wave speed was also observed in the large-scale MD simulations carried out by Abraham *et al.* in 1994 [22]

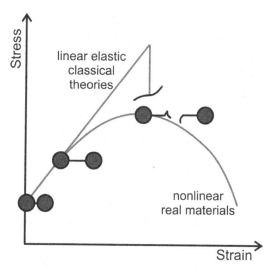

Fig. 15. The concept of hyperelastic softening close to bond breaking, in comparison to the linear elastic, bond-snapping approximation.

(see Fig. 14). The instability issue is further complicated by recent experiments which show stable crack propagation at speeds even beyond the shear wave speed in rubber-like materials [79]. The existing theoretical concepts seem insufficient to explain the various, sometimes conflicting, results and observations made in experimental and numerical studies.

The dynamical crack instability has been a subject of numerous theoretical investigations in the past decades, and several theoretical explanations have been proposed. First, there is Yoffe's model [78] which shows the occurrence of two symmetric peaks of normal stress on inclined cleavage planes at around 73% of the Rayleigh-wave speed. Later, Gao [80] proved that Yoffe's model is consistent with a criterion of crack kinking into the direction of maximum energy release rate. Eshelby [81] and Freund [82] made an argument that the dynamic energy release rate of a rapidly moving crack allows the possibility for the crack to split into multiple branches at a critical speed of about 50% of the Raleigh speed. Marder and Gross [26] presented an analysis which included the discreteness of atomic lattice [16], and found instability at a speed similar to those indicated by Yoffe's, Eshelby's and Freund's models . Abraham *et al.* [24] have suggested that the onset of instability can be understood from the point of view of reduced local lattice vibration frequencies due to softening at the crack tip [22, 24], and also discussed the onset of the instability in terms of the secant modulus [83].

Heizler *et al.*[84] investigated the onset of crack tip instability based on lattice models using linear stability analysis of the equations of motion including the effect of dissipation. These authors observed a strong dependence of the instability speed as a function of smoothness of the atomic interaction and the strength of the dissipation, and pointed out that Yoffe's picture [78] may not be sufficient to describe the instability. Gao [62, 63] attempted to explain the reduced instability speed based on the concept of hyperelasticity within the framework of nonlinear continuum mechanics. The central argument for reduced instability speed in Gao's model is that the atomic bonding in real materials tends to soften with increasing strain, leading to the onset of instability when the crack speed becomes faster than the local wave speed.

Despite important progresses in the past, there is so far still no clear picture of the mechanisms governing dynamical crack instability. None of the existing models have been able to explain all experimental and numerical simulations with a universal understanding applicable to a wide range of materials.

We hypothesize that hyperelasticity is the key to understanding the existing discrepancies among theory, experiments and simulations on dynamical crack instability. We will show that there exist two primary mechanisms which govern crack instability. The first mechanism is represented by Yoffe's model which shows that the local stress field near the crack tip can influence the direction of crack growth.   However, Yoffe's model is not sufficient. The second mechanism is represented by Gao's model [62, 63], which shows that hyperelastic softening drastically reduces the speed of energy transport directly ahead of the crack tip; hyperelasticity induces an anisotropic distribution of local wave speed near the crack tip and causes the crack to kink off its propagation path.

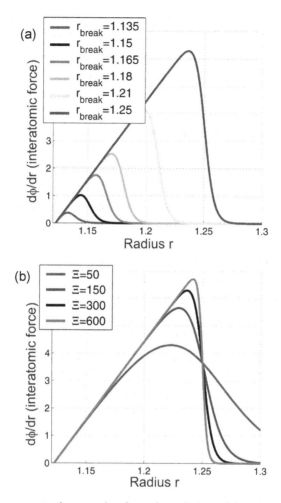

Fig. 16. Force versus atomic separation for various choices of the parameters r_{beak} and Ξ. Whereas r_{beak} is used to tune the cohesive stress in the material, Ξ is used to control the amount of softening close to bond breaking.

Figure 15 illustrates the concept of hyperelastic softening in contrast to linear elastic behavior. We demonstrate that Yoffe's model and Gao's model are two limiting cases of dynamical crack instability; the former corresponds to completely neglecting hyperelasticity (Yoffe) and the latter corresponds to assuming complete dominance of local hyperelastic zone (Gao). We use large scale MD simulations to show that the crack tip instability speeds for a wide range of materials behaviors fall between the predictions of these two models.

In the spirit of "model materials" as introduced in Section 3, we develop a new, simple material model which allows a systematic transition from linear elastic to strongly nonlinear material behaviors, with the objective to bridge different existing theories and determine the conditions of their validity. By systematically changing the large-strain elastic properties while keeping the small-strain elastic properties constant, the model allows us to tune the size of hyperelastic zone and to probe the conditions under which the elasticity of large strains governs the instability dynamics of cracks. In the case of linear elastic model with bond snapping, we find that the instability speed agrees well with the predicted value from Yoffe's model. We then gradually tune up the hyperelastic effects and find that the instability speed increasingly agrees with Gao's model. In this way, we achieve, for the first time, a unified treatment of the instability problem leading to a generalized model that bridges Yoffe's linear elastic branching model to Gao's hyperelastic model.

4.2 *Atomistic modelling*

Although simple pair potentials do not allow drawing conclusions for unique phenomena pertaining to specific materials, they enable us to understand universal, generic relationships between potential shape and fracture dynamics in brittle materials; in the present study we use a simple pair potential that allows the hyperelastic zone size and cohesive stress to be tuned. The potential is composed of a harmonic function in combination with a smooth cut-off of the force based on the Fermi-Dirac (F-D) distribution function to describe smooth bond breaking. We do not

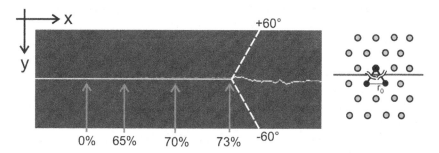

Fig. 17. Crack propagation in a homogeneous harmonic solid. When the crack reaches a velocity of about 73 percent of Rayleigh wave speed, the crack becomes unstable in the forward direction and starts to branch (the dotted line indicates the 60° plane of maximum hoop stress) [60].

include any dissipative terms. The force versus atomic separation is expressed as

$$\frac{d\phi}{dr}(r) = k_0(r - r_0)\left[\exp\left(r\frac{\Xi}{r_{break}} - \Xi\right) + 1\right]^{-1} \quad (17)$$

Assuming that the spring constant k_0 is fixed, the potential has two additional parameters, r_{break} and Ξ. The parameter r_{break} (corresponding to the Fermi energy in the F-D-function) denotes the critical separation for breaking of the atomic bonds and allows tuning the breaking strain as well as the cohesive stress at bond breaking. In particular, we note that

$$\sigma_{coh} \sim r_{break}. \quad (18)$$

The parameter Ξ (corresponding to the temperature in the F-D-function) describes the amount of smoothing at the breaking point.

In addition to defining the small-strain elastic properties (by changing the parameter k_0), the present model allows us to control the two most critical physical parameters describing hyperelasticity, (i) cohesive stress (by changing the parameter r_{break}), and (ii) the strength of softening close to the crack tip (by changing the parameter Ξ).

Figure 16 depicts force versus atomic separation of the interatomic potential used in our study. The upper part shows the force versus

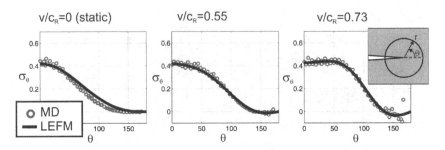

Fig. 18. Comparison between hoop stresses calculated from molecular-dynamics simulation with harmonic potential and those predicted by linear elastic theory [14] for different reduced crack speeds v/c_R [29, 60]. The plot clearly reveals development of a maximum hoop stress at an inclined angle at crack speeds beyond 73% of the Rayleigh-wave speed.

separation curve with respect to changes of r_{break}, and the lower part shows the variation in shape when Ξ is varied. For small values of Ξ (around 50), the softening effect is quite large. For large values of Ξ (beyond 1,000), the amount of softening close to bond breaking becomes very small, and the solid behaves like one with snapping bonds. The parameter r_{break} allows the cohesive stress σ_{coh} to be varied independently. This model potential also describes the limiting cases of material behavior corresponding to Yoffe's model (linear elasticity with snapping bonds) and Gao's model (strongly nonlinear behavior near the crack tip).

Yoffe's model predicts that the instability speed only depends on the small-strain elasticity. Therefore, the instability speed should remain constant at 73% of the Rayleigh-wave speed, irregardless of the choices of the parameters r_{break} and Ξ. On the other hand, Gao's model predicts that the instability speed is *only* dependent on the cohesive stress σ_{coh} (and thus r_{break}):

$$v_{inst}^{Gao} = \sqrt{\frac{\sigma_{coh}}{\rho}} \propto \sqrt{\frac{r_{break}}{\rho}} \qquad (19)$$

(note that ρ denotes the density, as defined above). According to Gao's model, variations in the softening parameter Ξ should *not* influence the crack instability speed.

In most of earlier computational studies, the analysis was performed only for a single interatomic potential, such as done for a LJ potential by Abraham *et al.* [22]. It remains unclear how the nature and shape of the atomic interaction affects the instability dynamics. Here we propose systematic numerical studies based on continuously varying potential parameters r_{break} and Ξ. Our investigations will focus on the predictions of Gao's model versus those of Yoffe's model.

4.3 *Atomistic modeling results: Hyperelasticity governs crack tip instabilities*

We have performed a series of numerical experiments by systematically varying the potential parameters r_{break} and Ξ. We start with harmonic systems serving as the reference and then increase the strength of the hyperelastic effect to study the dynamics of crack tip instability in hyperelastic materials.

4.3.1 *Harmonic potential – the linear elastic reference system*

We find that cracks in homogeneous materials with linear elastic properties (harmonic potential, achieved by setting Ξ to infinity) show a critical instability speed of about 73% of the Rayleigh-wave speed, independent of the choice of r_{break}. The crack surface morphology is shown in Fig. 17. This observation is in quantitative agreement with the key predictions of Yoffe's model.

We find that the occurrence of the instability can be correlated with the development of a bimodal hoop stress as proposed by Yoffe, as is shown in the sequence of hoop stress snapshots as a function of increasing crack speed depicted in Fig. 18. We conclude that in agreement with Yoffe's prediction, the change in deformation field governs the instability dynamics in the harmonic systems.

4.3.2 *Hyperelastic materials behavior in real materials*

It is observed that crack dynamics changes drastically once increasingly stronger softening is introduced at the crack tip and linear elastic Yoffe model fails to describe the instability dynamics.

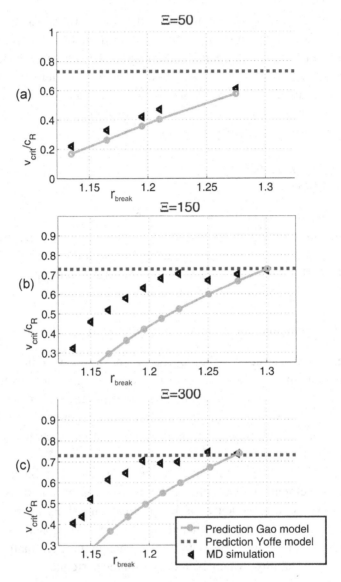

Fig. 19. The critical instability speed as a function of the parameter r_{break}, for different choices of Ξ. The results show that the instability speed varies with r_{break} and thus with the cohesive stress as suggested in Gao's model [62, 63], but the Yoffe speed [78] seems to provide an upper limit for the instability speed. The critical instability speeds are normalized with respect to the local Rayleigh-wave speed, accounting for a slight stiffening effect as shown in Figure 4.

The predictions by Yoffe's model are included in Fig. 19 as the red line, and the predictions by Gao's model are plotted as the blue points. As seen in Fig. 19, we observe that for any choice of r_{break} and Ξ, the instability speed lies in between the prediction by Gao's model and that by Yoffe's model. Whether it is closer to Gao's model or to Yoffe's model depends on the choice of r_{break} and Ξ.

For small values of r_{break} and Ξ, we find that the instability speed depends on the cohesive stress, which is a feature predicted by Gao's model.

Further, the instability speed seems to be limited by the Yoffe speed (as can be confirmed in Fig. 19 for $\Xi = 150$ and $\Xi = 300$ for large values of r_{break}). Whereas the observed limiting speeds increase with r_{break} for $r_{break} < 1.22$, they saturate at the Yoffe speed of 73% of Rayleigh-wave speed for larger values of $r_{break} \geq 1.22$ (see Fig. 19, bottom curve for $\Xi = 300$). In this case, the instability speed is independent of r_{break} and independent of Ξ. This behavior is reminiscent of Yoffe's deformation controlled instability mechanism and suggests a change in governing mechanism for the instability speed: The instability speed becomes governed by a Yoffe-like deformation field mechanism for $r_{break} \geq 1.22$, whereas the instability seems to be influenced by the cohesive-stress for $r_{break} < 1.22$. We find a similar transition for different choices of Ξ ranging from $\Xi = 50$ to $\Xi = 1,500$, with different values of r_{break} at which the transition is observed.

These results indicate that the instability speed depends on the strength of softening (parameter Ξ) and on the cohesive stress close to bond breaking (parameter r_{break}).

4.4 *The modified instability model*

We observe that the first derivative of the instability speed with respect to the cohesive stress in our MD simulations agrees reasonably well with Gao's model. However, the observed instability speed differs from Gao's model by a constant value which seems to depend on the softening parameter Ξ. We measure the deviation from Gao's model by a shift parameter $v_{shift} = v_{inst}^{MD} - v_{inst}^{Gao}$ which is a function of Ξ. The curve is shown in Figure 20.

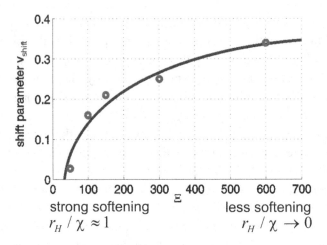

Fig. 20. Change of shift parameter v_{shift} as a function of the smoothing parameter Ξ. Small values of Ξ lead to cases where hyperelasticity dominates the instability onset ($r_H / \chi \approx 1$), and larger values of Ξ correspond to the case when the importance of hyperelasticity vanishes and the instability onset is governed by changes in the deformation field ($r_H / \chi \to 0$)

The physical interpretation of v_{shift} is that it accounts for the relative importance of hyperelastic softening close to the crack tip: Gao's model corresponds to the *limiting case* when the softening region is large and completely dominates the energy flow, and it therefore constitutes the *lower limit* for the instability speed. Indeed, we find that for very strong softening, that is, when $\Xi \to 0$, v_{shift} vanishes (Figure 20). In contrast, v_{shift} assumes larger values in the case of vanishing softening as $\Xi \to \infty$. The physical significance of this parameter can be understood from the perspective of the characteristic energy length scale of dynamic fracture

$$\chi \sim \frac{\gamma E}{\sigma^2} \qquad (20)$$

proposed earlier (see Section 3) [30]. The characteristic energy length scale χ describes the region from which energy flows to the crack tip to drive its motion. If the size of the softening region is comparable to χ, *i.e.*,

$$\frac{r_H}{\chi} >> 1, \tag{21}$$

hyperelasticity dominates energy flow, thus $v_{shift} \to 0$, and the predictions of Gao's model should be valid. In contrast, if the size of the softening region is smaller than χ, i.e.

$$\frac{r_H}{\chi} \approx 0, \tag{22}$$

hyperelasticity plays a reduced role in the instability dynamics and the purely hyperelastic model becomes increasingly approximate, so that v_{shift} takes larger values and eventually Yoffe's model of deformation field induced crack kinking begins to play a governing role. The shift parameter $v_{shift}(\Xi)$ should therefore depend on the relative size of the hyperelastic region compared to the characteristic energy flow

$$v_{shift} = f\left(\frac{r_H}{\chi}\right). \tag{23}$$

With the new parameter v_{shift}, the instability speed can be expressed as

$$v_{inst} = v_{shift}\left(\frac{r_H}{\chi}\right) + \sqrt{\frac{\sigma_{coh}}{\rho}} \tag{24}$$

Based on the results of our numerical experiments, we propose that eq. (19) and the Yoffe model should be combined to predict the onset of instability at a critical speed. The critical crack tip instability speed is then given by

$$v_{inst}^{mod} = \min\left(v_{YOFFE}, v_{shift}\left(\frac{r_H}{\chi}\right) + \sqrt{\frac{\sigma_{coh}}{\rho}}\right) \tag{25}$$

where $v_{YOFFE} \approx 0.73c_R$ is a constant, independent on the hyperelastic properties. We refer to this model as the modified instability model. Fig. 21 compares the predictions by eq. (25) with the MD simulation results (dashed curve).

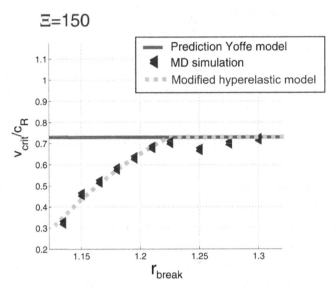

Fig. 21. Comparison of the *modified instability model* (see eq. (25)) for $\Xi = 150$ with the MD simulation results, showing the transition from an energy flow controlled instability mechanism to a deformation field controlled mechanism.

4.5 *Stiffening materials behavior: Stable intersonic mode I crack propagation*

A generic behavior of many rubber-like polymeric materials is that they stiffen with strain. What happens to crack instability dynamics in such materials? Recent experiments [79] have shown intersonic mode I crack propagation in rubber-like materials with elastic stiffening behaviors. According to the existing theories, such high speed crack propagation should not be possible in homogeneous materials. We hypothesize that local stiffening near the crack tip may lead to a locally enhanced Yoffe speed and enhanced energy flow, so that the onset of crack tip instability is shifted to higher velocities.

Our simulations are based on a simple model in which we change the large-strain spring constant and small-strain spring constant, similar to studies described in the previous section except for the smoothing part

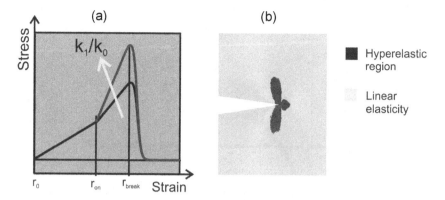

Fig. 22. Subplot (a): Schematic of stiffening materials behaviour as assumed in Section 4.5. Subplot b): Extension of the hyperelastic stiffening region for the case of $r_{break} = 1.17$, $r_{on} = 1.1375$ and $k_1/k_0 = 4$. Despite the fact that the stiffening hyperelastic region is highly localized to the crack tip and extends only a few atomic spacings, the crack instability speed is larger than the Rayleigh-wave speed.

near bond snapping. The model is depicted schematically in Fig. 22(a). Upon a critical atomic separation r_{on} (we choose $r_{on} = 1.1375$), the spring constant of the harmonic potential is changed and switched to a new "local" large-strain value.

Knowledge of r_{on} allows for clear definition of the extension of the hyperelastic zone near the crack tip (for details see reference [30]), as can be verified in Fig. 8 and Fig. 22(b). As described earlier, here we also use the F-D-function to smoothly cut off the potential at r_{break}. We discuss two different choices of the ratio $k_1/k_0 = 2$ and $k_1/k_0 = 4$.

We observe that if there exists a hyperelastic stiffening zone, the Yoffe speed is *no longer* a barrier for the instability speed and stable crack motion beyond the Yoffe speed can be observed. This behavior is shown in Figure 23. We find that the stronger the stiffening effect, the more rapid the increase of instability speed with increasing value of r_{break} (see the curves for $k_1/k_0 = 2$ and $k_1/k_0 = 4$ in Figure 23).

Note that all points fall together once r_{break} becomes comparable to r_{on}, corresponding to the case when no hyperelastic zone is present and the potential is harmonic with smooth F-D bond breaking (here the potential shape is identical because bonds rupture before onset of the

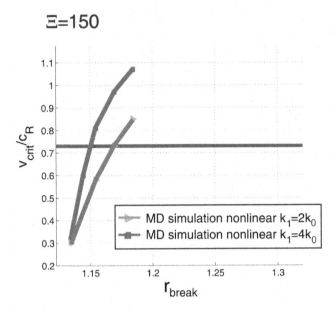

Fig. 23. MD simulation results of instability speed for stiffening materials behaviour, showing stable super-Rayleigh crack motion as observed in recent experiment (see the legends in the lower left). Such observation is in contrast to any existing theories, but can be explained based on the hyperelastic viewpoint.

hyperelastic effect). As can be clearly seen in Fig. 23 for $k_1 / k_0 = 4$, the instability speed can even be super-Rayleigh and approach intersonic speeds.

This is inconsistent with the classical theories but can be understood from a hyperelasticity point of view. This observation suggests that the stiffening materials behavior tends to have a stabilizing effect on straight crack motion.

4.6 *Discussion and conclusions*

Our simulation studies strongly suggest that the large-deformation elastic properties play a critical role in the instability dynamics of cracks.

Our results indicate that the onset of instability can be understood as a competition between energy flow governed instability (Gao's model

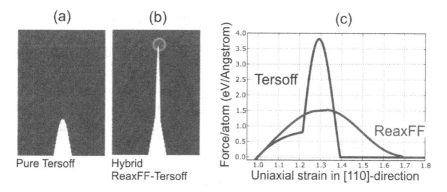

Fig. 24. Importance of accurate large-strain properties of bond breaking during fracture in silicon [85-87]. The results suggest that the large-strain properties close to bond breaking is critical in determining how and under which conditions cracks propagate. Subplot (a) shows results without correct large-strain elasticity, whereas subplot (b) depicts the results with correct large-strain elasticity at bond breaking. The difference of bond breaking is shown in subplot (c), indicating the incorrect behavior of the Tersoff potential at bond breaking.

[62, 63]) and stress field governed instability (Yoffe's model [78]), as summarized in eq. (12). To the best of our knowledge, Eq. (25) describes for the first time this transition between Yoffe's and Gao's model of instability dynamics (see Fig. 21). We hypothesize that the transition between the two mechanisms depends on the relative importance of hyperelasticity around the crack tip, as described by the ratio of the size of the hyperelastic region and the characteristic energy length scale r_H / χ. In most experiments and computers simulations, materials show a strong softening effect. Therefore, hyperelastic softening near the crack tip may explain the reduced instability speed observed in experiments.

In the case of a locally stiffening hyperelastic region, we find that the critical speeds for crack instability could be higher since the *local hyperelastic zone* with local higher wave speed allows for faster energy transport and higher local wave speeds; in this case, the small-strain Yoffe speed is no longer a barrier due to local stiffening. This concept provides a feasible explanation of recent experimental observations of stable mode I cracking at crack velocities beyond the shear wave speed in homogeneous materials that show a strong stiffening hyperelastic effect (see Fig. 23).

The notion that large-strain elastic properties close to bond breaking are important has recently also been demonstrated in studies of fracture in silicon [86]. Figure 24 shows a comparison of fracture dynamics by changing the description of bond breaking from an empirical Tersoff-type [88] to a first principles description using reactive potentials [89]. Only if a correct quantum mechanical derived description of bond breaking is used, the MD model is capable to reproduce important experimental features of fracture of silicon.

The work reported here, together with the results on the maximum crack speed (see Section 3), strongly suggest that hyperelasticity can play governing roles in dynamic fracture.

5. Dynamic crack propagation along interfaces between dissimilar materials: Mother-, daughter- and granddaughter cracks

Cracking along interfaces has received significant attention in the past decades, both due to technological relevance for example in composite materials, as well as because of scientific interest. In this section, we review recent progress in large-scale atomistic studies of crack propagation along interfaces of dissimilar materials [37, 66, 90].

We consider two linear-elastic material blocks bound together with a weak potential whose bonds snap early upon a critical atomic separation, as shown schematically in Fig. 25. This approach confines crack motion along the interface. In the two blocks, atoms interact with harmonic potentials with different spring constants adjacent to the interface. An initial crack is introduced along the interface and subjected to tensile (section 5.2) and shear (section 5.3) dominated displacement loading along the upper and lower boundaries of the sample.

Under tensile loading, we observe that upon initiation at a critical load, the crack quickly approaches a velocity a few percent larger than the Rayleigh-wave speed of the soft material. After a critical time, a secondary crack is nucleated a few atomic spacings ahead of the crack. This secondary crack propagates at the Rayleigh-wave speed of the stiff material. If the elastic mismatch is sufficiently large (e.g. ten as in our

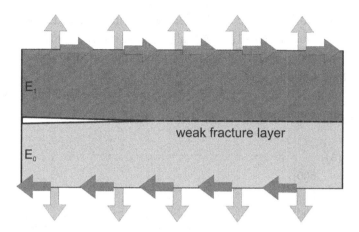

Fig. 25. Geometry used in the studies of crack dynamics at interfaces. We study both mode I (tensile) and mode II (shear) dominated loading.

study), the secondary crack can be faster than the longitudinal wave speed of the soft material, thus propagating supersonically. At this stage, supersonic crack motion is clearly identified by two Mach cones in the soft material. Our study suggests that such mother-daughter transition mechanism, which has been previously reported for mode II crack motion in homogeneous materials [13, 31, 38], may also play an important role in the dynamics of interfacial cracks under tensile loading. We also include some studies of unconstrained crack motion along interfaces of dissimilar materials, and demonstrate that the crack tends to branch off into the softer region, in agreement with previous study by Xu and Needleman [91, 92].

Under shear dominated loading, upon initiation of the crack, we observe that the crack quickly approaches a velocity close to the Rayleigh-wave speed of the soft material. After cruising at this speed for some time, a secondary crack is nucleated at a few atomic spacings ahead of the crack. This secondary crack, also referred to as the daughter crack [37, 38, 93], propagates at the longitudinal-wave speed of the soft material. Shortly after that, a tertiary crack, referred to as the granddaughter crack, is nucleated and begins to move at the longitudinal wave speed of the stiff material. The granddaughter crack is supersonic with respect to the soft material and is clearly identifiable by two Mach cones in the soft material [37]. Our results indicate that the limiting speed of shear dominated cracks along a bi-material interface is the longitudinal wave speed of the stiff material, and that there are two intermediate limiting speeds (Rayleigh and longitudinal wave speeds of the soft material) which can be overcome by the mother-daughter-granddaughter mechanism.

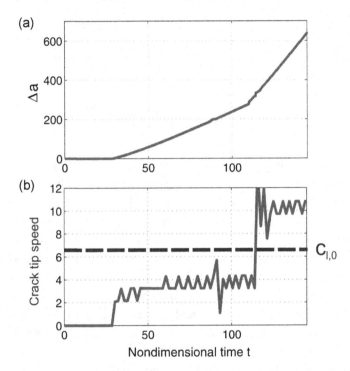

Fig. 26. Crack tip history and crack velocity history for a mode I crack propagating at an interface with X = 10. Subplot (a) shows the crack tip history, and subplot (b) shows the crack tip velocity over time. A secondary daughter crack is born propagating at a supersonic speed with respect to the soft material layer.

5.1 *Atomistic modeling of cracking along a bimaterial interface*

To model a crack along an interface between two elastically dissimilar materials, we study two crystal blocks described by harmonic interatomic potentials with different spring constants $k_1 > k_0$. For

σ_{xx} σ_{yy} σ_{xy}

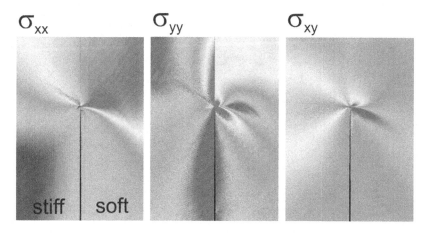

Fig. 27. Stress fields σ_{xx}, σ_{yy} and σ_{xy} for a crack at an interface with elastic mismatch $\Xi = 10$ before the secondary crack is nucleated. The red color corresponds to large stresses, and the blue color to small stresses.

simplicity, we consider two-dimensional triangular lattices with isotropic elastic properties. The harmonic potential is defined as in eq. (1).

We choose $k_0 = 36 / \sqrt[3]{2} \approx 28.57$ with the corresponding Rayleigh-wave speed $c_R \approx 3.4$, shear wave speed $c_s \approx 3.68$ and longitudinal wave speed $c_l \approx 6.36$. The difference in Young's modulus of the two materials can be related to the ratio

$$\Xi = \frac{k_1}{k_0}. \tag{26}$$

The wave speeds c_I (with $I = (R, s, l)$) of the two materials thus differ by a factor $\sqrt{\Xi}$:

$$c_{I,1} = \sqrt{\Xi} \cdot c_{I,0} \tag{27}$$

Atomic bonds across the weak interface have spring constant k_0 and break upon a critical separation $r_{break} = 1.17$.

The simulation slab is subjected to an applied tensile and/or shear strain rate along the upper and lower boundaries, as shown in Fig. 25. The system size is given by $l_x \approx 2,298$ and $l_x \approx 4,398$. Further details on modeling fracture based on harmonic potentials, as well as a detailed analysis on the elastic and fracture properties can be found elsewhere

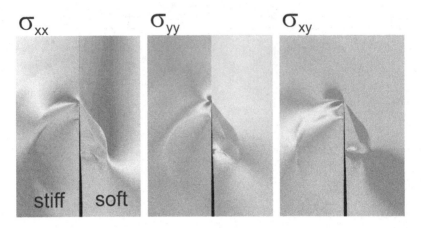

Fig. 28. Stress fields σ_{xx}, σ_{yy} and σ_{xy} for a crack at an interface with elastic mismatch $\Xi = 10$ after the secondary crack is nucleated. In all stress fields, the two Mach cones in the soft material are seen. The red color corresponds to large stresses, and the blue color to small stresses.

[13, 30, 31, 38] (see also discussion in previous sections). In homogeneous materials, it has been shown that the simulated results of limiting crack speeds, the energy flow fields and the stress fields near rapidly propagating cracks are in good agreement with continuum mechanics predictions (see results discussed in Section 3).

The geometry of our atomistic model, the loading as well as the crystal orientation of crack propagation is described in Fig. 25. The upper (left) block of the simulation slab is relatively stiff with higher Young's modulus and higher wave speeds than the lower (right) block.

5.2 Tensile dominated cracks at bimaterial interfaces: Mother-daughter mechanisms and supersonic fracture

In all cases studied in this paper, the cracked bi-crystal is loaded under tensile loading with a strain rate $\dot{\varepsilon}_{xx} = 0.000,1$ (the orientation of the loading is shown schematically in Figure 25) [37, 90]. In some cases, the loading is stopped as indicated in each case. Although there is usually no pure mode I interfacial cracks due to complex stress intensity factors,

E_{pot}

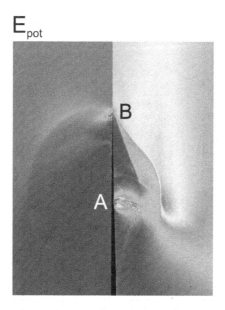

Fig. 29. The potential energy field for a crack at a biomaterial interface with elastic mismatch $\Xi = 10$. Two Mach cones in the soft solid can clearly be observed. The mother (A) and daughter crack (B) are marked in the figure. The mother crack is represented by a surface wave in the soft material after nucleation of the secondary crack.

the stress field near the crack tip is expected to be mode I dominated under pure tensile loading.

5.2.1 *Mode I dominated interfacial cracks: Constrained simulations*

All studies reported in this Section are carried out with an elastic mismatch of $\Xi = 10$. Figure 26(a) shows the crack tip history, and Figure 26(b) shows the crack tip velocity over time. The crack nucleates at time $t \approx 35$, and quickly approaches the Rayleigh wave speed of the soft material $v \rightarrow c_{r,0} \approx 3.4$. As the loading is increased, the crack speed increases slightly and becomes super-Rayleigh with respect to the soft material. We observe a large, discontinuous jump in the crack velocity at $t \approx 110$, when a secondary crack is nucleated that quickly approaches the Rayleigh speed of the stiff material $v \rightarrow c_{r,1} \approx 10.8 > c_{l,0} \approx 6.36$.

(1)   (2)

(3)   (4)

Fig. 30. Atomic details of nucleation of the secondary crack under tensile dominated loading. The plot shows the shear stress field σ_{xy} near the crack tip. Atoms with the energy of a free surface are drawn as larger, bluish atoms. The plot suggests that a maximum peak of the shear stress ahead of the crack tip leads to breaking of atomic bonds and creation of new crack surfaces. After the secondary crack is nucleated, it coalesces with the mother crack and moves supersonically through the material.

The secondary crack is nucleated approximately at a distance $\Delta a \approx 11$ ahead of the mother crack and rapidly propagates at Mach 1.7 with respect to the soft material. Nucleation of the secondary crack under tensile loading is only found under high strain rate loading (here we use $\dot{\varepsilon}_{xx} = 0.000,1$). If the strain rate is too low, the crack moves at a sub-Rayleigh speed (soft material) until the solid has separated, without nucleation of secondary cracking. The mechanism of nucleation of a secondary crack is reminiscent of the mother-daughter mechanism predicted by the Burridge-Andrew mechanism [94, 95] and observed in MD simulations for intersonic mode II cracks in homogeneous solids [13, 31, 38].

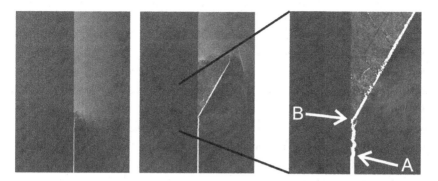

Fig. 31. Crack dynamics in homogeneous material without a weak layer (case (ii)). Unlike in the other cases, in this simulation we allow atomic bonds to break anywhere in the system. We observe that before the daughter crack is nucleated, the crack starts to wiggle and is thus limited in speed. Further, we observe that the crack branches off into the soft material and continues to propagate there. The plot on the right hand side shows a zoom into the crack tip region, showing the initial crack surface roughness (A) and the dominant branch (B) into the soft material. We observe a tendency of branching into the soft region from the wiggling stages (point A, for instance).

These results suggest that a mode I dominated interface crack under very large loading can propagate at the Rayleigh wave speed of the stiffer material, which can exceed the longitudinal wave speed in the soft material. This observation has not been reported in experiments so far, nor has it been predicted by theory. In experimental studies of mode I dominated cracks along bi-material interfaces, the crack speed has been reported to exceed the Rayleigh wave speed of the soft material, but has never been observed to attain supersonic velocities with respect to the soft material.

As the next step, we take a closer look at the deformation fields before and after nucleation of the secondary crack, taking advantage of the "computational microscope" that MD provides. Figure 27 shows the stress field before the secondary crack is nucleated. At the time the snapshots are taken, the crack propagates at a (slightly) super-Rayleigh speed with respect to the soft material. Since the crack motion is still subsonic, no shock front is observed. Figure 28 shows the stress field and the particle velocity field after nucleation of the secondary crack. The secondary crack propagates supersonically with respect to the soft material and the Mach cones in the right half space (soft material) are clearly visible. Figure 29 shows the potential energy field for a crack after the secondary, supersonic crack is nucleated. The two mach cones can clearly be seen. Similar mother-daughter mechanism for mode I dominated interfacial fracture has also been observed for smaller elastic mismatch ratios $\Xi = 2$, $\Xi = 5$, and $\Xi = 7$ (however, not in all cases crack motion is supersonic with respect to the soft material since the Rayleigh-wave speed of the stiff material may be smaller than the longitudinal wave speed of the soft material).

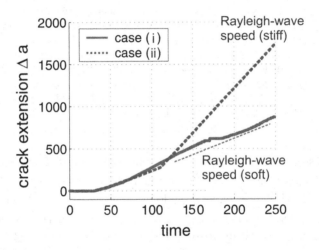

Fig. 32. Crack tip extension history as a function of time for the cases (i) and (ii) described in Section 4. The plot illustrates that once the daughter crack has been nucleated it does not branch off into either the weak or the soft material, but instead continues to propagate along the weak interface. If crack motion is unconstrained from the very beginning, it starts to branch off at low speeds (see also Figure 31) and eventually branches off into the soft material. In this case, a daughter crack is never nucleated.

What is governing nucleation of the secondary crack? MD simulations can be particularly helpful in investigating the atomistic details of the mechanism of nucleation of secondary cracks. In Figure 30, we plot the shear stress field σ_{xy} for several instants in time during nucleation of the secondary crack (note we chose quite small time intervals between the snapshots). Atoms with the energy of a free surface are colored blue and highlighted by larger spheres. The plot suggests that a peak of shear stress ahead of the crack tip may have caused breaking of atomic bonds. After the secondary crack is nucleated, it coalesces with the mother crack and moves supersonically with respect to the soft material. This mechanism via a peak in shear stress is quite similar to the understanding of nucleation of secondary cracks in shear loaded cracks in homogeneous systems as discussed in [38]. We note that the location of the maximum tensile stress σ_{xx} does not coincide with the location of nucleation of the secondary crack; rather, the latter was found to correlate with a peak in shear stress ahead of the mother crack. We there conclude that there exists a peak in shear stress ahead of a mode I dominated interfacial crack moving at the Rayleigh wave speed of the soft material and this peak shear stress leads to subsequent nucleation of a secondary crack which breaks the sound barrier at the soft Raleigh speed.

5.2.2 Unconstrained simulations

In this section, we relieve the constraint that crack propagation be restricted to the interface between the stiff and the soft material (glue). We report the results of three cases: (i) Crack motion is constrained along the interface until the secondary daughter crack is nucleated, and from then on crack motion is left unconstrained (we relieve the constraint at $t = 126$). (ii) Crack motion is unconstrained from the beginning of the simulation; the condition for breaking is given by $r_{\text{break}} = 1.17$ in both stiff and soft solid (therefore, the fracture surface energy is lower in the right half space than in the left half space). (iii) Crack motion is unconstrained from the beginning of the simulation; in contrast to case (ii), here we adjust r_{break} in the left material such that the fracture surface energy is equal to that in the right material so that the slab has a homogeneous fracture energy:

$$r_{break}^{left} = r_0 - \frac{1}{\sqrt{\Xi}}(r_0 - r_{break}^{right}) \tag{28}$$

and thus for $r_{break}^{right} = 1.17$ we assume that $r_{break}^{left} = 1.13749$.

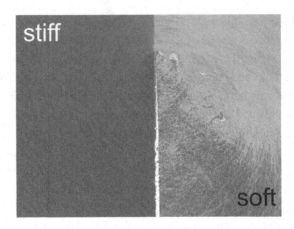

Fig. 33. Crack dynamics at a bimaterial interface with homogeneous fracture energy everywhere in the slab (case (iii)). In this study, the fracture surface energy in both the stiff and soft materials is equal. We observe that the crack continuously branches off into the soft solid and creates numerous small cracks. This indicates a general tendency of the crack to branch off into the soft material.

In case (i), we observe that the crack continues to move along its original direction and does not branch off into the soft or the stiff material until the bi-crystal is completely broken.

In case (ii), we observe that before the daughter crack is nucleated, the crack starts to wiggle and is thus limited in speed. Figure 29 shows the simulation snapshots of case (ii). We find that the crack branches off into the soft material and continues to propagate there. The plot on the right hand side in Figure 31 shows a zoom into the crack tip region, showing the initial crack surface roughness (A) and the large dominant branch (B) into the soft material.

Figure 32 plots the crack tip history for case (i) and case (ii). It clearly reveals that in case (ii) nucleation of a secondary daughter crack does not occur. However, the crack speed increases to speeds beyond the Rayleigh-wave speed of the soft material. In case (i), crack speed remains at the longitudinal wave speed of the stiff material, even after the constraint is relieved.

In case (iii), when the fracture surface energy is equal in the stiff and the soft materials, we observe a tendency of the crack to branch off into the soft region. Figure 33 shows a snapshot at late stages of simulation (here the loading rate was $\dot{\varepsilon}_{xx} = 0.000{,}02$, and loading never stopped). The branching into the soft region (right half) can clearly be recognized.

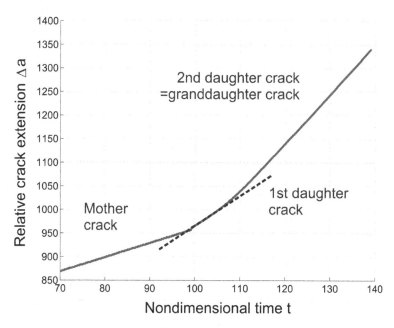

Fig. 34. Crack tip history for a shear loaded crack propagating along an interface with stiffness ratio $\Xi = 3$. This crack tip history plot reveals a mother-daughter-granddaughter mechanism: After a critical time, a secondary crack (the daughter crack) is nucleated ahead of the mother crack. Shortly afterwards, a tertiary crack (the granddaughter crack) is nucleated. The transition in crack speed is from the daughter to the granddaughter is more continuous compared to that from the mother to the daughter (see also the velocity history plot depicted in Fig. 35).

5.3 *Shear dominated cracks at bimaterial interfaces: Observations in MD simulation*

While noting that the basic observations here will hold for a range of elastic mismatch ratios, we mainly discuss the results at $\Xi = 3$. The primary crack is observed to initiate at time $t \approx 34$ and quickly approaches the Rayleigh speed of the soft material $v \rightarrow c_{R,0} \approx 3.4$.

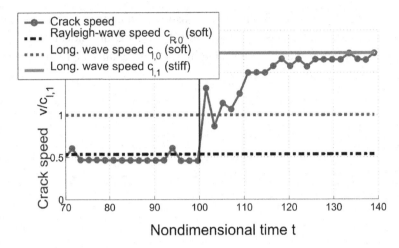

Fig. 35. Crack tip velocity history during the mother-daughter-granddaughter mechanism for a shear loaded crack propagating along an interface with stiffness ratio $\Xi = 3$. The plot is obtained by numerical differentiation of the crack tip history shown in Fig. 2. The crack speed changes abruptly at the nucleation of the daughter crack, and rather continuously as the granddaughter crack is nucleated.

Figure 34 shows the history of crack tip location, and Figure 35 that of the crack tip velocity [90]. Both curves are shown over a time interval close to the nucleation of secondary and tertiary cracks. We observe a discontinuous jump in the crack velocity at $t \approx 100$, as the daughter crack is nucleated in front of the mother crack. The daughter crack forms at a distance $\Delta a \approx 10$ ahead of the mother crack and quickly approaches the longitudinal wave speed of the soft material $v \rightarrow c_{l,0} \approx 6.36$. The mechanism of nucleation of the daughter crack is reminiscent of the mother-daughter mechanism of mode II cracks in homogeneous materials [31, 38].

Shortly after the daughter crack is nucleated, another change in propagation speed is observed at time $t \approx 108$ (see Figures 34 and 35). This is due to initiation of a tertiary granddaughter crack ahead of the daughter crack. The nucleation of the granddaughter crack occurs at a short time ($\Delta t \approx 8$) after nucleation of the daughter crack (Figs. 34 and 35).

(1)   (2)

(3)   (4)

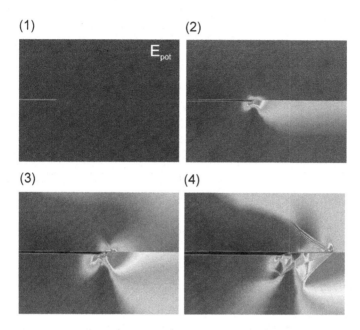

Fig. 36. The plot shows the potential energy field near a shear loaded interface crack with stiffness ratio $\Xi = 3$ (color code: red corresponds to high potential energy and blue corresponds to the potential energy of atoms in a perfect lattice). The plot shows a small section around the crack tip. The crack surfaces are highlighted red. In the upper left plot, the initial configuration with the starting crack is shown. As the loading is increased, the mother crack starts to propagate, eventually leading to secondary and tertiary cracks. Two Mach cones in the soft solid and one Mach cone in the stiff solid can be observed in the lower right figure, suggesting supersonic crack motion with respect to the soft material and intersonic motion with respect to the stiff material.

The transitions from mother to daughter and daughter to granddaughter cracks are quite sharp, and the crack motion stabilizes after the granddaughter crack is nucleated. Figure 36 shows a few snapshots of the potential energy field from the initial configuration until the birth of the granddaughter crack. A similar sequence is shown in Figure 37, depicting the σ_{yy} stress field.

In Figure 38, we show a blowup plot of the potential energy field close to the crack tip, with crack surfaces highlighted with "larger" atoms in red color. The mother (A), daughter (B) and granddaughter (C) cracks can be clearly identified.

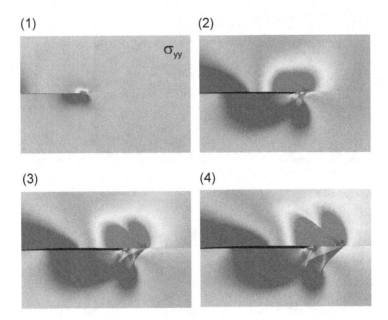

Fig. 37. The plot shows the σ_{yy} stress field near a shear loaded interface crack with stiffness ratio $\Xi = 3$ (color code: red corresponds to high stresses), for a similar sequence as shown in Figure 36. The Mach cones can clearly be seen in snapshot (4).

The result suggests that shear dominated cracks along a bimaterial interface can reach the longitudinal wave speed of the stiffer material and can become supersonic with respect to the wave speeds of the soft material. Similar observations have been made in experiments. Rosakis *et al.* [96] reported crack speeds exceeding the longitudinal wave speed of the softer material and, at least on one occasion, reaching the longitudinal wave speed of the stiffer material. In another experimental study by Wu and Gupta [97], a crack at Nb-saphire interface was found to approach the longitudinal wave speed of the stiff material. Our results suggest three critical wave speeds for shear dominated interface cracks: the Rayleigh-wave speed of the soft material, the longitudinal wave speed of the soft material and the longitudinal wave speed of the stiff material.

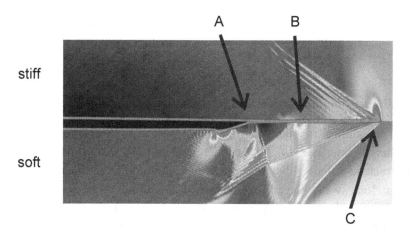

Fig. 38. The plot shows the potential energy field in the vicinity of the crack tip, shortly after nucleation of the granddaughter crack (color code: red corresponds to high potential energy and blue corresponds to the potential energy of atoms in a perfect lattice). The mother, daughter and granddaughter cracks can be identified. In the blow-up on the right, the mother (A), daughter (B), and granddaughter (C) are marked. The mother and daughter cracks are represented by waves of stress concentration in the soft material. For the daughter crack (B), the Mach cone associated with the shear wave speed in the soft material can clearly be observed. The granddaughter crack (C) carries two Mach cones in the soft material.

Atomistic studies are particularly suitable for studying the details of the processes close to the crack tip. Figure 39 shows a time sequence of the shear stress field very close to the crack tip during the nucleation of the secondary and tertiary cracks. The results suggest that a peak (concentration) in shear stress ahead of the crack causes nucleation of the daughter and granddaughter cracks. The shear stress peak ahead of the mother crack moves with the shear wave speed of the softer material (see Fig. 39, snapshot (2)). In Fig. 39, snapshot (3), the daughter crack has just nucleated, but there appears another peak in shear stress a few atomic spacings ahead of the daughter crack moving close to the shear wave speed of the stiffer material. In Fig. 39, snapshot (4), the granddaughter crack has appeared, as indicated by two shock fronts in the softer material. These observations are, at least qualitatively, in agreement with the mother-daughter mechanism observed for mode II cracks in homogeneous materials [31, 38]. The corresponding times of the snapshots are given in the caption of Fig. 39, and can be compared with the crack tip history and crack speed history shown in Figs. 34 and 35.

5.4 *Discussion and conclusions: dynamics of interfacial cracks*

The studies reported in this paper shows that cracking along a bimaterial interface is a rich phenomenon. We show that, barring hyperelastic effects, the limiting speed of an interfacial crack is the longitudinal wave speed of the stiffer material. A shear-dominated interface crack can propagate supersonically with respect to the softer material. We find that the crack speed changes discontinuously as the loading is increased, with more detailed analysis revealing a mother-daughter (tensile dominated loading) and mother-daughter-granddaughter mechanism (shear dominated loading) to achieve the ultimate limiting speed (see Figs. 25, 34 and 35). A clear jump in crack velocity can be observed when the daughter crack is nucleated. However, the nucleation of the granddaughter crack occurs with a more continuous velocity change.

A mother-daughter-granddaughter mechanism in shear dominated cracks has been reported in previous MD simulations of crack propagation along an interface between a harmonic material and an anharmonic material having the same elastic wave speeds under small deformation [13]. In the present study, we show that the mother-daughter-granddaughter mechanism not only occurs in nonlinear materials but also for linear elastic bimaterial interfaces. The fact that the mother-daughter-granddaughter mechanism occurs in both systems suggests that the phenomenon is robust and may occur under a wide range of conditions.

The elastic field of an interface crack can be very different from that of a homogeneous crack [98-101]. If crack propagation is supersonic with respect to one of the materials, multiple shock fronts are expected. The shock fronts shown in Figures 28, 29, 36-38 indicate that crack propagation is supersonic with respect to the soft material. A more detailed comparison of continuum theory and MD simulation is left to future work due to the known difficulties related to the oscillatory character of the continuum mechanics solutions.

We also simulated the dynamics of interfacial cracks with unconstrained crack branching under mode I dominated loading. The results of these studies are depicted in Figures 31 and 32. Figure 33 seems to reveal that there exists a tendency for cracks to branch off into the softer half space. This is in accord with observations made by other computational approaches at the continuum level by Xu and Needleman

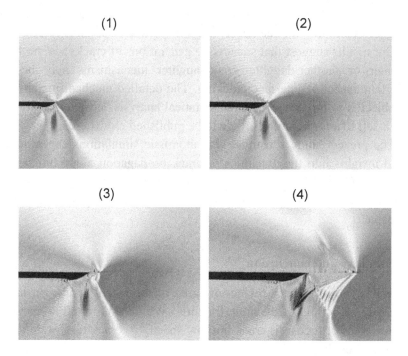

Fig. 39. Shear stress field during the transition from the mother crack (subplot (a)) to propagation of the granddaughter crack (see snapshot (1)). The analysis of the deformation field suggests that the large shear deformation ahead of the mother crack causes nucleation of the daughter crack. The peak in shear stress can be observed in snapshot (2), and appears by the dark red color. In snapshot (3), the daughter crack has nucleated, and there is still a peak shear stress ahead of the crack leading to initiation of the granddaughter crack (snapshot (4)). Here the granddaughter crack has appeared propagating at the longitudinal wave speed of the stiff material, while leaving two shock fronts in the soft material. Snapshots (1-4) are taken at times t=82.2, t=98.76, t=104.4, and t=115.2 respectively.

[91, 92]. However, we observe that if the constraint of interfacial crack motion is turned off *after* the secondary crack is nucleated, the crack maintains its supersonic speed without branching into either the soft or the stiff half space.

We find that the tendency to branch off into the soft material can not be explained by a difference in fracture surface energy. In a focused simulation study (case (iii)), we have adjusted the fracture surface energy so that it is equal in both the soft and the stiff regions and still observe a tendency of the crack to branch off into the soft region. This tendency of branching into the soft material may explain why supersonic cracking along interfaces has not been observed in experiments so far.

Our results suggest that successive generations of cracks via mother-daughter or mother-daughter-granddaughter mechanisms may be an important aspect of interface cracking. The detailed mechanisms should be subjected to more rigorous mathematical analysis and interpretations. This is left to a forthcoming article to be published elsewhere.

The present study illustrates that atomistic simulations can provide useful insights into the dynamics of crack propagation along bimaterial interfaces. Combining large-scale atomistic simulations with continuum mechanics is a powerful approach in modern studies of dynamic fracture mechanics.

6. Summary and conclusion

We have presented a review of some of the recent results on large scale MD modelling of dynamic fracture. We have reported progresses in three areas of focus: (i) crack limiting speed, (ii) crack tip instabilities, and (iii) crack dynamics at interfaces.

The main conclusion is that hyperelasticity, the elasticity of large deformation, is crucial in order to form a clear picture of the failure process near a rapidly moving crack tip: Both the maximum crack speed and the dynamic instability are strongly influenced by the large-strain elastic properties. Phenomena such as intersonic mode I fracture (Figure 9) and supersonic mode II fracture (Figure 10) can not at all be addressed by classical, linear elastic theories. The observation of stable cracks in homogeneous materials at speeds beyond the Rayleigh-wave speed, as shown in Figure 23, can only be understood from a hyperelastic point of view. Further, we have shown that hyperelastic softening can significantly reduce the critical instability speed in cracks, as shown for instance in Figure 19. Classical linear elastic theories such as the Yoffe-model [78] fail when significant softening is present at the crack tip. Our new concept of a characteristic energy length scale χ helps to understand the relative significance of hyperelasticity in dynamic fracture (see Figure 13), both for the limiting speed of cracks (Figure 12) and for the instability dynamics (Figure 20).

Further, we find that interfaces and geometric confinement can play an important role in the dynamics of cracks. Crack propagation constrained along interfaces can significantly change the associated maximum speeds of crack motion. This is illustrated for instance by the studies using the concept of a weak fracture layer where the Rayleigh-wave speed of cracks can be attained by cracks (see Figure 6), versus the studies of cracks in homogeneous materials where the crack starts to wiggle at 73% of the theoretical limiting speed (see Figure 17). If cracks propagate along interfaces of elastically dissimilar materials, the maximum crack speed can significantly change and new mechanisms of crack propagation such as daughter and granddaughter cracks appear, as illustrated in Figures 24, 32 and 33, leading to supersonic fracture relative to the soft material.

The results discussed in this article suggest that the definition of wave speeds according to the small-strain elastic properties is questionable in many cases, and a more complete picture of dynamic fracture should include local wave speeds near the cohesive failure of materials near a crack tip.

Acknowledgments

The simulations were carried out at the Garching Supercomputer Center of the Max Planck Society and at the MARS Linux cluster at the Max Planck Institute for Metals Research. We acknowledge continuing discussions with Dr. Farid F. Abraham on modeling dynamic fracture with molecular-dynamics simulations.

References

1. Field, J.E., *Brittle fracture-its study and application.* Contemp. Phys., 1971. **12**: p. 1-31.
2. Rice, J.R. and R.M. Thomson, *Ductile versus brittle behavior of crystals.* Phil. Mag., 1974. **29**: p. 73-97.
3. Argon, A., *Brittle to ductile transition in cleavage failure.* Acta Metall., 1987. **35**: p. 185-196.
4. Vashishta, P., R.K. Kalia, and A. Nakano, *Large-scale atomistic simulations of dynamic fracture.* Comp. in Science and Engrg., 1999: p. 56-65.
5. Kadau, K., T.C. Germann, and P.S. Lomdahl, *Large-Scale Molecular-Dynamics Simulation of 19 Billion particles.* Int. J. Mod. Phys. C, 2004. **15**: p. 193.
6. Buehler, M.J., et al., The dynamical complexity of work-hardening: a large-scale molecular dynamics simulation. Acta Mechanica Sinica, 2005. **21**(2): p. 103-111.
7. Zhou, S.J., et al., Large-scale molecular-dynamics simulations of three-dimensional ductile failure. Phys. Rev. Lett., 1997. **78**: p. 479-482.
8. Zhou, S.J., et al., Large-scale molecular-dynamics simulations of dislocation interactions in copper. Science, 1998. **279**: p. 1525-1527.
9. Buehler, M.J., et al., Atomic Plasticity: Description and Analysis of a One-Billion Atom Simulation of Ductile Materials Failure. Comp. Meth. in Appl. Mech. and Engrg., 2004.
10. Abraham, F.F., How fast can cracks move? A research adventure in materials failure using millions of atoms and big computers. Advances in Physics, 2003. **52**(8): p. 727-790.
11. Bulatov, V., et al., Connecting atomistic and mesoscale simulations of crystal plasticity. Nature, 1998. **391**: p. 669-672.
12. Abraham, F.F., et al., Simulating materials failure by using up to one billion atoms and the world's fastest computer: Work-hardening. P. Natl. Acad. Sci. USA, 2002. **99**(9): p. 5783-5787.
13. Abraham, F.F., et al., Simulating materials failure by using up to one billion atoms and the world's fastest computer: Brittle Fracture. P. Natl. Acad. Sci. USA, 2002. **99**(9): p. 5788-5792.
14. Freund, L.B., *Dynamic Fracture Mechanics.* 1990: Cambridge University Press, ISBN 0-521-30330-3.
15. Broberg, K.B., *Cracks and Fracture.* 1990: Academic Press.
16. Slepyan, L.I., *Models and Phenomena in Fracture Mechanics.* 2002: Springer, Berlin.
17. Cheung, K.S. and S. Yip, A molecular-dynamics simulation of crack tip extension: the brittle-to-ductile transition. Modelling Simul. Mater. Eng., 1993. **2**: p. 865-892.

18. Marder, M. and J. Fineberg, *How things break*. Phys. Today, 1996. **49**(9): p. 24-29.

19. Marder, M., *Molecular Dynamics of Cracks*. Computing in Science and Engineering, 1999. **1**(5): p. 48-55.

20. Holland, D. and M. Marder, *Ideal brittle fracture of silicon studied with molecular dynamics*. Phys. Rev. Lett., 1998. **80**(4): p. 746.

21. Ashurst, W.T. and W.G. Hoover, *Microscopic fracture studies in 2-dimensional triangular lattice*. Phys. Rev. B, 1976. **14**(4): p. 1465-1473.

22. Abraham, F.F., et al., *Instability dynamics of fracture: A computer simulation investigation*. Phys. Rev. Lett., 1994. **73**(2): p. 272-275.

23. Allen, M.P. and D.J. Tildesley, *Computer Simulation of Liquids*. 1989: Oxford University Press.

24. Abraham, F.F., et al., *A Molecular Dynamics Investigation of Rapid Fracture Mechanics*. J. Mech. Phys. Solids, 1997. **45**(9): p. 1595-1619.

25. Fineberg, J., et al., *Instability and dynamic fracture*. Phys. Rev. Lett., 1991. **67**(4): p. 457-460.

26. Marder, M. and S. Gross, *Origin of crack tip instabilities*. J. Mech. Phys. Solids, 1995. **43**(1): p. 1-48.

27. Sharma, A., R.K. Kalia, and P. Vashishta, *Large multidimensional data vizualization for materials science*. Comp. in Science and Engrg., 2003: p. 26-33.

28. Rountree, C.L., et al., Atomistic aspects of crack propagation in brittle materials: Multimillion atom molecular dynamics simulations. Annual Rev. of Materials Research, 2002. **32**: p. 377-400.

29. Buehler, M.J., F.F. Abraham, and H. Gao, *Stress and energy flow field near a rapidly propagating mode I crack*. Springer Lecture Notes in Computational Science and Engineering, 2004. **ISBN 3-540-21180-2**: p. 143-156.

30. Buehler, M.J., F.F. Abraham, and H. Gao, *Hyperelasticity governs dynamic fracture at a critical length scale*. Nature, 2003. **426**: p. 141-146.

31. Abraham, F.F. and H. Gao, *How fast can cracks propagate?* Phys. Rev. Lett., 2000. **84**(14): p. 3113-3116.

32. Swadener, J.G., M.I. Baskes, and M. Nastasi, *Molecular Dynamics Simulation of Brittle Fracture in Silicon*. Phys. Rev. Lett., 2002. **89**(8): p. 085503.

33. Li, J., et al., Atomistic mechanisms governing elastic limit and incipient plasticity in crystals. Nature, 2002. **418**: p. 307-309.

34. Cleri, F., et al., Atomic-scale mechanism of crack-tip plasticity: Dislocation nucleation and crack-tip shielding. Phys. Rev. Lett, 1997. **79**: p. 1309-1312.

35. Buehler, M.J., H. Gao, and Y. Huang, *Continuum and Atomistic Studies of a Suddenly Stopping Supersonic Crack.* Computational Materials Science, 2003. **28**(3-4): p. 385-408.

36. Gao, H., et al., *Flaw tolerant bulk and surface nanostructures of biological systems.* Mechanics & Chemistry of Biosystems, 2004. **1**(1): p. 37-52.

37. Buehler, M.J. and H. Gao, *A mother-daughter mechanism of supersonic crack growth of mode I dominated cracks at interfaces.* Accepted for publication in: Int. J. of Strength, Fracture and Complexity, 2005.

38. Gao, H., Y. Huang, and F.F. Abraham, *Continuum and atomistic studies of intersonic crack propagation.* J. Mech. Phys. Solids, 2001. **49**: p. 2113-2132.

39. Buehler, M.J. and H. Gao, Dynamical fracture instabilities due to local hyperelasticity at crack tips. under submission, 2005.

40. Abraham, F.F., et al., Dynamic fracture of silicon: Concurrent simulation of quantum electrons, classical atoms, and the continuum solid. MRS Bulletin, 2000. **25**(5): p. 27-32.

41. Abraham, F.F., et al., *Spanning the length scales in dynamic simulation.* Computers in Physics, 1998. **12**(6): p. 538-546.

42. Curtin, W.A. and R.E. Miller, *Atomistic/continuum coupling in computational materials science.* Modelling And Simulation In Materials Science And Engineering, 2003. **11**(3): p. R33-R68.

43. Shiari, B., R.E. Miller, and W.A. Curtin, *Coupled atomistic/discrete dislocation simulations of nanoindentation at finite temperature.* Journal Of Engineering Materials And Technology-Transactions Of The Asme, 2005. **127**(4): p. 358-368.

44. Tadmor, E.B., M. Ortiz, and R. Phillips, *Quasicontinuum analysis of defects in solids.* Phil. Mag. A, 1996. **73**: p. 1529.

45. Shenoy, V., et al., An Adaptive Methodology for Atomic Scale Mechanics - The Quasicontinuum Method. J. Mech. Phys. Sol., 1999. **73**: p. 611-642.

46. Knap, J. and M. Ortiz, *An analysis of the quasicontinuum method.* J. Mech. Phys. Sol., 2001. **49**(9): p. 1899-1923.

47. Dupuy, L.M., et al., Finite-temperature quasicontinuum: Molecular dynamics without all the atoms. Physical Review Letters, 2005. **95**(6).

48. Wagner, G.J. and W.K. Liu, *Coupling of atomistic and continuum simulations using a bridging scale decomposition.* Journal Of Computational Physics, 2003. **190**(1): p. 249-274.

49. Plimpton, S., *Fast parallel algorithms for short-range molecular-dynamics.* Journal of Computational Physics, 1995. **117**: p. 1-19.

50. Vashishta, P., R.K. Kalia, and A. Nakano, *Multimillion atom molecular dynamics simulations of nanostructures on parallel computers.* Journal of Nanoparticle Research, 2003. **5**: p. 119-135.

51. Buehler, M.J. and H. Gao, *Ultra large scale atomistic simulations of dynamic fracture.* Handbook of Theoretical and Computational Nanotechnology, ed. W. Schommers and A. Rieth. 2006: American Scientific Publishers (ASP).

52. Abraham, F.F., Computational statistical mechanics-methodology, applications and supercomputing. Advances in Physics, 1986. **35**(1): p. 1-111.

53. Chenoweth, K., et al., Simulations on the thermal decomposition of a poly(dimethylsiloxane) polymer using the ReaxFF reactive force field. Journal Of The American Chemical Society, 2005. **127**(19): p. 7192-7202.

54. Ercolessi, F. and J.B. Adams, Interatomic potentials from 1st principle-calculations - the force matching method. Europhys. Lett., 1994. **28**(8): p. 583-588.

55. Baskes, M.I., *Determination of modified embedded atom method parameters for nickel.* Materials Chemistry and Physics, 1997. **50**(2): p. 152-158.

56. Abraham, F.F., *Dynamics of brittle fracture with variable elasticity.* Phys. Rev. Lett., 1996. **77**(5): p. 869.

57. Abraham, F.F., et al., *Ab-initio dynamics of rapid fracture.* Modelling Simul. Mater. Sci. Eng., 1998. **6**: p. 639-670.

58. Abraham, F.F. and H. Gao, *Anamalous Brittle-Ductile Fracture Behaviors in FCC Crystals.* Phil. Mag. Lett., 1998. **78**: p. 307-312.

59. Washabaugh, P.D. and W.G. Knauss, A reconcilation of dynamic crack velocity and Rayleigh-wave speed in isotropic brittle solids. Int. J. Fracture, 1994. **65**: p. 97-114.

60. Buehler, M.J. and H. Gao, Dynamical fracture instabilities due to local hyperelasticity at crack tips. Accepted for publication in: Nature.

61. Buehler, M.J. and H. Gao, *Biegen und Brechen im Supercomputer.* Physik in unserer Zeit, 2004. **35**(1).

62. Gao, H., *A theory of local limiting speed in dynamic fracture.* J. Mech. Phys. Solids, 1996. **44**(9): p. 1453-1474.

63. Gao, H., Elastic waves in a hyperelastic solid near its plane-strain equibiaxial cohesive limit. Philosphical Magazine Letters, 1997. **76**(5): p. 307-314.

64. Fineberg, J., et al., *Instability in the propagation of fast cracks.* Phys. Rev. B, 1992. **45**(10): p. 5146-5154.

65. Sharon, E. and J. Fineberg, Confirming the continuum theory of dynamic brittle fracture for fast cracks. Nature, 1999. **397**: p. 333.

66. Buehler, M.J., Atomistic simulations of brittle fracture and deformation of ultra thin copper films, in Chemistry. 2004, University of Stuttgart: Stuttgart.

67. Tsai, D.H., *Virial theorem and stress calculation in molecular-dynamics.* J. of Chemical Physics, 1979. **70**(3): p. 1375-1382.

68. Zhou, M., Equivalent continuum for dynamically deforming atomistic particle systems. Phil. Mag. A, 2002. **82**(13).

69. Zimmermann, J.A., Continuum and atomistic modeling of dislocation nucleation at crystal surface ledges. 1999, Stanford University.

70. Buehler, M.J., H. Gao, and Y. Huang, Continuum and Atomistic Studies of the Near-Crack Field of a rapidly propagating crack in a Harmonic Lattice. Theoretical and Applied Fracture Mechanics, 2004. **41**: p. 21-42.

71. Buehler, M.J., A. Hartmaier, and H.J. Gao, Atomistic and continuum studies of crack-like diffusion wedges and associated dislocation mechanisms in thin films on substrates. Journal Of The Mechanics And Physics Of Solids, 2003. **51**(11-12): p. 2105-2125.

72. Griffith, A.A., *The phenomenon of rupture and flows in solids.* Phil. Trans. Roy. Soc. A, 1920. **221**: p. 163-198.

73. Broberg, K.B., *Dynamic crack propagation in a layer.* Int. J. Solids Struct., 1995. **32**(6-7): p. 883-896.

74. Cramer, T., A. Wanner, and P. Gumbsch, Energy dissipation and path instabilities in dynamic fracture of silicon single crystals. Phys. Rev. Lett., 2000. **85**: p. 788-791.

75. Cramer, T., A. Wanner, and P. Gumbsch, *Crack Velocities during Dynamic Fracture of Glass and Single Crystalline Silicon.* Phys. Status Solidi A, 1997. **164**: p. R5.

76. Gumbsch, P., S.J. Zhou, and B.L. Holian, *Molecular dynamics investigation of dynamic crack instability.* Phys. Rev. B, 1997. **55**: p. 3445.

77. Deegan, R.D., et al., *Wavy and rough cracks in silicon.* Phys. Rev. E, 2003. **67**(6): p. 066209.

78. Yoffe, E.H., *The moving Griffith crack.* Phil. Mag., 1951. **42**: p. 739-750.

79. Petersan, P.J., et al., *Cracks in rubber under tension break the shear wave speed limit.* Phys. Rev. Letters, 2004. **93**: p. 015504.

80. Gao, H., Surface roughening and branching instabilities in dynamic fracture. J. Mech. Phys. Solids, 1993. **41**(3): p. 457-486.

81. Eshelby, J.D., Elastic field of a crack extending non-uniformly under general anti-planbe loading. J. Mech. Phys. Solids, 1969. **17**(3): p. 177-199.

82. Freund, L.B., Crack propagation in an elastic solid subjected to general loading, II. Nonuniform rate of extension. J. Mech. Phys. Solids, 1972. **20**: p. 141-152.

83. Abraham, F.F., *Unstable crack motion is predictable.* Advances in Physics, 2005. **53**: p. 1071-1078.

84. Heizler, S.I., D.A. Kessler, and H. Levine, *Mode I fracture in a nonlinear lattice with viscoelastic forces.* Phys. Rev. E, 2002. **6**: p. 016126.

85. Buehler, M.J., et al., The Computational Materials Design Facility (CMDF): A powerful framework for multiparadigm multi-scale simulations. Mat. Res. Soc. Proceedings, 2006. **894**: p. LL3.8.

86. Buehler, M.J., A.C.T.v. Duin, and W.A. Goddard, Multi-paradigm multi-scale modeling of dynamical crack propagation in silicon using the ReaxFF reactive force field. Submitted for publication to Phys. Rev. Lett., 2005.

87. Buehler, M.J., A.C.T.v. Duin, and W.A. Goddard, Multi-paradigm multi-scale modeling of dynamical crack propagation in silicon using the ReaxFF reactive force field. Mat. Res. Soc. Proceedings 904E, 2006: p. BB4.28.

88. Tersoff, J., Empirical interatomic potentials for carbon, with applications to amorphous carbon. Phys. Rev. Lett., 1988. **61**(25): p. 2879-2883.

89. Duin, A.C.T.v., et al., *ReaxFF SiO: Reactive Force Field for Silicon and Silicon Oxide Systems.* J. Phys. Chem. A, 2003. **107**: p. 3803-3811.

90. Buehler, M.J. and H. Gao, A mother-daughter-granddaughter mechanism of supersonic crack growth of shear dominated intersonic crack motion along interfaces of dissimilar materials. Journal of the Chinese Institute of Engineers, 2004. **27**(6): p. 763-769.

91. Xu, X.P. and A. Needleman, Numerical simulations of dynamic interfacial crack growth allowing for crack growth away from the bond line. International Journal of Fracture, 1996. **74**(3): p. 253-275.

92. Xu, X.P. and A. Needleman, *Numerical simulations of dynamic crack growth along an interface.* International Journal of Fracture, 1996. **74**(4): p. 289-324.

93. Rosakis, A.J., O. Samudrala, and D. Coker, *Cracks faster than the shear wave speed.* Science, 1999. **284**(5418): p. 1337-1340.

94. Burridge, R., Admissible speeds for plain-strain self-similar cracks with friction but lacking cohesion. Geophys. J. Roy. Astron. Soc., 1973. **35**: p. 439-455.

95. Andrews, D.J., *Rupture velocity of plane strain shear cracks.* J. Geophys. Res., 1976. **81**: p. 5679-5687.

96. Rosakis, A.J., *Intersonic crack propagation in bimaterial systems.* J. Mech. Phys. Solids, 1998. **6**(10): p. 1789-1813.

97. Wu, J.X. and V. Gupta, *Observations of Transonic Crack Velocity at a Metal/Ceramic Interface.* Journal of the Mechanics and Physics of Solids, 2000. **48**(3): p. 609-619.

98. Zak, A.R. and M.L. Williams, *Crack point singularities at a bi-material interface.* J. Appl. Mech., 1963. **30**: p. 142-143.

99. Williams, M.L., *The stresses areound a fault or crack in dissimilar media.* Bull. Seismol. Soc. America, 1959. **49**: p. 199-204.

100. Liu, C., J. Lambros, and A.J. Rosakis, Highly transient elastodynamic crack growth in a bimaterial interface: higher order asymptotic analysis and experiments. J. Mech. Phys. Solids, 1993. **41**: p. 1887-1954.

101. Liu, C., Y. Huang, and A.J. Rosakis, Shear dominated transonic interfacial crack growth in a bimaterial-II. Asymptotic fields and favorable velocity regimes. J. Mech. Phys. Solids, 1993. **43**(2): p. 189-206.

Chapter 2

Dynamic Crack Initiation Toughness

Daniel Rittel

Faculty of Mechanical Engineering, Technion
32000 Haifa, Israel
merittel@technion.ac.il110

The dynamic crack initiation toughness is a measure of material resistance to fracture under transient loading conditions. As of today, there are no standardized techniques for the assessment of this property. This chapter provides a detailed presentation of recently developed experimental techniques, largely based on the author's experience. Each technique is thoroughly and critically discussed to provide the necessary background for the practitioner. Selected results are described in an attempt to establish a relationship between microstructure aspects and mechanical characteristics. Future directions are pointed out.

1. Introduction

Since the early days of Griffith, fracture mechanics has made considerable progress, and from a practical point of view, one considers that the resistance of a material to crack propagation (fracture) under quasi-static conditions is represented by the so-called fracture toughness – K_{IC}. Recommended experimental procedures are available for the selection of the specimen, loading conditions and processing of the results[1]. However, quasi-static loading conditions are not always encountered and structures must frequently withstand transient loading conditions, subsequently referred to as dynamic loading. The subject of dynamic fracture mechanics has for long aroused much scientific interest, and basic solutions are available for two selected cases: the dynamically initiated crack and the rapidly propagating crack[2, 3]. The dynamically propagating crack has traditionally drawn much more attention, in both its analytical and experimental aspects with emphasis on the maximum achievable crack speed and the issue of crack bifurcation. By contrast, despite its important practical aspect, dynamic crack initiation has drawn much less attention, so that the experimental aspects of the dynamic initiation toughness have not been widely investigated.

Therefore, compared to the vast amount of literature on the quasi-static fracture toughness as a material property, relatively little is known on the dynamic initiation toughness as a potential material property. Also less known are aspects such as how to measure it, and eventually how to introduce it into design criteria.

This chapter addresses the dynamic crack initiation toughness – K_{IC}^d – from an experimental point of view. Its purpose is to familiarize the practicing engineer with the basic concepts and experimental techniques. Each technique will be presented along with its theoretical background. Two main aspects will be discussed: the mode I and mode II initiation toughness with its peculiarities such as the so-called failure mode transition. The techniques and results presented here are largely drawn upon the author's experience. Therefore, this chapter does not mean to present a complete and exhaustive review of the available techniques and their evolution, but rather to focus on simple and easily implemented techniques for the characterization of the dynamic fracture properties of materials. The following discussion will be restricted to *stationary* cracks in *linear elastic materials*, allowing for the development of small scale yielding at the tip of the crack.

We will first briefly present a path-independent integral based method which was developed as a new framework for mode I dynamic fracture investigations, with applications to mixed-mode fracture. This technique was soon greatly simplified, as will be shown in detail, to yield the one-point impact technique that is currently used on a routine basis for dynamic initiation toughness measurements. Selected results on the dynamic initiation toughness of various materials, and its rate dependence, will then be presented to establish a preliminary correlation with fracture micromechanisms. It will then be shown that mode II measurements are the natural extension of the above-mentioned technique, with a special mention of the failure mode transition phenomenon which is of very high importance. Thermomechanical issues related to dynamic crack initiation will be briefly introduced. The chapter will be concluded by a discussion on future directions.

2. Mode I initiation toughness: Compact compression Specimen and H integral

Path independent integrals are particularly interesting to circumvent a detailed characterization of the immediate vicinity of a crack-tip. Path-independence can be practically taken advantage of in two selected cases: a vanishingly small path around the crack-tip, and a large path encompassing the specimen boundaries, where measurements are usually made. An elastodynamic path independent integral (the H integral) was originally devised by Maigre and Bui[4], which is a particular case of the dynamic weight functions[2]. For the special case of mode I loading, the integral writes as:

$$H(t) := \frac{1}{2} \int_S \left\{ T^{\exp} * \frac{\partial u^{ref}}{\partial a} \right\} d\Sigma = \frac{1-\nu^2}{E} \left\{ K_{Id}^{\exp} * K_{Id}^{ref} \right\} \qquad (1)$$

This expression combines, through convolution products (*), experimental (*exp*) and reference *(ref)* quantities. The left hand side of the expression represents the specimen's outer boundaries where the experimental load T^{exp} is applied to any linear elastic structure containing a stationary crack of length a. The right hand side of eqn. (1) contains two stress intensity factors: $K_{Id}^{\exp}$ that is to be determined, and a reference stress intensity factor, K_{Id}^{ref}. The reference displacement u, its derivative with respect to the crack length a, and the corresponding stress intensity factor are all calculated numerically. All the components of eqn. (1) are time dependent. The meaning of eqn. (1) is that, given a cracked structure subjected to dynamic loading conditions, the stress intensity factor $K_{Id}^{exp}(t)$ can be determined by solving the integral equation, i.e. through a deconvolution procedure. This approach emphasizes boundary measurements instead of direct crack-tip monitoring.

A special specimen was later devised by Rittel et al.[5-6] which was especially adapted to the concept. As shown in Fig. 1, the Compact Compression Specimen (CCS) contains a crack that opens under compressive loads applied on its boundaries.

This specimen can easily be inserted between the bars of a split Hopkinson (Kolsky) pressure bar[7]. While a detailed description of the Kolsky bar is beyond the scope of the present chapter (for additional details, see, e.g. Meyers[8]), we will just remind that this setup applies stress wave loading to a specimen (of any shape). While the bars (incident and transmitter) remain elastic, thus justifying the use of elastodynamic 1-D wave theory, the specimen may either crush or

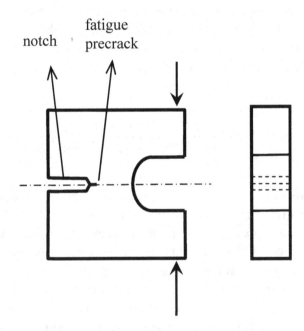

Fig. 1. Schematic representation of the compact compression specimen (CCS). This specimen allows crack opening as a result of compressive loads on its branches. The degree of mode-mix depends on the symmetry of the applied loads.

fracture. As a result of the striker's impact on the incident bar, an incident strain pulse - ε_{inc} - progresses towards the specimen, is partly reflected at the bar-specimen boundary - ε_{ref}, and partly transmitted to the transmitter bar after passing through the specimen - ε_{tra}. Each bar is instrumented with a set of strain gauges in order to measure those signals. The experimental setup is shown in Fig. 2.

The interfacial loads and displacements that are applied to the specimen by means of the bars are given by:

$$F_{in} = EA\left(\varepsilon_{inc} + \varepsilon_{iref}\right)$$
$$F_{out} = EA\left(\varepsilon_{tra}\right) \tag{2}$$

where E and A stand for Young's modulus and diameter of the bars respectively.

One should note that, as a result of the transfer time of the signals through the specimen, F_{in} and F_{out} are generally not equal during the initial period of the loading. This point will be discussed later. Therefore,

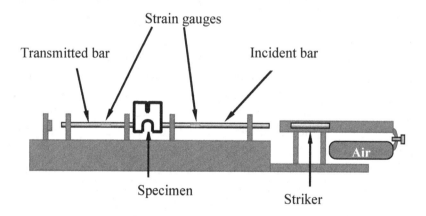

Fig. 2. Schematic representation of a Kolsky (split Hopkinson) bar, with the CCS sandwiched between the bars. A compressed air gun fires the striker against the incident bar, which induces a compressive stress pulse. The latter travels towards the specimen, gets partly reflected and partly transmitted through the specimen to the transmitter bar. The transient strains are measured away from the specimen, using strain gauges cemented to the bars.

once the structure and applied loads (boundary conditions) are selected and determined, the dynamic crack initiation toughness is defined as the value of the stress intensity factor at the onset of crack propagation:

$$K_{IC}^d = K_{Id}(t) \; @t = t_{fracture} \tag{3}$$

The fracture time is determined by single wire fracture gauges that are connected to a timing circuit. A typical experiment lasts a few tens of microseconds. It should be emphasized that accurate assessment of the fracture time is crucial, since the typical stress intensity rate is of the order of $\dot{K}_{Id} = 10^6 \, MPa\sqrt{m}/s$. As a result of the linearity of the problem, the approach can be extended to mixed-mode loading, using the symmetric $0.5(F_{in}+F_{out})$ and anti-symmetric $0.5(F_{in}-F_{out})$ components of the loading with appropriate mode I and mode II boundary conditions, respectively[9]. The SCS was used in a series of validation experiments with various brittle materials.

A key observation was that the brittle specimens fractured at a time where $F_{in} \neq 0$, while F_{out} was still equal to 0. This indicates that fracture is purely inertial, and the crack propagates upon the first passage of the stress wave prior to exiting the specimen at the transmitter bar-specimen interface. This observation opens the way for the testing of *unsupported* specimens, since fracture occurs before the wave has reached the supports. In spite the symmetry of the CCS, the lack of symmetry of the applied loads at fracture preclude the development of a pure mode I fracture. Consequently, the specimen may experience a mixed mode loading with slant initial crack trajectory. However, this situation can be controlled through the use of one point loaded structures (e.g. beams), as discussed next.

Keeping in mind these points, the ensuing developments focused on the application of the so-called one-point impact technique[10].

3. The one-point impact technique

An alternative to the deconvolution technique mentioned in the previous section consists of calculating the response $\hat{k}$ of the cracked structure to a unit impulse. As before, the stress intensity factor will be given by the convolution of the applied load with the unit impulse response:

$$K_{Id} = F * \hat{k} \qquad (4)$$

The dynamic initiation toughness will be again given by eqn. (3). This approach is less involved than the (unstable) deconvolution process, although it requires one calculation per given crack length. A great advantage of the method is that one can now enforce pure mode I conditions by selected (symmetrical) beam specimens, for instance. The experimental setup is schematically represented in Fig. 3. A significant advantage of this setup is that the beam is not supported, so that the numerical calculation of $\hat{k}$ is greatly simplified when compared with 3 point bend experiments (e.g. Charpy) for which the boundary conditions at the supports may be ill-defined.

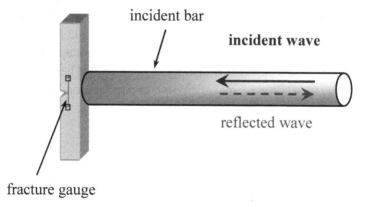

Fig. 3. The one-point impact technique. A single instrumented bar is brought in contact with the fracture (beam) specimen. A striker is fired against the bar, which induces a compressive signal that loads the specimen. The latter is unsupported and fractures by inertia. The fracture time is measured by means of a single wire fracture gauges cemented on each side of the specimen. For this geometry, pure mode I loading is ensured.

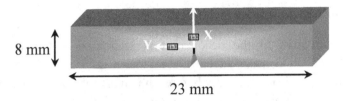

Fig. 4.1. Dimensions [mm] of the short beam specimens used by Weisbrod and Rittel[11]. Note that the dimensions of the beam dictate the use of numerical analysis techniques to determine the dynamic stress intensity factor.

Although (fatigue) cracked beam specimens are commonly used, the beams' dimensions are not restricted by any specific geometrical assumption. Weisbrod and Rittel[11] investigated the fracture properties of short beams made of tungsten heavy alloy (Fig. 4.1), using the one-point impact technique.

Typical experimental signals are shown in Fig. 4.2, as they are collected on the incident bar (ε_{in} and ε_{ref}), and close to the crack-tip by the fracture gauge.

The validity of the approach was assessed by comparing the measured and calculated strains ahead of the crack-tip. Miniature strain gauges were mounted close to the crack-tip to determine $\varepsilon(t)$ at (r, θ). At the same time, the strain was calculated from $K_{Id}(t)$ based on the applied load and eqn. (4). As shown in Fig. 5, the simple assumption of a 1 term LEFM, K-dominated strain, yielded an excellent approximation of the measured strain in several distinct experiments. These experiments showed the feasibility of the one-point impact approach for dynamic initiation toughness measurements.

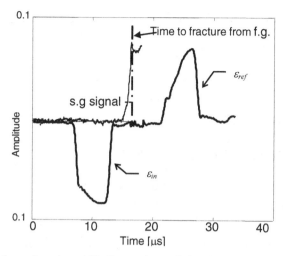

Fig. 4.2. Typical raw (not time shifted) experimental signals, including the incident and reflected pulses. A miniature strain gauge has been cemented on one side of the specimen and a single wire fracture gauge on the other side. Both indicate quite similar fracture time. (Reprinted from [11]).

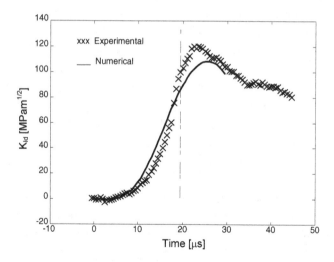

Fig. 5. Typical evolution of the $K_I(t)$ in a one-point impact experiment of a short beam. The SIF is calculated from the boundary conditions (eqn. 4) and measured using a miniature strain gauge. Note the excellent agreement between the two signals. The dashed vertical line indicates fracture time, beyond which the determination of $K_I(t)$ is no longer valid. (Reprinted from [11]).

We will now discuss two important aspects of the experiment, namely the accuracy of fracture gauge readings, and the duration of the contact phase between an unsupported specimen and the incident bar in typical experiments.

4. Specific issues related to the one-point impact technique

4.1 *Determination of the fracture time*

As mentioned before, one should tend to achieve the best possible accuracy in determining the onset of crack propagation. There are basically only two available methods to determine the onset of fracture. The first relies on single wire fracture or miniature strain gauges, and the other on high speed photography (visual detection). One should keep in mind that while we are currently limited to surface monitoring of the specimen, (dynamic) fracture is a three-dimensional process during which a finite time is required for microcracks that initiate at mid-thickness of the specimen, to coalesce into a main crack, visible on its surface. Unfortunately, more sophisticated techniques such as acoustic emission, flash X-ray or potential drop techniques have not yet lifted the limitation related to surface monitoring. Other techniques, essentially optical, rely on direct crack-tip monitoring through high speed photography, and they still require the crack to emerge on the free surface of the specimen. Therefore, one can conclude that the fracture time, determined using either technique, is the upper bound corresponding to the final stage of fracture, e.g. gross specimen separation. The reader interested in a quantitative estimate of the time delays that characterize the process is referred to a paper by Aoki and Kimura[12], who analyzed the method of caustics in dynamic fracture.

In a set of experiments, a steel specimen was tested, which was both instrumented with a fracture gauge on one side and monitored by high speed photography on its other side. The low test temperature (specimen extracted from liquid nitrogen) induced a brittle fracture mode. Figure 6.1 shows the evolution of the recorded load and the fracture time indicated by the fracture gauge. It can be noted that fracture occurs beyond the maximum load peak and that after some time, the load drops to zero (this point will be addressed later). With the time origin taken as the arrival of the stress wave on the specimen, the measured fracture time, from the fracture gauge, was found to be 32.3µs. The estimated fracture time, from the photographic record was 34.0 µs (Fig. 6.2). An excellent agreement was observed between the two crack detection methods[13]. Therefore, while additional effort is indeed required to refine our detection capabilities of early damage, fracture gauges can be considered as a cheap and reliable method to assess the onset of fracture.

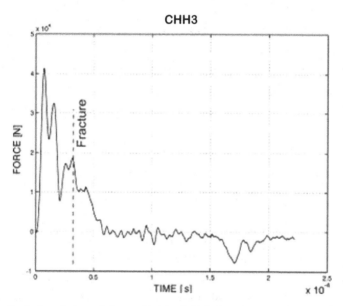

Fig. 6.1. Typical variation of the applied load, F(t), in a one-point impact experiment. Fracture is detected by means of the single wire fracture gauge. Note that the onset of fracture starts past the maximum load at t=32.3µs. The load drops to zero at t≈60µs, indicating completion of the fracture process. (Reprinted from [13]).

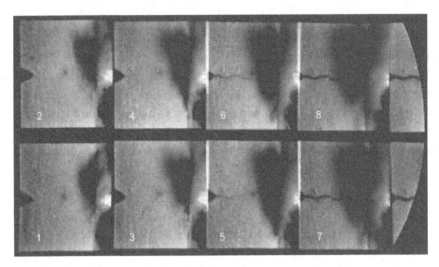

Fig. 6.2. High speed photographic recording of the experiment shown in Fig. 6.1. the crack originates between frames 4 and 5 corresponding to a fracture time of 34μs, in excellent agreement with the fracture gauge reading of 32.3μs. (Reprinted from [13]).

4.2 *Contact time between the specimen and incident bar*

It is well established that an impacted and unsupported structure will not immediately start to propagate upon impact. Rather, a certain amount of time will elapse, during which waves travel back and forth in the structure, causing its step-like rigid body motion. It is therefore important to have an idea of the duration of the contact phase between the specimen and the bar, as compared to the fracture time. In other words, one would like to be sure that the specimen fractures while it is still in contact with the bar, and not as the result of its later impact on some restraining structure or wall. However, contrary to 1-D wave propagation analysis, the 2-D case of the beam specimen must be investigated experimentally and/or numerically. In this context, impact experiments were performed on steel (Charpy) specimens that were loaded in a 1-point configuration[13]. The specimens, made of ductile steel at room temperature, were monitored using high speed photography and the duration of the contact phase was assessed visually. Here, the specimens did not fracture, but rather took off the bar after a given time. Since the incident and reflected signals were recorded during the experiments, the applied load was calculated (eqn. 2) so that when it dropped to zero, the hypothesis was that the specimen separated from the bar. One should note that a zero load condition means that the incident and reflected signals are equal, which alternatively means that the end of the bar acts as a free surface (total reflection). The separation time, thus determined was found to be 80μs (Fig 7.1). On the other hand, the high speed sequence of photographs clearly shows a tiny gap that opened between the specimen and the bar at about 82.5μs (Fig. 7.2). This experiment clearly shows that the contact phase duration can be reliably estimated from the load-time data, even if this point is obvious from a theoretical point of view. One can now compare the fracture time of 34 μs with the duration of the contact phase of 80 μs in this series of experiments, and the obvious conclusion is that the brittle specimen fracture in the 1-point impact test occurs during the contact phase with the loading bar.

However, this conclusion should not be generalized to *any* 1-point impact test. Indeed, the contact phase depends, among other factors on the geometry and the material of the test specimen. Intuitively, one would expect that the higher the wave velocity of the material, the shorter the contact period. One should also pay attention to the slenderness of the beam specimen, which may cause temporary loss of contact as a result of the mode shapes of the specimen. Fortunately, these points can be reliably investigated using finite element simulations. Figure 8 shows a numerical simulation of the 1-point impact test, including the various stages of deformation of the beam specimen until it definitely separates from the bar[14]. From these simulations, the contact time can be reliably estimated and was also found to match the experimental observations. In addition, such simulations may be extremely useful for the design of one-point impact experiments with specimens of various geometries, when one wants to compare the contact time with that required for the stress wave to reach and interact with the crack-tip.

D. Rittel

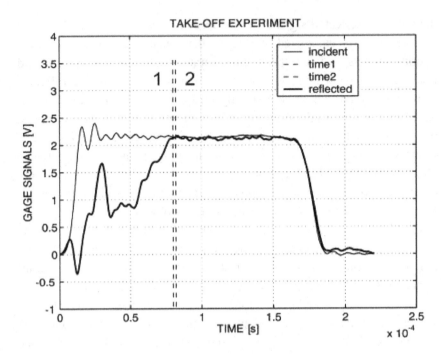

Fig. 7.1. Typical incident and reflected signals. The time at which they reach equal levels indicates a free end condition (specimen take-off or fracture), in excellent agreement with high speed photographic recordings. (Reprinted from [13]).

The practitioner should therefore keep in mind that the experiment yields redundant information by means of the fracture gauges and the strain pulses. The most obvious indication that fracture indeed occurs during the contact phase is obtained from the fact that fracture time is noticeably shorter than the time at which contact cessation is recorded, i.e. load drop to zero.

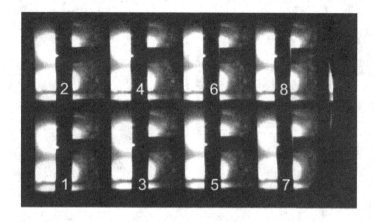

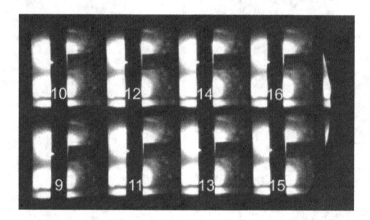

Fig. 7.2. Two separate sequences of high speed photographs in a one-point impact experiment of a Charpy specimen at room temperature. The specimen did not fracture. The rigid body kinematics is evidenced from the growing gap that develops at the specimen-bar interface. The first evidence of loss of contact appears between frames 4-5, corresponding to t = 82.5 μs. (Reprinted from [13]).

D. Rittel

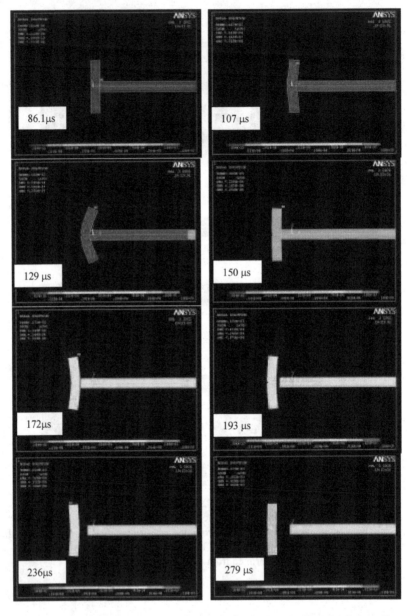

Fig. 8. Finite element simulation of the one-point impact tests. The simulation allows a precise estimate of the specimen-bar contact phase as well as the various vibration modes of the specimen, in order to optimize the specimen geometry. (Reprinted from [14]).

One last, but not the least addressed question concerns the accuracy of the experimental determination of the dynamic crack initiation toughness using the above mentioned methods.

4.3 *A cross-technique comparison*

As mentioned in the introduction, the determination of the dynamic crack initiation toughness is not a standardized practice. So, every researcher develops his own methodology. As of today, the author is not aware of any round-robins aimed at assessing the accuracy of the measurements. However, a first comparison of that kind was recently carried out by Rittel and Rosakis[15], in which two radically different approaches were compared. The selected material was a bulk metallic glass that is well known for its limited ductility to fracture and tendency to fail by (adiabatic) shear band formation. In one series of experiments, plate specimens were dynamically loaded (mode I) by an impact pendulum, while the crack-tip displacement field was monitored (imaged) in real time using the Coherent Gradient Sensing (CGS) method and high speed photography. The evolution of the stress intensity factor was subsequently determined until the onset of fracture, as evidenced from the photographic recording. The fracture toughness was thus determined according to eqn. (2). An independent series of experiments was carried out on the same material, using small beam specimens and the one-point impact technique, as described above. Fig. 9 shows the dynamic crack initiation toughness, as determined from the 2 methods, as a function of the stress intensity rate (by analogy with the strain rate in constitutive tests). As can be noted, the 2 techniques yielded rather similar results, in terms of both K_{IC}^d and its dependence on $\dot{K}_{Id}$. The good agreement between the 2 techniques opens the way towards more standardized tests on one hand, and also allows for a more reliable comparison of results originating from different sources on the other. One should nevertheless keep in mind that the main bottleneck in these experiments remains the accurate determination of the fracture time, regardless of the experimental approach.

D. Rittel

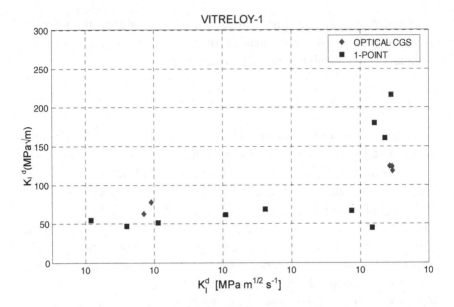

Fig. 9. Dynamic initiation toughness of a bulk metallic glass, as determined by 2 different experimental techniques: CGS and one-point impact. Note the excellent agreement of the two techniques. (Reprinted from [15]).

5. Characteristic experimental results

Having described and discussed some approaches to the experimental determination of the dynamic crack initiation toughness, we will now present selected results about this material property, in an attempt to clarify the underlying physical mechanisms. While Fig. 9 showed the rate sensitivity of K_{IC}^d for a bulk metallic glass, Figs. 10 and 11 show results obtained for 2 more material systems. The first is a glassy polymer (commercial polymethylmethacrylate - PMMA) that was tested using the CCS technique[16]. The second material is porous ceramic skeleton (TiC) infiltrated with a steel binding phase[17].

Fig. 10 shows that the dynamic initiation toughness of PMMA is noticeably higher than the quasi-static fracture toughness. Dynamic loading conditions produce values that are twice the quasi-static values, even if at the higher stress intensity rates, the crack was observed to

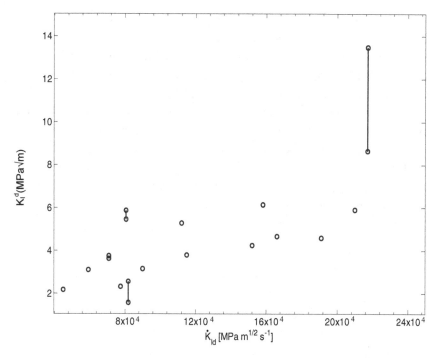

Fig. 10. Dynamic initiation toughness of commercial PMMA as a function of the stress-intensity rate. Two corresponding points correspond to the 2 fracture gauges on each side of the specimen. At the highest rate, fracture is unsymmetrical and exhibits mode-mix. (Reprinted from [16]).

extend at a slant angle, thus indicating some mode-mix. In addition, fracture was not always symmetrical and different values were recorded on each side of the specimen, indicating asymmetrical initiation conditions.

Figure 11 shows the distribution of the quasi-static and dynamic fracture toughness of the TiC-steel cermet, for which the steel matrix has been subjected to various heat treatments. While additional information about this material can be found in Rittel et al.[17], it is interesting to note at that stage that the dynamic fracture toughness is again markedly higher than the quasi-static, and it may triple its quasi-static value, on the average. Therefore, Figs. 9-11 all point to the same observation: the dynamic fracture toughness is affected by the stress intensity rate, in the

D. Rittel

sense that higher loading rates correspond to higher toughness values. This observation cannot be generalized to all the classes of materials, since in certain cases the dynamic initiation toughness may be lower than its static counterpart or even remain unaffected by the rate of loading, as reported in the literature. However, if one considers the brittle materials in particular, the present results point to the puzzling fact that the apparent material resistance to crack propagation is higher under dynamic conditions. While the physical mechanism responsible for this trend is not elucidated, it seems obvious that strain rate dependent mechanical properties (e.g. plasticity) can be ruled out since they play a minor role in the brittle fracture process. The following section presents some results and clues on this issue.

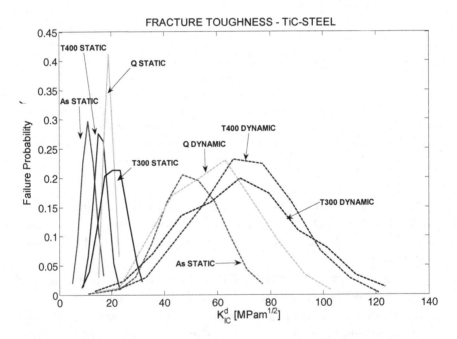

Fig. 11. Static and dynamic fracture toughness of a TiC-steel cermet. The steel matrix underwent various heat treatments, as indicated (As-received, Quenched, Tempered at 300 and 400C). Note the significantly higher dynamic fracture toughness. (Reprinted from [17]).

6. On the rate sensitivity of the dynamic initiation toughness

In the absence of a well-defined (universal) explanation for the rate sensitivity of the dynamic crack initiation toughness, one can tentatively explain the observed trend on the basis of simple micromechanical considerations, namely damage, for which the specific mechanism is material dependent. In other words, the failure mechanisms must be characterized using optical and scanning electron microscopy as well as other materialographic techniques.

In the above mentioned work on commercial PMMA[16], it was noted that the region immediately adjacent to the crack front was significantly rougher for the dynamically loaded specimens, while it was relatively smooth for those which were loaded quasi-statically. In addition, away from the crack initiation area, the crack propagation region was observed to be relatively smooth to an extent that one could not differentiate static from dynamic specimens. This observation points to a distinct signature of the dynamic initiation process. It was therefore suggested that the dynamic initiation process differs from the quasi-static one in the sense that, whereas the latter proceeds by a self-similar extension of the crack-front, the former seems to consist of the nucleation on multiple microcracks, responsible for the observed roughness. In such a case, a finite time is required for these microcracks (damage) to coalesce with the main crack-front, during which the macroscopic stress intensity factor may increase. As of today, there is no analytical model for this complicated phenomenon. Yet, the simple argument that final failure is delayed in time by a given micromechanism, may well be responsible for the high fracture toughness values that are measured.

In the second case of the composite TiC-steel cermet[17], metallographic sections, perpendicular to the fracture plane, were prepared in the vicinity of the crack-origin. As shown in Fig. 12, the quasi-statically loaded specimen shows no apparent damage below the main fracture plane. By contrast, several cleavage microcracks can be noted in a limited radius around the crack-tip. These microcracks are confined to the ceramic phase, as was shown by transmission electron microscopy[18].

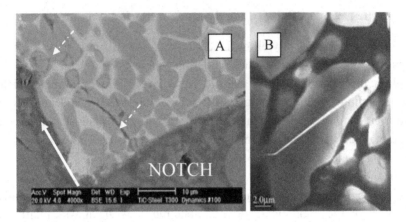

Fig. 12.1. Longitudinal cross-section through the fracture plane. Static fracture. The dark gray particles are TiC skeleton and the clear gray phase is the steel matrix. Note the lack of apparent damage below the main fracture plane. (Reprinted from [17]).

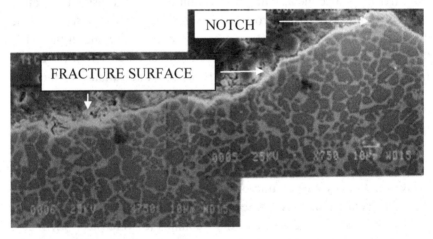

Fig. 12.2. (A) Longitudinal cross-section through the fracture plane. Dynamic fracture. Note the extensive cleavage cracking of the TiC phase. (Reprinted from [17]). (B) ibid. at higher magnification using transmission electron microscopy. (Reprinted from [18]).

One can therefore identify two phases of dynamic *damage* evolution: cleavage damage in the ceramic phase which dissipates energy that would otherwise be available for crack-propagation, and the second phase which, consists of bridging the ceramic cleavage cracks across the steel ligaments to complete the final failure process. Additional evidence of energy dissipation through cleavage microcracking can be found in the work of Rittel et al.[19]. This work, which was carried out on bainitic steel, showed a correlation between the fracture energy, the qualitatively assessed roughness of the fracture surfaces and mostly the extent of cleavage microcracking beneath the main fracture plane.

Concerning the above-mentioned bulk metallic glass[15], it was observed that the static and dynamic *failure* mechanisms are quite different. As shown in Fig. 13, the quasi-static fracture specimen fails essentially by cleavage with some dimples. This failure mechanism is deemed to consume less energy than the ductile mechanism (dimples) observed in the dynamically fractured specimens. Qualitatively, the two failure mechanisms correspond to the trend observed in the initiation toughness values recorded for these specimens.

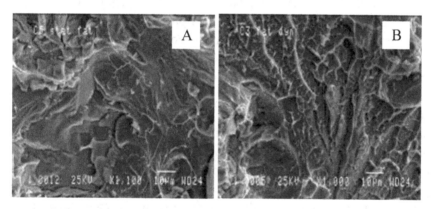

Fig. 13 Static (A) and dynamic (B) fracture surfaces of a bulk metallic glass. Note the difference in failure mechanisms, showing mostly cleavage and some dimples in (A), as opposed to profuse dimples in (B). (Reprinted from [15]).

The present observations are brought to illustrate the point that the dynamic initiation toughness and its rate sensitivity are dictated by micromechanical damage and failure considerations. The damage and failure mechanisms should be systematically characterized for each case, so that distinct classes can ultimately be identified, thus allowing for prediction of the rate sensitivity of the dynamic initiation toughness.

The next subject to be addressed concerns dynamic fracture under mode II loading.

7. Dynamic mode II loading

7.1 *Failure mode transition*

Dynamic fracture under mode II conditions has been less investigated by the engineering community than mode I. By contrast, the geophysics community has a strong interest in mode II because of its occurrence during earthquakes (the reader can find a comparative review in [20]). In parallel, the formation of adiabatic shear instabilities (shear bands) has drawn a large analytical and experimental interest, although outside of the fracture mechanics framework. However, seminal work by Kalthoff et al.[21] has shed new light on dynamic mode II fracture by revealing the so-called "failure mode transition". This phenomenon, which was first discovered for maraging steel specimens, consists of a dramatic change of fracture path of dynamically loaded mode II cracks, as a function of the impact velocity. In the range of "low impact velocities" (to be determined for each material, and probably experimental setup), the crack proceeds roughly along a 70° inclined path. The selection of this path is determined by a maximum normal stress criterion, as e.g. in a brittle fracture case. However, when the impact velocity increases beyond a certain value ("high impact velocity"), the crack extends parallel to itself at a 0° angle. The crack is therefore driven under shear loading, thus responding to a different criterion that is more adequate to ductile failure. And indeed, the 0° angle was found to correspond to adiabatic shear failure, thus tying fracture mechanics and adiabatic shear banding. The transition in crack-path is called the failure mode transition. Following this observation, both analytical and experimental work was carried out on other materials. Kalthoff's experiments were repeated by Ravi-Chandar on polycarbonate[22], confirming the failure mode transition for this polymer. Zhou and coworkers[23] performed similar experiments on maraging steel and a titanium alloy, which were modeled by finite element simulations[24]. The latter work included the thermomechanical coupling ahead of the crack/notch tip, to replicate the observed transition. Rittel et al.[25] investigated the failure mode transition in commercial polycarbonate, for dynamically loaded sharp fatigue cracks, and confirmed the existence of a failure mode transition in this material, in accord with Ravi-Chandar[22]. However, one noticeable difference is the fracture mechanics framework adopted in [25], within which the authors sought for a critical mode II stress intensity factor that triggers the failure mode transition. The main advantage of this approach is that by contrast to a critical impact velocity which is much dependent on specimen and setup, a critical K_{II} is of a more universal character. However, the actual criterion for the failure mode transition is not yet universally established and much additional experimental work is needed to clarify this issue.

Therefore, at this stage, the reader might want to note that depending on the loading conditions, 2 different failure modes may develop, each of them answering a different criterion. The most important point is that a similar situation does not occur in mode I loading, so that it is important to investigate the possible occurrence of the transition for a given material that will be subjected to mode II loading conditions. Consequently, the rest of this paragraph will be dedicated to mode II loading, from an experimental point of view.

7.2 One-point impact experiment

fracture gauge

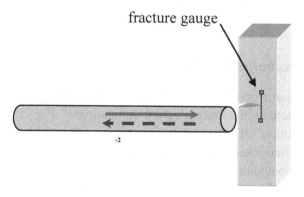

Fig. 14. Typical setup for mode II experiments, using one-point impact

The one-point experiment shown in Fig. 14 is again quite convenient to perform dynamic mode II experiments, provided the stress-intensity factors are well characterized. As before, the main advantage of this setup is that the boundary conditions are those of a free specimen.

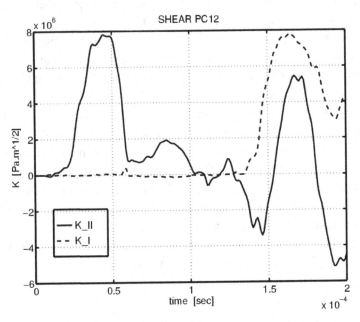

Fig. 15. Typical evolution of the dynamic SIF's for a side impacted polymeric plate (numerical calculation). Note that up to t≈120 µs, the crack experiences almost pure mode II loading. (reprinted with permission from [25])

7.3. *The stress intensity factors*

The stress intensity factors of the side impacted plate specimens have been calculated, using finite element simulations[25]. Fig. 15 shows the typical evolution of the stress intensity factors for a crack that experiences frictionless contact conditions.

The main outcome of this calculation is that during a significant part of the loading history, the crack experiences pure mode II conditions. At a later stage, corresponding to multiple wave reflections in the specimen, the situation evolves into a mixed-mode loading condition. It should be mentioned that the influence of friction (Coulomb) was investigated preliminarily in the same work, and the numerical calculations show that friction of the crack flanks has a very minor influence on the history of the SIF's. The main influence of the contact between the crack flanks is to limit closure effects that would otherwise develop, as shown by Lee and Freund[26]. One should note that dominance of an almost pure initial mode II component is rather remarkable, when one thinks of quasi-static testing for which pure mode II is not so easily achieved!

7.4 *Determination of the fracture time*

Similar to mode I tests, fracture is determined using single wire fracture gauges that are cemented close to the crack-tip. However, two possible cases may be encountered. In the first case, due to low impact velocity, the crack propagates at a given angle and actual fracture occurs, so that the fracture gauges provide an accurate estimate of the onset of crack propagation. In the second case, due to a high velocity impact, the actual specimen fragmentation occurs at a later stage that is preceded by adiabatic shear band formation and fracture along the original crack direction. In that case, the initial lack of fragmentation will most often correspond to some tunneling of the crack/shear band underneath the fracture gauge, so that there is no clear indication of the fracture time. This lack of crack opening precludes the determination of the corresponding stress intensity factors. This problem may be circumvented using miniature strain gauges cemented in the vicinity of the crack-tip, although their accurate position requires a preliminary assessment of the angles at which the strains will be best measured and also the radius at which to position these gauges to ensure K dominance with respect to the anticipated plastic zone size. For such experiments, high speed photographic recording provides valuable information, either in the form of photo elastic patterns or through direct observation. In both cases, numerical simulations must be carried out to verify the actual stress field ahead of the crack-tip.

For slant crack propagation, additional information can be gained by measuring the initial crack angle. If a fracture criterion is assumed, such as a maximum normal stress, the latter can be calculated from the simulated history of the stress intensity factors. The outcome is the evolution of the corresponding angle as a function of time[27], so that the observed angle will indicate the fracture time. However, the reader should keep in mind that measurement of the initial kink angle may be a delicate task since this angle may not have an identical value on each side of the specimen, or may even vary within the crack plane itself such that a good engineering judgment might be required. The information provided by the initial kink angle value is of redundant nature if valid gauge readings are obtained. If this is not the case, the kink angle will provide an approximate estimation of the fracture time which would not be available otherwise.

8. Effects of thermomechanical couplings in fracture

Since the fundamental contribution of Taylor and Quinney[28], it is a well known fact that the mechanical energy invested during a mechanical test is partly dissipated into heat and partly stored in the deformed microstructure as the so-called stored energy of cold work[29]. When high rates of deformation prevail, heat cannot leave the specimen by conduction and convection, so that the temperature of the specimen changes. For instance, this phenomenon is particularly evident in the case of adiabatic shear banding. One can distinguish two heat sources: the first one, related to elastic deformations, is reversible and relates to the volume change. For the vast majority of materials, hydrostatic expansion will cause a temperature drop, while compression will cause a temperature rise. This thermoelastic effect causes very small temperature changes of the order of 1K in smooth specimens of metallic materials and slightly more in polymers. The second heat source, thermoplastic coupling, is linked to the dissipated energy, and it will always cause a temperature rise that is much more significant than that related to thermoelastic effects. Therefore, at this stage, it is evident that dynamic crack initiation (and propagation) is certainly not an isothermal phenomenon. While the issue of thermomechanical coupling is not the main concern of this chapter, it should nevertheless be mentioned for the sake of completeness of the physical picture.

The crack-tip is characterized by a strong stress and strain gradient, and geometrically, the crack-tip surrounding can be divided into a plastic zone surrounded by an elastic zone, the extent of which is dictated by fracture mechanics considerations and the mechanical properties of the selected material. Rice and Levy[30] addressed the thermoplastic coupling effect at the tip of dynamically loaded cracks and provided estimates of the temperature rise, on the generally accepted premise that thermoelastic effects are negligible. From a practical point of view, this result shows that the crack-tip material is hotter than the surroundings so that the dynamic initiation toughness is affected and may thus differ from its static counterpart. While this does not account for inertial effects on the micromechanisms of dynamic crack initiation, it would be quite desirable to perform experiments aimed at quantifying the effects of temperature on the quasi-static fracture toughness, and compare it to the dynamic toughness to establish some sort of time-temperature equivalence for this property.

While it is commonly admitted that thermoelastic effects are negligible with respect to thermoplastic effects, early work by Fuller et al.[31] showed that this may not always be the case. These authors used infra-red sensing techniques to measure the temperature ahead of a propagating crack in a brittle polymers, and their results clearly show a thermoelastic cooling effect that precedes the temperature rise as the crack traverses the field of view of the detector. This point was further investigated by Rittel[32] for commercial polymethylmethacrylate. However, these authors investigated the crack initiation process, directly related to the initiation toughness. The experimental technique consisted of embedding small thermocouples in the close vicinity of the crack-tip. The reliability of transient temperature sensing using embedded thermocouples was thoroughly assessed beforehand, both experimentally[33] and analytically[34]. A typical temperature recording until crack propagation is shown in Fig 16. The temperature ahead of the crack-tip remains constant until the stress wave impinges upon the crack. A striking observation is that the temperature drops, as expected, but the drop itself is not negligible and may reach a few tens of degrees. Looking

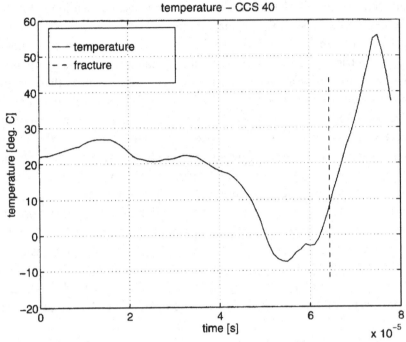

Fig. 16. Transient record of the temperature ahead of the crack-tip of a dynamically loaded PMMA specimen. Note the significant temperature drop, that occurs prior to crack extension, as indicated by the fracture gauge. (Reprinted from [32])

at the indication of the fracture gauge, one may note that the temperature drop actually precedes the onset of crack propagation.

To rationalize this observation, one must first assume that plasticity effects are negligible in this material so that the crack-tip surroundings remain essentially elastic. In addition, assuming that LEFM applies, the expression for the temperature change is given by:

$$T(t) = T_0 \left(1 + \frac{3\alpha}{\rho c_p} p(t) \right) \qquad (5)$$

where T_0 is a reference temperature, α is the linear expansion coefficient, ρ the density, c_p the heat capacity and p stands for the hydrostatic pressure. Equation (5) is valid for an elastic solid under adiabatic conditions. It predicts that the temperature change is of a singular nature, like the crack-tip stress field. This prediction makes no physical sense at the very crack-tip, just like it makes no sense regarding the stresses or the strains. One may nevertheless insert the material properties of this polymer and investigate the nature of the temperature drop as a function of the distance from the crack-tip. The outcome of this exercise is that up to distances of the order of several hundred microns, the temperature drop is not negligible and certainly superior to 1K[31, 32]. Here it should be noted that the distances in question are quite physical, and certainly of a size that is currently dealt with in conventional fracture mechanics. The same exercise can also be carried out for metallic materials, and again, there will be a noticeable temperature drop, closer to the crack-tip [35].

At this stage, the discussion of thermomechanical couplings in fracture might be summarized by noting again that dynamic crack initiation is not an isothermal phenomenon. Two heat sources (sinks) compete, namely thermoelastic and thermoplastic effects. While thermoplastic effects, if present, are deemed to exceed by far thermoelastic effects, these should not be neglected a-priori, without estimating the plastic zone size and the extent of the K dominated elastic zone. Indeed, real materials always tend to display some crack-tip plasticity (even if limited), but the latter is affected by the high strain rate which acts to reduce the plastic zone size by increasing the flow properties of the material. The same high strain rates do not affect the elastic properties of the material, so that when elasticity becomes dominant, crack-tip cooling effects may be significant. This issue requires additional experimental and analytical/numerical work. It should also be noted that, from a practical point of view, a local temperature drop may in some cases trigger a ductile to brittle transition. However, again, the thermal effects may be difficult to single out since the operating fracture micromechanisms seem to be different under dynamic loading conditions, as discussed above.

9. Discussion and conclusion

The preceding paragraphs have presented experimental procedures for the determination of the dynamic crack initiation toughness.

Irrespective of the crack loading mode, it was shown that one-point impact experiments provide valuable information which compares to that obtained with more sophisticated techniques involving high speed photography. A major advantage of the technique is that the specimen lays unsupported so that simple and reliable numerical models can be built, which are useful for data reduction and interpretation.

When compared to other techniques, such as direct impact of the specimen by a projectile, the one-point impact test has the advantage that the applied boundary conditions (loads and displacements) are well defined and measured. The validity of the hybrid experimental-numerical approach was established by several independent tests. For example, the crack-tip stress fields were simultaneously measured and calculated, and an excellent agreement was noted, thus validating the overall procedure used to determine the SIF's.

A key issue is that of the accurate determination of the onset of crack propagation at which the critical fracture conditions are met. The use of single wire fracture gauges was emphasized throughout this chapter, and their reliability was established by comparing gauge readings and photographic recordings. Specimen design considerations were discussed with respect to the duration of the loading phase. It was shown that compact specimens will remain in contact with the loading apparatus for a significant time, whose extent can be modified through suitable geometrical design and numerical simulations.

Typical results were shown for various materials, including those obtained through a limited round-robin test. One can observe that the dynamic fracture toughness, whether a material property or not, is definitely a rate dependent quantity. From our experience with quasi-brittle materials, the dynamic fracture toughness increases markedly with the rate of loading. It appears that this trend may well be dictated by the operating fracture micromechanisms at the very crack-tip. Some of these mechanisms may include multiple cracking and localized damage, whose purpose is to dissipate energy around the crack-tip, thus "shielding" it from the main fracture process.

An additional physical effect that is likely to influence dynamic crack initiation is the existence of thermomechanical couplings that will cause local shifts in the crack-tip temperature. It seems that additional work is required to comprehend the extent to which coupling effects influence the dynamic initiation toughness, while separating them as much as possible from the specific failure micromechanisms.

Mode II dynamic fracture was shown to be analogous, in experimental terms, to mode I fracture. Yet a noticeable difference lies in the selection of the failure mode, crack trajectory and micromechanisms, that is not observed for mode I fracture.

To conclude this chapter, it seems that while the experimental techniques and accuracy will certainly improve in the future, the following two points ought to be investigated further. The first calls for a multiplication of round-robin tests to cross-validate the various techniques, unify them and produce a reliable database of the dynamic fracture properties of materials. This concerns not only mode I but also mode II fracture (and of course mixed-mode) for which additional progress is required to pinpoint the onset of crack propagation and define the appropriate fracture mechanics framework.

The second important aspect of dynamic crack initiation concerns its micromechanical aspects. As of today, little, if none, is really known on the physics of early crack propagation from a micromechanical point of view, acting fracture mechanisms and so on. This information is essential to identify an adequate criterion from a fracture/damage mechanics point of view. As digital high speed cameras become increasingly available with improved performance (higher frame rate and higher resolution), it appears that new micromechanical experiments can be thought of by developing high speed microscopes. Such microscopes will have the capability to monitor the immediate vicinity of the crack-tip, thus revealing the physical details of dynamic crack nucleation, local stress fields and dynamics of microcrack coalescence, most of which remains to be investigated.

References

1. ASTM STANDARDS, *Standard test method for plane-strain fracture toughness of metallic materials*. E399-90 ((1993).
2. L.B. Freund, *Dynamic Fracture Mechanics*. 1990, Cambridge.
3. K. Bertram Broberg, *Cracks and Fracture*. 1999, London: Academic Press.

4. H.D. Bui and H. Maigre, *facteur d'intensité dynamique des contraintes tiré des grandeurs mécaniques globales.* Comptes rendus de l'Académie des Sciences de Paris.T306 *Série* II: p 1213-1216. (1988).

5. D. Rittel, H. Maigre, and H.D. Bui, *A new method for dynamic fracture toughness testing.* Scripta Metallurgica et Materialia. 26: p. 1593-1598. (1992).

6. H.D. Bui, H. Maigre, and D. Rittel, *A new approach to the experimental determination of the dynamic stress intensity factor.* International Journal of Solids and Structures. 29(23): p. 2881-2895. (1992).

7. H. Kolsky, *An investigation of the mechanical properties of materials at very high rates of loading.* Proceedings of the Royal Society of London. 62-B: p. 676-700. (1949).

8. M.A. Meyers, *Dynamic behavior of materials.* 1994, New York, NY: J. Wiley and Sons.

9. H. Maigre and D. Rittel, *Mixed-mode quantification for dynamic fracture initiation: application to the compact compression specimen.* International Journal of Solids and Structures. 30(23): p. 3233-3244. (1993).

10. J.H. Giovanola, *Investigation and application of the one-point-bend impact test.* ASTM STP 905. 1986, Philadelphia, PA: ASTM.

11. G. Weisbrod and D. Rittel, *A method for dynamic fracture toughness determination using short beams.* International Journal of Fracture. 104(1): p. 91-104. (2000).

12. S Aoki and T. Kimura, *Finite element study of the optical method of caustic for measuring impact fracture toughness.* Journal of the Mechanics and Physics of Solids. 41(3): p. 413-425. (1993).

13. D. Rittel, A. Pineau, J. Clisson and L. Rota, *On testing of Charpy specimens using the one point bend impact technique.* Experimental Mechanics. 42(3): p. 247-252. (2002).

14. Y. Raitman and D. Rittel, *Private communication.* (2000).

15. D. Rittel and A.J. Rosakis, *Dynamic fracture of beryllium-bearing bulk metallic glass systems: a cross-technique comparison.* Engineering Fracture Mechanics. 72(12): p. 1905-1919. (2005).

16. D. Rittel and H. Maigre, *An investigation of dynamic crack initiation in PMMA.* Mechanics of Materials. 23(3): p. 229-239. (1996).

17. D. Rittel, N. Frage, and M.P. Dariel, *Dynamic mechanical and fracture properties of an infiltrated TiC-1080 steel cermet.* International Journal of Solids and Structures. 42(2): p. 697-715. (2004).

18. W.D. Kaplan, D. Rittel, M. Lieberthal, N. Frage and M.P. Dariel, *Static and dynamic mechanical damage mechanisms in TiC-1080 steel cermets.* Scripta Metallurgica et Materialia. 51(1): p. 37-41. (2004).

19. D. Rittel, B. Tanguy, A. Pineau and T. Thomas, *Impact fracture of a ferritic steel in the lower shelf regime.* Intl. Journal of Fracture 117: p 101-112. (2002).

20. B.N. Cox, H. Gao, D. Gross and D. Rittel, *Modern topics and challenges in dynamic fracture.* Journal of the Mechanics and Physics of Solids. 53: p. 565-596. (2005).

21. J.F. Kalthoff., Shadow optical analysis of dynamic fracture. Optical Engineering 27: 835-840. (1988).

22. K. Ravi-Chandar, *On the failure mode transitions in polycarbonate under dynamic mixed mode loading*. International Journal of Solids and Structures. 32(6/7): p. 925-938. (1995).

23. M. Zhou, A.J. Rosakis, and G. Ravichandran, *Dynamically propagating shear bands in impact-loaded prenotched plates. I- Experimental investigations of temperature signatures and propagation speed.* Journal of the Mechanics and Physics of Solids. 44(6): p. 981-1006. (1996).

24. M. Zhou, A.J. Rosakis, and G. Ravichandran ,*Dynamically propagating shear bands in impact-loaded prenotched plates. II- Numerical Simulations.* Journal of the Mechanics and Physics of Solids. 44(6): p. 1007-1032. (1996-b).

25. D. Rittel and R. Levin, *Mode-mixity and dynamic failure mode transitions in polycarbonate.* Mechanics of Materials. 30(3): p. 197-216. (1998).

26. Y.J. Lee and L.B. Freund, *Fracture initiation due to asymmetric impact loading of an edge cracked plate.* Journal of Applied Mechanics. 57: p. 104-111. (1990).

27. D. Rittel and H. Maigre, *A study of mixed-mode dynamic crack initiation in PMMA.* *Mechanics Research Communications.* 23(5): p. 475-481. (1996).

28. G.I. Taylor and H. Quinney, *The latent energy remaining in a metal after cold working.* Proceedings of the Royal Society of London. A(143): p. 307-326. (1934).

29. M.B. Bever, D.L. Holt, and A.L. Titchener, *"The stored energy of cold work" in Chalmers, B., Christian, J.W. and Massalski, T.B. (eds).* Progress in Materials Science. 17. (1973).

30. J.R. Rice and N. Levy in *The physics of strength and plasticity*, 1969, A.S. Argon ed., MIT Press, Cambridge

31. K.N.G. Fuller, P.G. Fox, and J.E. Field, *The temperature rise at the tip of a fast-moving crack in glassy polymers.* Proceedings of the Royal Society of London. A 341: p. 537-557. (1975).

32. D. Rittel, *Experimental investigation of transient thermoelastic effects in dynamic fracture.* International Journal of Solids and Structures. 35(22): p. 2959-2973. (1998).

33. D. Rittel, *Transient temperature measurement using embedded thermocouples.* Experimental Mechanics. 38(2): p. 73-79. (1998).

34. Y. Rabin and D. Rittel, *A model for the time response of solid-embedded thermocouples.* Experimental Mechanics. 39(1): p. 132-136. (1999).

35. O. Bougaut, *On crack-tip cooling during dynamic crack initiation.* International Journal of Solids and Structures. 38(15): p. 2517-2532. (2001).

Chapter 3

The Dynamics of Rapidly Moving Tensile Cracks in Brittle Amorphous Material

Jay Fineberg

The Racah Institute of Physics, The Hebrew University of Jerusalem, Givat Ram, Jerusalem 91904, Israel
jay@vms.huji.ac.il

The dynamics of fast fracture in brittle amorphous materials are reviewed. We first present a picture of fracture in which numerous effects commonly observed in dynamic fracture may be understood as resulting from an intrinsic (micro-branching) instability of a rapidly moving crack. The instability, when a single crack state undergoes frustrated microscopic crack branching, occurs at a critical propagation velocity. This micro-branching instability gives rise to large velocity oscillations, the formation of non-trivial fracture surface structure, a large increase in the overall fracture surface area, and a corresponding sharp increase of the fracture energy with the mean crack velocity. We present experimental evidence, obtained in a variety of different materials, in support of this picture. The dynamics of crack-front interactions with localized material inhomogeneities are then described. We demonstrate that the loss of translational invariance resulting from this interaction gives rise to both localized waves that propagate along the crack front and the acquisition of an effective inertia by the crack. Crack-front inertia, when coupled with the micro-branching instability, leads to an understanding of the chain-like form of the micro-branch induced patterns observed both on and beneath the fracture surface.

1. Introduction

1.1 *The equation of motion for a rapidly propagating crack*

The detailed dynamics of a rapidly propagating crack have been the object of intensive study since the early scaling theories of Mott[1,2], which were initiated during the second World War. These theories generalized the concept of energy balance, which was first introduced by Griffith[3] in 1920 to describe the onset of fracture. Griffith proposed that fracture occurs when the potential elastic energy per unit area released by a unit extension of a crack is equal to the fracture energy, Γ, defined as the amount of energy necessary to create a new unit of fracture surface. This concept was generalized by Mott and, later Dulaney and Brace[1, 2], to include a global accounting of the kinetic energy within the elastic medium that is released by a propagating crack. Utilizing dimensional anaylsis, these first theories predicted that, in a two-dimensional medium of infinite extent, a crack should continuously accelerate as a function of its instantaneous length to a limiting, but finite, asymptotic velocity. Stroh[4] noted that this asymptotic limit could be reached by either asymptotically increasing the amount of energy stored in the material or, equivalently, by reducing Γ to zero. With $\Gamma \to 0$, a crack propagating at its asymptotic velocity is equivalent to a disturbance moving across two free surfaces. As the speed of a disturbance moving along a free surface is the Ralyeigh wave speed, V_R, of the medium, Stroh concluded that this must be the asymptotic velocity of a dynamic crack. This intuitive idea was later[5-7] shown to be rigorously correct, when a quantitative theory for dynamic fracture was developed.

The scaling arguments mentioned above yielded the general form of an equation of motion. The first quantitative description of the motion of a rapidly moving crack in a 2D linearly elastic material was, however, derived by considering the form of the stress singularity surrounding the tip of a crack. As first noted by Irwin and Orowan[8, 9], the stress field, σ_{ij}, at the tip of a crack is singular in a linearly elastic material. The form of the singularity at a distance r and angle θ from the tip of a crack moving at velocity, v, is given by[5]:

$$\sigma_{ij} \to K_I / (2\pi r)^{1/2} f_{ij}(v, \theta) \tag{1}$$

where the stress intensity factor, K_I, measures the strength of the singularity.

In a linearly elastic material there is no intrinsic dissipation of energy. The only location where energy can be dissipated is at the singular point at the crack's tip. The energy release rate, G, is defined as the energy flux, per unit of extension, that is flowing into the tip of a crack. Irwin showed that G is related to K_I by:

$$G = (1-v^2)/E \cdot A(v) \, K_I^2 \qquad (2)$$

where v and E are, respectively, the Poisson ratio and Young's modulus of the medium. $A(v)$ is a function of the instantaneous crack velocity, v. Energy balance, as originally invoked by Mott, is now used locally (along a path surrounding the crack tip) to obtain an equation of motion for a crack by equating G with the fracture energy, Γ.

To obtain this equation of motion a tacit assumption is used; that *all* of the dissipation in the system occurs within a "small" region surrounding the tip of the crack. The dissipative region is small in the sense that all of the dissipative mechanisms, inherent in the fracture process, occur in a region for which the surrounding stress field is entirely dominated by the singular stress field described in Eq. 1. This condition, called the *condition of small-scale yielding*, is, in essence, the definition of a brittle material. The small zone surrounding the tip in which all nonlinear and dissipative processes occur is defined as the *process zone*. Thus, the fracture energy, Γ, need not necessarily be simply the energy cost to break a single plane of molecular bonds – but embodies all of the complex nonlinear processes that result from the action of the putatively singular stress field near the crack tip. The condition of small-scale yielding is at the foundation of the theoretical basis for describing the motion of a rapidly moving crack. This condition ensures the equivalence of the global energy balance criterion used by Mott with the locally formulated equation of motion inherent in:

$$G = (1-v^2)/E \cdot A(v) \, K_I^2 = \Gamma \qquad (3)$$

Using Eq. (3), an explicit equation of motion was obtained by Eshelby, Kostrov and Freund[5-7] for the (rather general) case where all the externally imposed stresses can be mapped onto the fracture plane. This was accomplished by explicitly calculating the velocity-dependent piece of $K_I(v)$ as well as $A(v)$. Surprisingly, (for the types of loading considered) Freund demonstrated that $K_I(L,v)$ has the separable form

$K_I(L,v)=K_I(L)k(v)$. $K_I(L)$, which can be explicitly calculated, is solely dependent on the external loading configuration and the instantaneous value of the crack length, L. In addition, both $k(v)$ and $A(v)$ are *universal* functions of only the instantaneous velocity v of the crack. Freund[5] also found that, to a good approximation $k(v) \cdot A(v) \sim 1-v/V_R$ thereby yielding the following equation of motion:

$$G = (1-v^2)/E \cdot A(v)\ K_I^2 \sim (1-v^2)/E \cdot (1-v/V_R)\ K_I(L)^2 = \Gamma \qquad (4)$$

$$\blacktriangleright v/\ V_R = 1 - E\Gamma/[(1-v^2)\ K_I(L)^2] \qquad (5)$$

1.2 *Discrepancies with theoretical predictions*

Eq. 5 should describe the motion of any dynamic crack. Experimental tests of this equation of motion, however, were both disappointing and confusing. One firm prediction of this equation was that the limiting speed of a Mode I crack should be the Rayleigh wave speed, V_R. Early experiments[10-12] in brittle amorphous materials consistently showed that the velocity of a crack never really approached V_R, with the maximal observed velocities[13] between 0.5 and $0.6V_R$. At very high velocities, in fact, large increases in the energy release rate, G, only resulted in incremental increases in propagation velocities, as shown in fig. 1. Dynamic cracks, however, in either crystalline materials[14-17] or along a weakened plane in amorphous materials[15] were easily seen to propagate a velocities approaching $0.9V_R$.

In light of the puzzling stubbornness of a crack in amorphous materials to be pushed beyond $0.5-0.6V_R$, it was perhaps surprising to find that Eq. 5 appeared to be in good quantitative agreement with experiments at relatively low velocities. Bergquist[18] found, in experiments conducted on the brittle acrylic PMMA, that the equation of motion worked very well in experiments on crack arrest, where the propagation velocities were less than $0.3V_R$.

J. Fineberg

There were a number of ideas proposed to explain these puzzling results. One of these, a velocity dependence of the fracture energy, Γ, is "hidden" within the equation of motion derived in Eq. 5. Although the fracture energy is assumed to be a material-dependent parameter, it need not be constant. If $\Gamma(v)$ were a rapidly increasing function of the crack velocity, v, then many of the apparent discrepancies of the equation of motion could be explained. Measurements of $\Gamma(v)$, in fact, do show that beyond 0.3-$0.4V_R$, the fracture energy is indeed a rapidly increasing function of the crack velocity. Some representative measurements of the energy release rate, G, as a function of the crack velocity, are presented in Fig. 1. Although this idea may formally solve the discrepancies between the measurements and the predicted equation of motion, one still has no fundamental idea of *why* $\Gamma(v)$ has this high rate dependence for crack velocities exceeding 0.3-$0.4V_R$. Thus, the strong effective rate

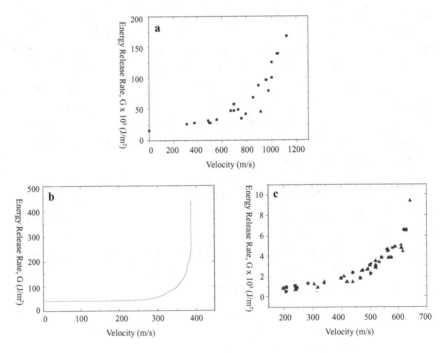

Figure 1. The velocity dependence of the fracture energy, $G(v)$ for (a) AISI 4340 steel from [19], (b) Homalite-100 from [12], and (c) PMMA from [20].

dependence of the fracture energy does not really present a solution to understanding the problem of the dynamics of a rapidly propagating crack. This behavior, instead, simply demonstrates that Eq. 5 is not necessarily inconsistent with observed crack dynamics.

Additional discrepancies between Eq. 5 and experiment were suggested by a series of experiments on the brittle acrylic, Homalite-100, by Ravi-Chandar and Knauss[21, 22]. In these experiments, high-speed (5 µs between exposures) photographs of the caustic formed at the tip of a crack initiated at high loading rates were performed. The velocity of the crack, as deduced from the position of the crack tip in the photographs, was compared with the instantaneous value of the stress intensity factor, which was derived from the size of the caustic. These authors found that at low velocities (below ~$0.3V_R$) a change in the value of the stress intensity factor resulted in an "instantaneous" change in the crack's velocity, exactly as the theory predicts. On the other hand, at higher velocities significant changes in the stress intensity factor produced no discernible corresponding change in the crack's velocity. These experiments have been interpreted as indicating that the stress intensity factor is not a unique function of crack velocity. This is in direct contradiction with the equation of motion, described by Eq. 5, which predicts a one to one correspondence between the instantaneous values of the crack velocity and stress intensity factor.

Eq. 5 was derived on the assumption of a single propagating crack. An alternative picture of the fracture process, that crack propagation is due to the coalescence of microvoids or preexisting defects situated in the crack's path, was then proposed[23]. The observation of parabolic markings on the fracture surface of many brittle polymers prompted the suggestion that the propagation of a dynamic crack takes place by the nucleation, expansion and subsequent coalescence of microscopic voids, or defects, in the crack's path. In this picture, defects ahead of the crack will start to propagate due to the intense stress field that exists in the near vicinity of the crack tip. The parabolic markings left on the fracture surface were shown to be consistent with the interaction of a planar crack front with small spherical voids that initiate ahead of the front and expand toward it[24]. The idea that the dynamics of rapid cracks needed to be described by a mechanism other than an equation of motion for a single crack was further enforced by experiments in Homolite 100. Using high speed photography[21], these experiments showed that cracks, at high speeds, are surrounded by an ensemble of small propagating cracks. As defects exist in most materials, the view of crack propagation as a process of micro-void coalescence would suggest that crack propagation via interacting micro-voids should, in general, occur as a randomly activated process and not by the dynamics predicted by Eq. 5.

2. A quantitative comparison of the equation of motion with experiment

We will now suggest an alternative explanation for the fracture process that is based on a series of experiments in brittle amorphous materials. These experiments suggest that the motion of a rapidly moving crack is well-described by the equation of motion derived in Eq. 5 until the crack reaches a critical velocity, or, equivalently, a critical value of the energy release rate. At that point, an intrinsic instability is excited in which a single propagating crack is no longer stable. We will then show that, beyond this critical velocity, the character of the instability entirely governs the details of the fracture process.

2.1 *Description of the experimental system*

Let us briefly digress to describe the experimental system that was used to obtain many of the results that will be described in this section. This description is important to both understand how the results were obtained as well as to point out some of the limitations (and advantages) of other experimental techniques.

The experiments described in the following section were conducted in thin, quasi-2D sheets of brittle, amorphous materials. Materials used were either the brittle acrylic poly-methyl-methacrylate (PMMA), soda-lime glass, brittle polyacrylamide gels, or rock. The experimental system is shown schematically in Fig. 2, where we define the thickness direction as Z, the direction of applied loading as the Y direction and the direction of crack propagation as the X direction. Details of this apparatus can be found in [13, 25-27]. Stress was applied to the sample via steel bars, which were ground to be straight to within 10μm tolerances, and glued to the opposing faces of the sample at each of its vertical boundaries. In this loading configuration, samples were loaded via uniform displacement of the vertical boundaries with the fracture was initiated at a constant displacement. Prior to loading, a small "seed" crack is introduced at the edge of the sample, midway between the vertical boundaries. Tensile stress (Mode I loading) was applied quasi-statically until arriving at the desired stress. Crack propagation was initiated either by allowing fracture to start spontaneously or by the gentle application of a razor blade in order to sharpen the initial crack tip.

The sample geometry could be varied to provide either steady-state crack propagation at a constant energy density within the sample, or a continuously changing velocity throughout the experiment. Steady-state propagation was achieved by using a thin strip configuration with the ratio of its vertical (Y) to horizontal (X) dimensions between 0.25 - 0.5.

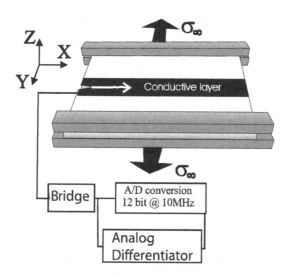

Figure 2. A Schematic picture of the apparatus used to measure crack velocities. A thin conductive coating, whose resistance varies with the instantaneous length of the moving crack, is evaporated onto one or both of the sample faces. As the crack progresses, both the change and value of the sample's resistance are monitored at 12bit resolution at a 10MHz rate. This yields both the location of the crack tip and its instantaneous velocity along the measurement plane to better than a 0.1mm precision throughout the experiment.

When the crack tip is sufficiently far from the horizontal boundaries of the system, this geometry approximates an infinitely long strip with approximate "translational invariance" in the direction of propagation. This state is realized when the crack reaches a length of about half of the vertical size of the system. At this point, advance of the crack by a unit length frees an amount of energy equal to the (constant) energy per unit length stored in the plate far ahead of the crack. Under these conditions, a crack arrives at a state of constant mean velocity with the energy release rate given by $G = \sigma^2 L/(2E)$ where σ is the applied stress at the vertical boundaries, L the vertical size of the system and E the Young's modulus of the material. In this geometry we can directly measure G to within an 8% accuracy.

The crack velocity was measured by first coating the side(s) of the sample with a thin (0.1-1µm) resistive layer. Upon fracture initiation, a propagating crack will cut this coating, thereby changing the sample's resistance. As the crack propagates across the sample, we measure its resistance change by digitizing the voltage drop across the sample to 12 bit accuracy at a rate of 10 MHz. Thus, in PMMA, the location of the crack tip can be established to a 0.1mm spatial resolution at 0.1µsec intervals. The use of an analog differentiating circuit circumvents the need for numerical differentiation of the signal and enables a velocity resolution of better than 25 m/s. The relation between the crack length and sample resistance was calibrated after each experiment.

This experimental apparatus has a number of distinct advantages over other methods used to measure crack velocities. The first of these is an unprecedented accuracy in both the instantaneous location and velocity of a rapidly propagating crack tip throughout the entire duration of a given experiment. This is in contrast to most velocity measurement methods based on high-speed photography, which are generally limited either by the achievable frame-rates or the total number of photographs that can be acquired. Another advantage of this method is that it enables simultaneous high accuracy measurements on two different XY planes (on both sample faces). This is a decided advantage if the propagating crack front (i.e. the leading edge of the crack) undergoes dynamics which are non-uniform in the Z direction. An application of this will be later presented.

Disadvantages of this method are that it only gives the location of the *leading edge* of a crack on a given XY plane. This is fine as long as a crack has not undergone large-scale crack branching. If the lengths of the branched cracks remain small compared to the overall crack length, the method is still accurate – but the information that it provides is the velocity and location of the *leading* crack. If one is interested in the detailed dynamics between two or more bifurcating cracks, one must revert to methods such as high-speed photography.

2.2 *The validity of the equation of motion for a dynamic crack*

Let us now consider the equation of motion derived in Eq. 5. We have seen that there had been indications that it may be valid for relatively slow crack velocities[18, 21], whereas for higher crack velocities there were questions raised regarding its applicability. Sharon et al.[26] were the first to conduct a detailed and quantitative examination of Eq. 5, using the experimental apparatus described in the previous section.

In these experiments, the detailed dynamics of rapidly propagating cracks were measured in both PMMA and soda-lime glass, two brittle amorphous materials. As the experiments were designed as a quantitative test of the validity of Eq. 5, they were conducted using relatively large samples of both materials, whose geometry was conducive to precise numeric calculation of the static stress intensity factor, $K_I(L)$, for each value of the instantaneous crack length, L. As Eq. 5 neglects the effects of elastic waves reflected from the system's boundaries, the measurements were solely considered for times shorter than those necessary for waves, reflected from the system boundaries, to return to the crack tip. To quantitatively examine the predictions of Eq. 5, the experiments were designed to compare the values of the fracture energy, $\Gamma(v)$, for widely different experimental conditions. By inverting Eq. 5, the functional dependence of Γ with the crack velocity, v, is predicted to be:

$$\Gamma = (1\text{-}v^2)/E \cdot (1\text{-}v/V_R)\, K_I(L)^2 \qquad (6)$$

where $K_I(L)$ is a numerically calculated value[a] that is a quadratic function of the applied stress, σ. The instantaneous crack velocity, v, and crack tip location, L, are quantities that are obtained to high resolution for any given experiment, usng the measurement system described previously.

If Eq. 6 is valid, we would expect the value of $\Gamma(v)$ to be independent of the conditions of a specific experiment. Thus, values of $\Gamma(v)$ derived from Eq. 6, using values of L and v which are obtained by performing measurements for a variety of different experimental conditions, should all collapse onto a single curve for each material tested.

The results of these experiments were surprising. As we see in Fig. 3, the experiments indeed collapse onto single curves for both glass and PMMA at crack velocities that are less than $0.4V_R$ in each material. This indicates that the equation of motion derived in Eq. 5 is indeed in excellent quantitative agreement with experiments for crack velocities below $0.4V_R$. The collapse of the data for both glass and PMMA is significant, in that the micro-structure of these two materials is entirely different. These results are also consistent with the earlier results of Bergqvist[18] and Ravi-Chandar et al.[21] which showed that for relatively low crack velocities, cracks appear to behave in a way that is, at least, consistent with Eq. 5.

What occurs for velocities that are greater than $0.4V_R$? The results of the same experiments are presented in Fig. 4 for the entire range of measured crack velocities. These results are compared[26] to independently measured values of Γ for PMMA that were performed by Sharon et al[20].

---

[a] In this comparison we use values of $K_I(L)$ that were calculated from the static stress intensity factor of a stationary crack of length L for the precise geometry of our experimental system. Strictly speaking, the theory predicts that one should use the integral: $\int p/\sqrt{x}\, dx$ from the initial crack length, L_0 to L, where p(x) are the tractions applied to the crack faces[6, 28]. These two formulations coincide when $L_0 \sim 0$, but differ for finite values. As the experiments were performed for small values of L_0, the first formulation better approximates the effects of the close proximity of the lateral boundary at $L=0$, whereas the formulation used in [6] is more appropriate for large values of L_0.

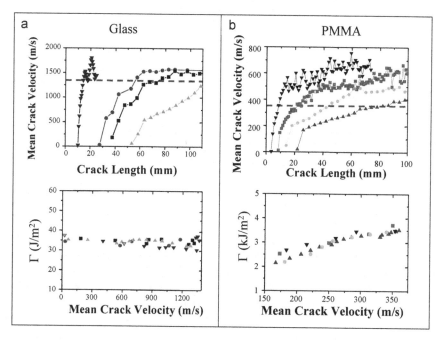

Figure 3. Fracture energy, Γ, derived using Eq. 6, as a function of the mean velocity for soda-lime glass (a) and PMMA (b). Top: Velocities of cracks driven with different initial stresses and seed crack lengths as a function of their instantaneous lengths. The data were smoothed over a 1mm length, to filter out velocity fluctuations. The dashed lines indicate the highest values of v used to derive the corresponding $\Gamma(v)$ (lower plots). These are 350 m/s ($0.38V_R$) in PMMA and 1325 m/s ($0.40V_R$) in glass. The observed data collapse to a single (material-dependent) function, $\Gamma(v)$ is a quantitative validation of the equation of motion (Eqs. 5 and 6) derived for a single crack. Data were taken from [26]. Values of V_R for PMMA and soda-lime glass are, respectively, 930m/s and 3,300m/s.

Fig. 4 demonstrates that the excellent quantitative agreement between the Eq. 6 and directly measured values[20] of Γ breaks down for $v \sim 0.4V_R$. The error bars in the figure are not the result of systematic error but, instead, indicative of the increasing scatter that occurs when one attempts to use Eq. 6 to derive Γ for velocities above $0.4V_R$. These velocities are also those for which the stress intensity factor appears[21, 22, 29] to be a "non-unique" function of v. It is obvious that something interesting is happening at these velocities. An explanation for this and other observed

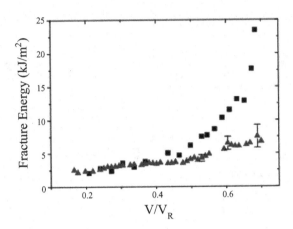

Figure 4. The measured fracture energy[20] (squares) compared with the values of Γ (triangles) derived[26] from Eq. 6. Shown are data from PMMA as a function of the mean crack velocity, where the velocities are scaled by the Rayleigh wave speed, V_R. The error bars are not the result of systematic error, but rather indicative of the increasing scatter in the results when, after $0.4V_R$, derivation of Γ from Eq. 6 is attempted.

(and otherwise unexplained) characteristics of dynamic fracture will be presented in the next section.

3. Instability in dynamic fracture

3.1 *The onset of instability in dynamic fracture*

In the previous section we saw that behavior inconsistent with Eq. 5 occurred for velocities beyond $0.4V_R$. Let us now look carefully at the dynamics that occur for crack velocities above this range.

In Figure 5 we present the dynamics and photographs of the resulting fracture surfaces in typical experiments in both PMMA[25] and brittle polyacrylamide gels[30]. In both of these experiments, the cracks accelerated rapidly up to a maximal mean velocity of about $0.6V_R$. In the initial stages of fracture the velocity smoothly accelerates. At this stage, the fracture surface formed by the crack is smooth and mirror-like, with no apparent structure. At a critical velocity of $v_c = 340\text{m/s}$ ($0.36V_R$) in

PMMA[27, 31] and 1.7m/s ($0.34V_R$) in polyacrylamide[30] the crack undergoes a dynamic instability. Rapid oscillations appear in the instantaneous crack velocity and, at the same time, non-trivial structure appears on the fracture surface. As the crack velocity increases beyond v_c, both the size and amplitude of these fracture surface features increase, as the fracture surface becomes increasingly rougher and the velocity oscillations increase in amplitude. In PMMA, a roughly periodic structure, which is approximately in phase with the crack velocity oscillations is formed. This rib-like structure is generic to PMMA, but is not necessarily evident in other materials. In gels, the structure on the fracture surface[30] formed at v_c (shown in Fig. 5b) is similar to that formed in soda-lime glass[32]. We will return to a discussion of the details of the fracture surface structure later.

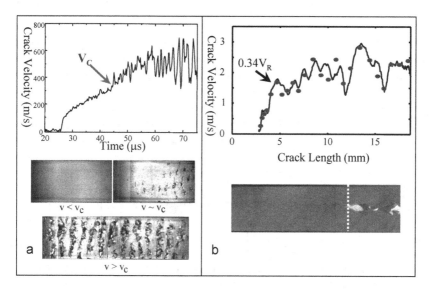

Figure 5. (a) (top) A typical plot of the crack velocity in PMMA as a function of time. Note that the character of the crack velocity changes qualitatively, once v is greater than a critical velocity, v_c= 340m/s = $0.36V_R$ (shown in figure). (bottom) Photographs of typical fracture surfaces created in PMMA prior to v_c, slightly above v_c, and well above v_c. (b) (top) Plot of the crack velocity as a function of its length in a polyacrylamide gel[30]. At a critical velocity (arrow) the crack undergoes the same instability as in PMMA (bottom). A photograph of the fracture surface formed. Note that at the critical velocity of v_c=$0.34V_R$ (dashed line) non-trivial structure is formed on the fracture surface.

3.2 *Frustrated micro-branches as the instability mechanism*

Let us now consider the mechanism that gives rise to the instability from "featureless" dynamics, that are well-explained by the equation of motion derived in Eq. 5, and the onset of both oscillating, non-trivial crack dynamics and the abrupt formation of non-trivial surface structure.

The answer to this question lies beneath the fracture surface. As we see in Figure 6, the onset of instability is marked by a change in the crack *topology*. Instead of a single propagating crack, as envisioned in the derivation of Eq. 5, we now observe that the main crack loses stability to a state composed of a large "mother" crack surrounded by an ensemble of small "daughter" cracks. Once formed, these daughter cracks are not very long-lived, but rapidly arrest, while the mother crack continues. The daughter cracks do not run parallel to the mother crack, but branch out from it, as shown in Fig. 6. Although the fracture surface formed by the

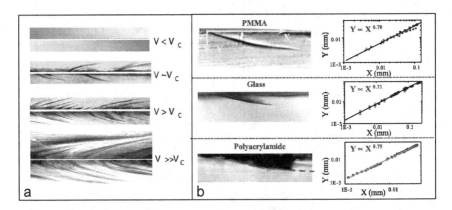

Figure 6. (a) Photographs in the XY plane of the two half-surfaces formed by the main, "mother" crack in PMMA for 4 different velocities. The (straight) path of the mother crack is indicated by the white region along the center of each photograph. At v_c micro-branches, frustrated microscopic crack branches, are formed beneath the fracture surface of the mother crack. The micro-branches become progressively longer and more complex with increasing velocity. The photographs, of length 0.75mm, are to scale. (b) Typical photographs (left) and XY profiles (right) of single micro-branches in PMMA (top) soda-lime glass (center) and a polyacrylamide gel (bottom). The functional form of micro-branches in each of these very different materials is nearly identical with a roughly power-law form of $Y \propto X^{0.7}$.

instability can have a very different appearance in different materials (see e.g. Fig. 5), Figure 6b demonstrates that the functional form of each daughter branch is material-independent. A single micro-branch has a roughly power-law shape in the XY plane of $Y \propto X^{0.7}$. This functional form has been observed in three very different materials: PMMA[33], soda-lime glass[34] and brittle polyacrylamide gel[30] (Fig. 6b). This approximate form is also observed, numerically, in simulations of dynamic crack propagation in a lattice model with randomly distributed quenched noise[35]. Recent analytical work has also predicted a functional form, consistent with these observations, for crack branches[36, 37].

Micro-branches first appear at $v = v_c$. As Fig. 6a (top) demonstrates, *no* micro-branches are observed for $v < v_c$. The materials in which this instability was observed were not "perfect" in the sense that numerous pre-existing microscopic voids and defects were interspersed within each material. Although these small defects certainly interact with the moving crack, no micro-branches are observed below the threshold velocity, v_c. An excellent example of repeated perturbations due to material inhomogeneities is in the parabolic markings formed by voids coalescing with the crack front[24] in brittle polymers. Despite a crack's interaction with these voids, which exist both before and after v_c, only beyond v_c were micro-branches observed to propagate.

The velocity fluctuations evident in Fig. 5 result from the repetitive birth and death of micro-branches. The fluctuations of the instantaneous crack velocity in a given XY plane are correlated[30, 33] with the appearance of micro-branches at the same spatial point. The velocity fluctuations can be understood in the following picture. As a single crack accelerates, the energy released from the potential energy stored in the medium is channeled into creating new fracture surface (i.e. the two crack faces). When the crack velocity reaches v_c, the energy flowing into the tip of the crack is now divided between the mother and daughter cracks, which are formed by branching events. Thus, *less* energy is directed into each crack and the velocity of the crack ensemble decreases. The daughter cracks, competing with the main crack, have a finite lifetime, presumably because the main crack can ''outrun'' and screen them from the surrounding stress field. The daughter cracks then die and the energy that had been diverted from the main crack returns, causing it to accelerate until the scenario repeats itself. The acoustic emissions broadcast by these repetitive accelerations are quite strong. In PMMA[38, 39] their onset corresponded to the previously measured value of v_c, whereas the acoustic emission onset at $0.42V_R$ was the first measurement of the onset of the micro-branching instability in glass[38]. This mechanism of deceleration and acceleration was also observed in finite element simulations[40, 41] of brittle amorphous materials.

3.3 *The influence of micro-branches on surface structure and fracture energy*

In PMMA, detailed studies of the dependence of the micro-branches with the mean crack velocity were carried out[20, 25, 27, 31, 39]. These studies showed that, beyond v_c, the character of the micro-branches was correlated with the mean velocity, or alternatively, with the energy release rate, G, driving the crack ensemble. Although consistent with the observed properties of micro-branches in other materials, such detailed measurements have yet to be performed in materials other than PMMA.

Let us first consider the correlation between micro-branches and the formation of the non-trivial structures on the fracture surface. A comparison of the velocity dependence of the RMS surface amplitude with the mean micro-branch length is presented in Fig. 7. In both figures data from a number of different experiments all collapse onto a common curve. The sharp transition, in both figures, to the micro-branching instability at v_c is evident. For $v > v_c$ both the mean surface amplitude and mean branch lengths increase linearly with v. A comparison of the scales in the two graphs shows that the micro-branch lengths are, at any given crack velocity, nearly two orders of magnitude larger than the typical surface structure size. This indicates that the micro-branches are the dominant objects formed by the instability. The structure formed on the fracture surface is simply the initial section of a micro-branch that is observed before the main body of the micro-branch "disappears" beneath the fracture surface.

Figure 7a indicates how the typical length of a micro-branch grows with the mean crack velocity. An important question is the amount of additional surface area that is being formed by these branches. Sharon

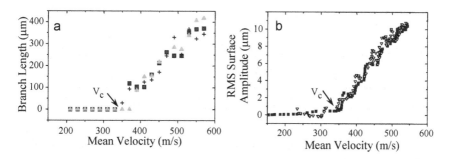

Figure 7. The mean micro-branch length (a) and the RMS surface amplitude (b) as a function of the mean crack velocity in PMMA. The critical velocity of 340 m/s in PMMA is indicated in both plots. The different symbols in both (a) and (b) are from different experiments. For details see [27, 33].

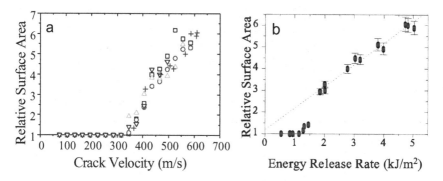

Figure 8. (a) The total surface area formed by both the mother and daughter cracks, formed in PMMA as a function of the crack's velocity. The relative surface area is the total surface area normalized by the surface area created by a single straight crack. Different symbols correspond to different experiments and conditions (for details see [20]). (b) The relative surface area as a function of the measured energy release rate[20] for the experiments presented in (a). The energy release rate as a function of crack velocity in these experiments was presented in Fig. 1c.

et al.[20] performed detailed (and rather arduous) measurements of the total surface area formed in PMMA by the micro-branches. The main results of these measurements are reproduced in Figure 8.

Below v_c the total fracture energy for PMMA was nearly constant and consistent with the value obtained by calculation of the crazing energy needed to separate two polymer surfaces[42]. As Fig. 8a shows, beyond v_c the total fracture surface produced in PMMA increases significantly with the mean crack velocity. At 600 m/s ($1.8v_c$), for example, 6 times more fracture surface is produced by a crack together with its surrounding micro-branches than is produced by a single straight crack. Beyond 600 m/s, measurements of the total fracture surface could not be performed by the optical method used in [20]. This was because the complexity of the branched surfaces (see e.g. the lower panel of Fig. 6a, which corresponds to a mean velocity of 580m/s) was not amenable to quantitative measurement by direct visualization.

The direct measurements of the energy release rate (Fig. 1c) as a function of the mean crack velocity together with the corresponding measurements of the fracture surface area increase (Fig. 8a) enabled, for the first time, a direct comparison of the increase of the fracture surface area with the energy release rate[20]. The results of this comparison, which are presented in Fig. 8b, were surprising. Beyond mean velocities of approximately $1.15v_c$, the energy release rate is a linearly increasing function of the fracture surface. Significantly, the slope of this graph is consistent with the value of the fracture energy corresponding to a single crack. This suggests a simple, geometric explanation for the observed increase of the fracture energy with mean crack velocity. *The energy cost to propagate a crack at any given velocity is simply equal to the energy to create a unit crack surface times the total surface created by both the main and daughter cracks.* The observed increase in fracture energy with the crack's mean velocity is simply a reflection of the amount of extra fracture surface created by the micro-branches surrounding the main crack. It is worth stressing here that, when viewed without the use of a microscope, the crack ensemble appears (by eye) to be a single crack. Even at the highest crack velocities obtained in laboratory experiments, where the total fracture surface was nearly an order of magnitude greater than that formed by a single crack, the crack ensemble simply appears to the eye as a single, relatively rough, "hackled" surface. The large amount of subsurface structure formed by this crack is not trivially apparent.

To the author's knowledge, no measurements analogous to those presented in Fig. 8 have been performed in materials other than PMMA. It still remains to be shown that the above explanation for the origin of the sharp increase in fracture energy observed in other materials can be explained by this simple mechanism. It does not seem unreasonable, however, that this simple geometrical explanation for the increase in the effective fracture energy with velocity is generally applicable. This is because the evolution of the surface structure with velocity (i.e. beyond v_c the fracture surface roughness increases along the fracture surface of accelerating cracks, giving rise to the phenomenological "mirror-mist-hackle" progression of the fracture surface appearance) as well as the appearance and evolution of the micro-branching instability seems to be a general property of brittle amorphous materials. These important (although difficult) measurements, however, have yet to be performed in other materials.

3.4 *The value of the critical velocity for the onset of micro-branching*

There is, as yet, no conclusive quantitative theoretical description for the either the onset or subsequent evolution of the micro-branching instability. Experiments performed in glass[34, 38] and brittle polymers (PMMA or Homalite 100)[13, 31, 39, 43] suggest a value for the critical value, v_c, of close to $0.4V_R$. The most precise values for v_c are available in PMMA and representative measurements[27, 31] are presented in Figure 9. The values in the figure were the crack velocities measured at the onset of the characteristic surface structure shown in Fig. 5. The values of v_c in Fig. 9 were plotted as a function of the crack length at which the instability onset occurred. This roughly reflects the mean acceleration that the cracks in the different experiments had undergone. The experiments were conducted in sheets of PMMA that were formed in different thicknesses with different manufacturing processes. The experiments were also conducted with a variety of different loading conditions. The measured values of v_c/V_R were constant, within experimental resolution, with a value of $v_c/V_R = 0.36 \pm 0.02$. The value of $v_c=0.4V_R$ was also inferred for Homalite[13, 43], and $v_c=0.42V_R$ was measured by both the onset of acoustic emissions[38] and the onset of non-trivial fracture surface structure[34] in the fracture of soda-lime glass.

More recent measurements of v_c in brittle polymer gels[30] suggest that the value of v_c may not be constant, but may depend on the crack's acceleration in the following sense. For values of acceleration that are

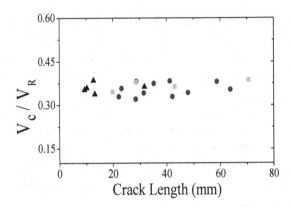

Figure 9. The critical velocity, v_c (normalized by V_R), for the initiation of non-trivial structure (as e.g. in Fig. 5b) on the fracture surface as a function of the crack length at the onset of the instability. The data were obtained in a number of different experiments on PMMA. Symbols denote different experimental conditions and sample geometries (see [27, 31]).

comparable to those measured in the experiments on PMMA and glass, the critical velocity is approximately $v_c \sim 0.34 V_R$, which is compatible with the values of v_c measured in these other materials. In these experiments, (for a typical example see Fig. 5b) the crack accelerated to v_c over a distance of over 1 cm or so (approximately 10% of the sample width). These acceleration rates were typical for quasistatically loaded cracks, where the maximal imposed strain in the sample did not exceed 10% before the onset of propagation.

Higher acceleration rates, on the other hand, produced (on average) much higher values for v_c. These acceleration rates in the Livne et al[30] experiments were attained via highly dynamic loading. As Figure 10a demonstrates, the critical velocity is significantly delayed when these material are subjected to high loading rates, and is roughly a linearly increasing function of the crack's acceleration.

An interesting feature of Fig. 10a is the large *apparent* scatter in the values of v_c for a given value of the crack acceleration. This scatter is intrinsic, and is far larger than the measurement error. This can be seen in the histogram, presented in Fig. 10b, of the time interval between the moment when v surpasses $0.34 V_R$ until the appearance of the first micro-

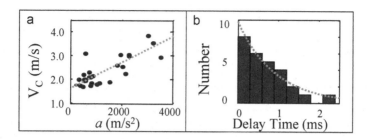

Figure 10. (a) The critical velocity, v_c, for the onset of micro-branches in a polyacrylamide gel increases systematically with the mean acceleration, a, over the range $0.34V_R < v_c < 0.75V_R$ (where $0.34V_R$ is the minimum observed value of v_c). The slope of a linear fit (line) yields a characteristic "activation" time of $\tau=0.5$ms. (b) A histogram of the time intervals between $v=0.34V_R$ and the onset of the instability at v_c. The mean value of these activation times, 0.6 ms, is consistent with the value of τ derived from (a). This figure is reproduced from [30].

branch. This uncertainty was shown in [30] to result from the fact that the micro-branching instability, at least in this material, undergoes a sub-critical (hysteretic transition) bifurcation from a single crack state. A large amount of hysteresis in v was observed[30] for the reverse transition from a crack state with micro-branches to a single-crack state. In light of this clear hysteresis, the minimal observed value of v_c ($0.34V_R$) may simply be indicative of the intrinsic noise level within the system (e.g., vibrations induced by the loading). This picture would predict a distribution of the activation times which is similar to Fig. 10b. Once within the *bistable* region of velocities, where either a single or multi-crack state could exist, the instability may be triggered when random perturbations surpass a critical threshold for triggering the first micro-branch[44]. This would yield a finite probability to bifurcate at each time interval after crack velocity surpasses a minimum value of v_c and produce an exponential distribution similar to Fig. 10b.

The bistability of multi-crack and single crack states above a minimum critical velocity may provide at least a partial explanation for a number of "open" problems in dynamic fracture. The first of these is the observation of the non-uniqueness of the stress intensity factor at high loading rates[21]. The delay in the onset of micro-branching would certainly yield non-unique values of the measured values of the stress intensity factor for given values of v. This would occur because either a single crack state or a crack surrounded by an ensemble of micro-cracks would be equally probable for the high acceleration rates induced by high loading rates. The local stress fields or, equivalently, the stress intensity factors, would be expected to be significantly different for each of these highly different states. An additional explanation for the observed non-uniqueness of K_I measurements could be that the crack velocities, as defined by the distance between successive frames in high speed photographs, were not be sufficiently resolved. The mean velocities over the 5μs intervals, that were used to deduce non-unique values of $K_I(v)$, may not be sufficient to resolve the instantaneous values of v, when, beyond v_c the crack velocity undergoes large amplitude, high frequency oscillations (see Fig. 5) due to the micro-branching instability.

An additional observation that may be explained by the delayed onset of micro-branching with crack acceleration comes from geological observations of rock structures surrounding large meteor impacts[45, 46]. In these regions of intense energy fluxes, large-scale dynamic fracture of brittle rock occurs. The loading rates in these cases are extremely high, as the loading is due to the outgoing shock waves generated at impact. One might expect that under such extreme loading conditions, extensive micro-branching at all scales would occur. This, conceivably, would lead to extensive fragmentation of the rock down to very small scales. Although *mean* velocities approach V_R (more on this later) and crack branches at many scales are indeed observed, surprisingly, extensive fragmentation reaching the scales of the rock grains is *not* observed. The surprising amount of intact rock in these events could be explained by delay of the micro-branching instability (Fig. 10) at high loading rates. Under shock-wave loading of finite duration (approximately the meteorite's impact time), the acceleration of the resulting cracks would be immense. Let us now consider the distribution of activation times for micro-branching presented in Fig. 10b. The mean activation time for such a distribution is finite, although its precise value for brittle rock can only be surmised. Any finite activation time, τ, for micro-branching, however, would define a mean length scale, $\delta x \sim V_R \cdot \tau$, for a micro-branch size. In rock, a 10μs activation time would yield micro-branches whose mean length would be a few centimeters, a scale consistent with field observations[45, 46].

4. Crack Front Waves

4.1 *Introduction and theoretical background*

Non-trivial crack dynamics can be induced by the interaction of a crack with material defects. These interactions, under the proper conditions, will generate long-lived disturbances, coined "Crack Front Waves", that propagate along the leading edge of the crack. As we will see, these disturbances can both generate and be generated by micro-branching events.

In the work summarized in the preceding sections, only ideal (defect-free) quasi-2D materials were considered. In two dimensional materials, a crack's tip is a singular point progressing through a perfect two-dimensional sheet. As long as the material is homogeneous along the Z (thickness) direction, translational invariance exists in this direction and an effectively two-dimensional description of fracture is justified. Let us now consider a crack progressing through an initially homogeneous three-dimensional plate of finite thickness. The tip of a one dimensional crack now becomes a singular one-dimensional front ("crack front") defined by the leading edge of the two–dimensional fracture plane. A crack front is schematically shown in Figure 11. We'll now assume that

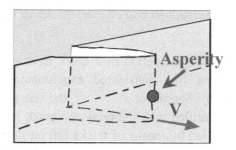

Figure 11. A schematic view of a crack front propagating with velocity v, encountering an asperity, a spatially localized perturbation whose fracture energy is either below or above that of the surrounding medium.

the crack front encounters an asperity, defined as a spatially localized inhomogeneity whose fracture energy is either higher or lower than that of the surrounding medium. The existence of the asperity breaks the system's translational invariance in the Z direction.

An asperity's effect on the subsequent evolution of the crack was first considered[47-49] in scalar (Mode III) models of fracture. These models suggested that a crack front will continually roughen under repeated interactions with asperities. The interaction of a Mode I crack with a single asperity has been theoretically considered by both Morrissey and Rice[50, 51] and Ramanathan and Fisher[52]. Ramanathan and Fisher, using Willis and Movchan's[53, 54] calculation of the change in G induced by a localized perturbation, analytically demonstrated that a single asperity interacting with the crack front will induce a new type of wave coined "crack front waves". Front waves (FW) induce velocity fluctuations within the fracture plane and propagate along the crack front with a velocity, V_{FW}, relative to the material's frame of motion. V_{FW} was predicted to lie between $0.94V_R$ and V_R.

FW were shown to be marginally stable for constant (velocity independent) values of the fracture energy, Γ, and predicted[52] to grow (decay) if Γ were a decreasing (increasing) function of the crack velocity, v. FW were observed by Morrissey and Rice[50, 51] in dynamic finite element calculations. These calculations also predicted that, under repeated interaction with asperities, linearly increasing roughness[55] of the crack front with time would occur at constant propagation velocities.

4.2 *Experimental observations and characteristics of Crack Front Waves*

The dynamics of the interaction of a crack front with an externally induced asperity were first considered experimentally in soda-lime glass[32, 56-60]. Work by Sharon et al.[32, 56, 57] showed that coherent, and long-lived waves, identified with the predicted crack front waves, could be quantitatively studied by means of tracks left on the fracture surface upon the interaction of a crack front with externally imposed asperities. Later experiments demonstrated the existence of these waves in both brittle polyacrylamide gels[30] and rock[45, 46]. In Figure 12 we present examples of tracks formed by front waves that were generated by either externally induced or intrinsic asperities on fracture surfaces of glass, volcanic obsidian, gels and rock.

Front waves that were considered theoretically resulted from perturbations to the crack front, where both the perturbations and resultant disturbances were confined to *within* the fracture plane. In experiments, the tracks observed on the fracture surface were out-of-plane disturbances. Do these result from the same phenomenon? One

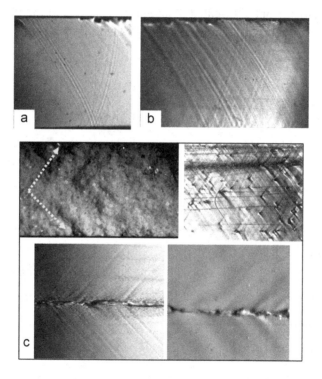

Figure 12. Tracks on the fracture surface of soda-lime glass as a result of the interaction of crack fronts with (a) a single asperity externally induced on the sample surface and (b) a number of such externally induced asperities32, 56, 57. In (a) and (b) the width of the fracture surface was 3mm. (c) Pairs of counter-propagating front waves induced by the interaction of crack fronts with intrinsic material inhomogeneities in (top left) rock46 and (top right) volcanic obsidian (adapted from [61]). FW generated by micro-branches are shown in glass32 (lower left) and polyacrilamide gel30 (lower right). In all of the frames shown the crack front was propagating from right to left.

firm theoretical prediction[52] is that V_{FW} should lie in the range $0.94V_R <$ $V_{FW} < V_R$, when V_{FW} is measured in the material's reference frame. The simple geometric construction[32, 56] presented in figure 13a shows that tracks formed by two intersecting FW form an angle, α, defined by

$$v = V_{FW} \cdot \cos(\alpha/2) \tag{7}$$

for crack fronts oriented normal to the propagation direction and

$$v = V_{FW} \cdot \cos(\alpha/2)/\cos(\beta) \tag{8}$$

when the crack front is oriented at an angle β relative to the propagation direction. An example of intersecting tracks with $\beta=0$ is shown in Figure 13b.

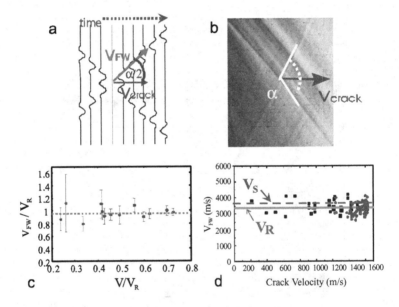

Figure 13. The front wave velocity can be measured by means of two intersecting front wave tracks by means of (a) the geometrical construction leading to Eq. 7, where front profiles of tracks left by two intersecting front-waves are drawn schematically at constant time intervals. (b) A photograph of intersecting tracks on the surface of soda-lime glass. Values of VFW derived using Eq. 7 and measured values of both v and the intersection angle α. Values of VFW= 0.95±0.03VR for gels30 (c) and VFW= 1±0.05VR for soda-lime glass56 (d) are in excellent agreement with predicted values. Values of VR and the shear wave velocity, VS are indicated.

Inverting Eqs. 7 and 8, V_{FW} can be obtained by measurement of the instantaneous values of v and front-wave intersection angles, α. The values of V_{FW} for a large number of such independent measurements in both soda-lime glass[56] and polyacrylamide gel[30] are reproduced in Figure 13c,d. The values of $V_{FW}= 1\pm0.05V_R$ for glass and $V_{FW} = 0.95\pm0.03V_R$ for gels are in excellent agreement with predictions.

In both gels and glass the measured FW velocities significantly (by two standard deviations) differ from the shear wave velocities. This is one way to differentiate between FW formation and the fracture surface features formed by Wallner lines[62]. Wallner lines result from the interaction of the crack front with shear waves generated either externally or, for example, by a point source near the crack front. It has been argued that FW may be confused with this mechanism[58-60]. A closer look at the characteristic features of the observed front waves indicates that front waves are, indeed, distinct entities.

Theoretically predicted FW's were considered as in-plane velocity perturbations of the crack front, whereas the observed tracks on the fracture surface suggest a more three dimensional character. Let us now look more closely at the velocity fluctuations carried by the FW tracks. This can be done by correlating the behavior of the crack velocity along the measurement plane with the arrival of the FW tracks at this plane. This is possible because the velocity measurements described in Fig. 2 are performed only along the specific XY planes located at the *faces* of the samples used. The surface amplitude and instantaneous crack velocity along this measurement plane are compared for both glass[32] and rock[63] in Figure 14a. In both of these materials, a high degree of correlation between the velocity fluctuations and local amplitudes of the front wave tracks is apparent. In addition, although the surface features have an extremely small amplitude (less than 1μm for glass and about 1mm in rock), they generate large velocity fluctuations (hundreds of m/s in glass and over 1km/s in rock). This suggests that the in-plane velocity fluctuations may be the dominant effect caused by the front waves. These large velocity fluctuations then couple to a relatively small out-of-plane component that generates the observed structure on the fracture surface.

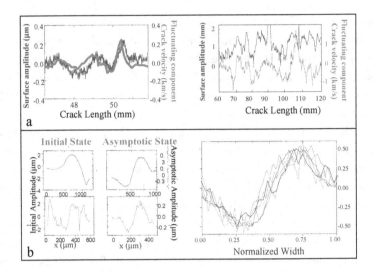

Figure 14. Characteristics of front waves. The amplitudes along the fracture surface formed by front waves have both a characteristic profile and are correlated with large fluctuations in the crack velocity.   (a) Comparisons of the fracture surface amplitudes (blue) and fluctuating components of the velocity (red) in soda-lime glass[32] (left) and rock[63] (right). Both the surface height and velocity measurements were performed on the same XY planes, at the sample faces. (b) Despite large variations in initial conditions, front wave profiles on the fracture surface converge to a unique characteristic profile[56]. (left) Two examples (in glass) of initial profiles together with their final asymptotic form. (right) A superposition of six asymptotic profiles taken from different experiments in which the initial scale of the perturbations driving the front waves varied over a factor of 6. The profiles collapse onto the same shape when normalized by their initial widths (for details of the figure see [56]).

The existence of this out-of-plane component enables us to perform quantitative measurements of front-wave properties. FW are relatively long-lived. Once generated, FW surface amplitudes initially decay exponentially with a characteristic scale, a, the size of the initial perturbation that generated them. The front wave amplitudes then level off at a small, but finite, values[32]. Once arriving at this amplitude, FW can propagate long distances. They undergo reflections at the free boundaries of the sample and have been observed to propagate for up to 7 successive reflections. One interesting FW property is highlighted in Figure 14b, where we examine the asymptotic profile of front waves measured on the fracture surface of glass[56]. These measurements show that the asymptotic form of the front waves is independent of the initial conditions that formed the wave. Profiles generated at a variety of different scales and initial conditions all converge to the same final form. Both the spatial extent of the asymptotic profile as well as the characteristic size of their initial decay FW scale with the size, *a*, of the initial perturbation that generated the FW.

4.3 *Shatter Cone formation in large meteorite impacts: Front Waves as geophysical probes*

Shatter cones are rock structures typically found surrounding the site of large meteorite impacts. Sagy et al.[46] provide a detailed description of these phenomena. Two photographs of shatter cones are presented in figure 15. Although these structures have been studied for many years, their characteristic forms have remained, until recently, an enigma. Shatter cones range in size from a few centimeters to a few meters. They are characterized by curved surfaces that are decorated with radiating striations. Shatter cones are often observed as part of a hierarchic structure; many generations of small shatter cones emanating from progressively larger ones. As shatter cones are formed by the shock wave

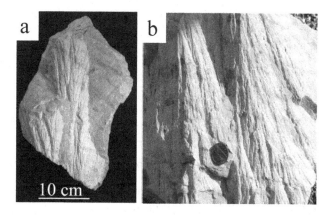

Figure 15. Examples of shatter cones, rock structures formed by the impact of large meteorites[45, 46].

generated by the impact of a large meteorite, their formation should be related to dynamic fracture.

The sharp, V-like striations that are characteristic of shatter cone surfaces, provided the key for understanding their origin. In figure 16 we present a close-up view of the surfaces of two shatter cones. In these photographs the characteristic striations decorating the shatter cone surfaces are easily apparent. Detailed measurements by Sagy et al[45] showed that, for each given rock sample, the angles formed by striation pairs belong to a relatively narrow distribution. They surmised that these characteristic striation pairs were front waves which resulted from the interaction of large-scale cracks generated by the meteorite impact with heterogeneities within the surrounding rock. Thus, as demonstrated in the figure, striation pairs may simply be viewed as counter-propagating front waves. Using Eq. 7, the angles formed by the striations can be used to quantitatively measure of the crack-front propagation speed at the instant of shatter cone formation. The very acute (10-15°) angles typical in

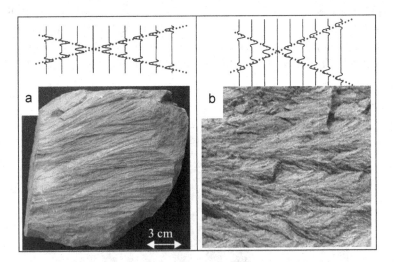

Figure 16. Photographs of the shatter-cone striations in two rock samples (bottom) are accompanied by a schematic picture (as in Fig. 13a) of the front-waves forming the striation angles (top). The striation angles in the two samples yielded propagation velocities of $0.96V_R$ (a) and $0.93V_R$ (b). Note (see Eq. 7) that, at these large propagation velocities, the striation angle is a very strong function of the ratio v/V_R, so that small differences in v are easily detectable.

shatter cones indicate that these structures were formed by cracks propagating at velocities approaching V_R, an observation consistent with the huge energy fluxes inherent in the impact events.

Field observations by Sagy et al[45, 46] found that the striation angles systematically increased with the distance from the impact site. As the angle, α, in Eq. 7 increases with decreasing v, this observation is also consistent with the front-wave explanation, as the energy flux, hence the mean crack velocities would be expected to decrease with distance.

Once the origin of the striations on the shatter cones was understood, the additional features characteristic of shatter cone structure could be explained by the same mechanism[46]. Shatter cones can simply be understood as "large" micro-branches. Their scale is small compared to the huge (km or more) scale of the crack front that formed them, but large compared to the 10-100μm scale micro-branches that are typically observed in laboratory experiments. The overall curved "spoon-like" structure of shatter cones is simply a reflection of the 3D form of micro-branches. Their characteristic hierarchical structure is indicative of the large-scale hierarchical branching that occurs at the huge values of the energy release rate generated by the meteorite impact. Qualitatively similar hierarchical branching structure can be seen at laboratory scales (see e.g. Fig. 6a) when cracks are driven to (relatively) high velocities.

5. Three Dimensional Effects and Crack Front Inertia

The equation of motion (Eq. 5) derived in the first section has the interesting property that the crack velocity, v, has no dependence on the crack's acceleration. In other words, the equation describing crack dynamics attributes no inertia to a moving crack. As Eq. 5 indicates, a local change in the fracture energy should cause an immediate corresponding change in the instantaneous velocity of a crack. In addition, the moment that a crack's tip passes an asperity's immediate vicinity, both G and v should instantly revert to their initial values, and no "memory effects" should be evident.

As we showed earlier, this equation works well as long as the system can be considered as a single crack propagating through an infinite two-dimensional medium. In the previous section, we demonstrated that when the translational symmetry in the Z direction is broken, crack front waves are formed. Here we will describe an additional consequence of the breaking of this symmetry by an asperity; the acquisition by the crack of an effective inertia. We will show that, once translational invariance is broken, a crack retains a "memory" of its prior history, and thereby exhibits inertial behavior that is not described in the framework of Eq. 5.

In figure 17 we demonstrate a crack's acquisition of inertia[32, 57]. In the figure, a profilometer measurement of the fracture surface is presented in which decaying oscillations are observed *ahead* of the point at which the crack front encountered an externally introduced asperity. The amplitudes of these oscillations decay exponentially in X, with the decay length scaling (Figure 17b) with the size, *a*, of the initial disturbance. Each of the oscillations generated a crack front wave, which propagated parallel to the FW generated at the asperity's location. Such parallel tracks are also observed in the photographs of the fracture surface presented in Fig. 12a.

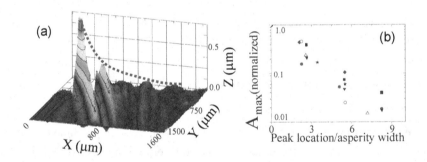

Figure 17. Aligned trains of propagating front waves are typically generated by the interaction of a crack front with a single asperity. (a) A high-resolution profilometer measurement of the fracture surface amplitude where a single asperity, situated at the point (X,Y)=(0,0), generated the front wave train shown. The crack front was propagating in the direction of increasing X. The dotted line traces the envelope of the maximum peak amplitudes, A_{max}, of the front waves generated *ahead* of the asperity. (b) A_{max} as a function of the distance ΔX from the initial asperity, where ΔX is normalized by the asperity width, *a*. Note[32, 57] that the initial amplitude of each FW in the train exponentially decays with a characteristic decay length of 1.8*a*.

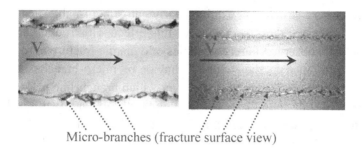

Micro-branches (fracture surface view)

Figure 18. Photographs of branch-lines along the fracture surface in a polyacrylamide gel (left) and soda-lime glass (right). The branch lines are formed by chains of aligned micro-branches. In both photographs the crack was propagating from left to right.

The effects of asperity-generated oscillations, as presented in Fig. 17, disappear rapidly as their amplitude decays exponentially. These oscillations were generated at velocities that were below the critical velocity, v_c, for the onset of the micro-branching instability. What occurs for $v>v_c$ when the crack front is perturbed? Figure 18 demonstrates the generation of branch-lines in both glass and gels beyond v_c. These chains of aligned micro-branches have also been observed in the fracture of rock[63].

Above y_c, as shown in Fig. 18, micro-branching events themselves serve as FW sources since, like an asperity, they effectively increase the local value of the fracture energy. Micro-branches, once excited, are not randomly dispersed throughout the fracture surface but, as shown in Fig. 18, are aligned along straight lines in the propagation direction[30, 32, 57, 63]. We will call these aligned lines of micro-branches "branch lines".

As demonstrated in Figure 19, the internal structure of a branch line has a rough periodicity[32, 57] in X. This periodicity is a consequence of the crack-front inertia presented in Fig. 17. Instead of the exponentially decaying "ring-down" experienced by the crack front for $v<v_c$, for $v>v_c$ each oscillation ahead of the crack becomes a trigger for the next micro-branch in the chain. As in Fig. 17, where the spatial interval between oscillations scaled with the asperity size, the distance between sequential branches scales with the width, ΔZ, of the preceding micro-branch. The observation (fig 19c) that both glass and gels, materials with entirely different micro-structures, exhibit nearly indistinguishable quantitative behavior indicates the universality of this effect. This rough periodicity has also been observed in recent studies of dynamic fracture in rock[63].

One difference in the scaling behavior of micro-branches between glass and gels is the range of scales over which branch-line periodicity is observed. In soda-lime glass, branch-line widths extend over 3 orders of magnitude to below μm scales. It is conceivable that these scales extend much lower than the micron scale, but branch-line visualization was limited by the (~1μm) diffraction limit. In polyacrylamide gels[30] a clear lower cut-off (Fig. 19a) at approximately 50μm exists for branch-line widths. We surmise that this cut-off is related to the polyacrylamide micro-structure and may possibly be related to the size of the process

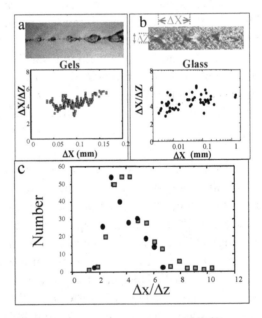

Figure 19. Branch-line scaling in both polyacrylamide gel[30] and glass[32, 57]. Close-up photographs of single branch-lines in gels (a-top) and glass (b-top). In both materials the distance, ΔX between successive branches scales with their width, ΔZ. The ratio, ΔX/ ΔZ, is plotted for hundreds of different measurements as a function of ΔX in both materials (a-bottom) and (b-bottom). Both the mean value of ΔX/ ΔZ ~ 3 and *distributions* of the measured values (c) are nearly indistinguishable for both of these very different materials. Data from gels (glass) are marked by squares (circles).

zone in these materials. In both glass and gels no upper scale cutoff for branch-line widths is observed, other than the overall sample widths.

As a consequence of the increased surface area created by micro-branches upon branch-line formation, the energy release rate, G, into the front can not be evenly distributed as a function of z. The total energy dissipated by a branch line at a given z location should be significantly larger than in the surrounding, featureless surface. This inhomogeneous distribution of G, which is perpetuated for the life of a branch line, indicates a nonlinear focusing of energy in the z direction.

It is important to note that crack-front inertia, the resulting branch-line structure and inhomogeneous distribution of G are all intrinsically three-dimensional effects. Theoretical descriptions which are based on the assumption of a two-dimensional medium, such as that leading to the derivation of the Eq. 5, can not describe these phenomena.

6. Discussion and conclusions

We have summarized, here, results of experimental work in which the detailed dynamics of rapid crack propagation in brittle amorphous materials were studied. These results were divided into three, perhaps overlapping, sections; the micro-branching instability, the interaction of a crack front with localized inhomogeneities, and the existence of crack front inertia.

6.1 *The origins of the micro-branching instability*

In the section describing the micro-branching instability, we demonstrated that many previously open problems in dynamic fracture may be understood as consequences of the micro-branching instability. This instability has been observed in the fracture of brittle acrylics, glasses, brittle gels, and rock. The generality of this instability suggests that it is an intrinsic feature of dynamic fracture, independent of the specific details of a material's underlying micro-structure. Some of the important aspects of dynamic fracture that can be understood as resulting from the micro-branching instability are:

- The origin of the fracture surface structure (e.g. the "mirror, mist, hackle" scenario) that is commonly observed in the fracture of the brittle materials
- The relevance and region of applicability of the equation of motion derived in the linear elastic fracture mechanics (LEFM) framework for a dynamic crack.
- A simple explanation for the strong dependence of the fracture energy with crack velocity for rapidly propagating cracks
- An explanation of the apparent non-uniqueness of the stress intensity factor that had been observed for high loading and propagation rates
- The micro-branching instability may provide a key towards understanding the origin of crack branching, if we assume that large-scale crack branching is due to the development and evolution of micro-branches.

The robust nature of the instability is demonstrated by its appearance in molecular dynamic simulations of fracture[64,65-68]. Abraham[64] in a 6-12 Lenard-Jones crystal and Zhou et al. using a Morse potential[68] demonstrated that, despite the explosive loading necessary to reach dynamic crack velocities in the small (10^6 atom) sample, the instability was observed at respective velocities of $0.32V_R$ and $0.36V_R$.

Despite the apparent universality of the micro-branching instability, our theoretical understanding of its origins is surprisingly unclear. Although the LEFM framework provides an excellent quantitative description of the dynamics of a *single* crack, there are few analytical frameworks that provide criteria for the onset of the instability. One such candidate is a classic calculation in the framework of LEFM by Yoffe[5, 69] showing that the hoop stress develops an off-axis maximum beyond approximately $0.6V_s$. Yoffe suggested that this may provide an explanation for crack branching, although both the high value of the critical velocity and the predicted 60° branching angle do not quantitatively agree with experiment. Gao, using a piecewise linear model for the behavior of the effective elastic moduli in the vicinity of the crack tip, suggested that v_c corresponds to a Yoffe-like instability, where the local wave speed is governed by local (near tip) value of the shear modulus[70, 71]. This work proposes that an explanation for the similar values of v_c for different materials is that the ratio of the yield stress to shear modulus is similar in many materials.

The necessity of considering non-linear continuum models for the process zone is highlighted by the fact that to even obtain straight-ahead propagation of a crack at any velocity, one must consider the material behavior away from the crack tip[72]. If only the stress fields at the tip are considered[73, 74] the entire problem is severely ill-posed. In these models there is essentially no velocity at which a single propagating crack is stable. A crack, at *all* velocities, becomes unstable to off-axis motion. Linear instability calculations in a number of cohesive zone models have stubbornly refused to yield evidence of a micro-branching instability[74-76]. In summarizing the work performed in this class of models[74] Langer and Lobkovsky stated that "... the general conclusion is that these cohesive-zone models are inherently unsatisfactory for use in dynamical studies. They are extremely difficult mathematically and they seem to be highly sensitive to details that ought to be physically unimportant".

A continuum framework that yields frustrated crack branching is that of phase field models[77-80] in which the process zone region is approximated by a local Ginzburg-Landau equation describing the "damage" inherent in the near-tip vicinity. These models, however, have yet to make a quantitative connection with real materials and are, at this time, still being actively developed. Explicit predictions by these models[79, 80] of sinusoidal oscillations of the crack path, whose period is dependent on the system size, have not been experimentally observed for rapid fracture.

Lattice models[13, 81-83] with and without explicit dissipation [13, 44, 82, 84-87] have qualitatively observed similar bifurcation sequences. In contrast to the phase-field and cohesive zone models, there is no need to explicitly simulate the process zone in these models. The process zone simply comes out of the microscopic description that is imposed on the "atoms". Marder has noted that, although these models are perhaps better representations of an ideal brittle crystal than of an amorphous material, it is possible that simply the intrinsic discreteness of these models is necessary for exciting the instability. This supposition is supported by finite element calculations of fracture that use a cohesive zone formulation in the vicinity of the crack tip[40, 41]. These calculations reproduced the appearance of many observed experimental features. It turns out[88], however, that even the existence of micro-branching is very dependent on the grid size used in the simulations. Micro-branching tends to disappear in the limit where the grid size goes to zero.

In all of the experiments, the onset of the instability is very close to $0.4V_R$ in materials with very little microscopic similarity. This suggests that this velocity is, in some sense, a universal critical velocity. No first principles explanation, however, exists for this number. Recent analytical work by Adda-Bedia for both mode III loading[89] and general loading[90] in which the Eshelby criterion (the Eshelby condition states that crack branching may occur if sufficient energy is available to propagate two cracks simultaneously) was shown to yield upper limits of 0.50-$0.52V_R$ for the instability, depending on the crack path criteria that was assumed. These calculations, however, were performed under the assumption that the system is truly two-dimensional. As we have seen, micro-branching first appears when the micro-branches are highly localized in the Z direction. Once micro-branching occurs, empirically, the energy flux in Z becomes highly non-uniform. Thus, the Eshelby condition does not provide a stringent limiting condition for instability – since micro-branches can bifurcate while taking up only a negligible percentage of the sample's thickness.

A hint to understanding why this instability seems to be so difficult to "crack" might be found in the recent experiments by Livne et al.[30] in brittle gels. Their observation of a hysteretic transition from a single-crack state to one in which micro-branching occurs could explain why perturbative approaches have consistently failed to yield an instability. In analogy to problems such as turbulent pipe flow, which has been shown to be linearly stable for all flow velocities, infinitesimally small perturbations might not be sufficient to trigger the instability. *Finite* sized perturbations might be necessary to drive the system into an unstable state. Additional experiments are required in order to determine whether this intrinsic bi-stability is a universal feature of dynamic fracture.

6.2 *Crack inertia and the three-dimensional nature of fracture*

The generation of inertia via the interaction of a crack front with a localized material inhomogeneity is an intrinsically three-dimensional effect. These effects are certainly not incorporated in current analytic theories of fracture in which massless cracks propagate within a two-dimensional medium. When coupled to the micro-branching instability, this crack front inertia generates the branch-line structure and scaling that, as we observed, appears to be a characteristic of dynamic fracture in glass, gels and rock. There are hints that this effect is also observed in numerical simulations of the in-plane behavior of a crack front that encounters a localized asperity[51]. In these simulations, Morrissey and Rice showed that a distinct overshoot of the crack front occurred immediately after the crack front's interaction with a localized asperity.

We believe that this effect is an important one, which may be linked with the generation of crack front waves. Sharon et al.[32] showed how a protrusion (indentation), breaking the front's translational invariance, can influence (be influenced by) other parts of the front via stress waves. As shown in [91] for a static crack front, the local deviation of the front creates a local decrease (increase) in K_I that acts as a "restoring force" which drives a crack front back towards its initial (flat) state. Any addition of inertial effects would cause the front to "overshoot" and thereby lead to ringing behavior similar to that presented in Fig. 17. Effects similar to these have recently been calculated by Dunham et al.[92] as a mechanism drive a mode II (shear) crack to super-shear velocity upon encountering an asperity.

The qualitatively different structure of the fracture surface in PMMA (see Fig. 5) might be related to the rapid decay of front waves in this material. Although very transient front wave tracks have been observed[57] in PMMA, the increase of the fracture energy with v in these materials (see the comparison of $\Gamma(v)$ in glass and PMMA in figure 3) may, as theory[52] predicts, lead to their rapid decay relative to gels and glass. The suppression of front waves in PMMA may also suppress the inertia induced in the crack front. We would then not expect branch-lines to be observed. Indeed, it seems that the length scale observed in the rib-like patterns on the fracture surface of PMMA can be ascribed to their molecular weight of its polymer building blocks[93, 94].

Acknowledgments: This research was supported in part by the National Science Foundation under Grant No. PHY99-0794

References

[1] N. F. Mott, Engineering **165**, 16 (1947).
[2] E. N. Dulaney and W. F. Brace, Journal of Applied Physics **31**, 2233 (1960).
[3] A. A. Griffith, Mechanical Engineering **A221**, 163 (1920).
[4] A. N. Stroh, Philosophical Magazine **6**, 418 (1957).
[5] L. B. Freund, *Dynamic Fracture Mechanics* (Cambridge, New York, 1990).
[6] J. D. Eshelby, Journal of the Mechanics and Physics of Solids **17**, 177 (1969).
[7] B. V. Kostrov, Appl. Math. Mech. **30**, 1077 (1966).
[8] G. R. Irwin, Journal of Applied Mechanics **24**, 361 (1957).
[9] E. Orowan, Weld. Res. Supp. **34**, 157 (1955).
[10] T. Kobayashi and J. W. Dally, in *Crack Arrest Methodology and Applications* (American Society for Testing Materials, Philadelphia, 1979), p. 189.
[11] J. W. Holloway and D. G. Johnson, Philosophical Magazine **14**, 731 (1966).
[12] J. W. Dally, Experimental Mechanics **19**, 349 (1979).
[13] J. Fineberg and M. Marder, Physics Reports-Review Section of Physics Letters **313**, 2 (1999).
[14] D. Hull and P. Beardmore, Int. J. of Fracture **2**, 468 (1966).
[15] P. D. Washabaugh and W. G. Knauss, International Journal of Fracture **65**, 97 (1994).
[16] J. J. Gilman, C. Knudsen, and W. P. Walsh, Journal of Applied Mechanics **6**, 601 (1958).
[17] J. E. Field, Contemporary Physics **12**, 1 (1971).
[18] H. Bergkvist, Engineering Fracture Mechanics **6**, 621 (1974).
[19] A. J. Rosakis, J. Duffy, and L. B. Freund, Journal of the Mechanics and Physics of Solids **32**, 443 (1984).
[20] E. Sharon, S. P. Gross, and J. Fineberg, Physical Review Letters **76**, 2117 (1996).
[21] K. Ravi-Chandar and W. G. Knauss, International Journal of Fracture **26**, 141 (1984).
[22] K. Ravi-Chandar and W. G. Knauss, International Journal of Fracture **25**, 247 (1984).
[23] K. B. Broberg, edited by K. Kawata and J. Shioiri (Springer-Verlag, Berlin, 1979), p. 182.
[24] K. Ravi-Chandar and B. Yang, Journal of the Mechanics and Physics of Solids **45**, 535 (1997).
[25] E. Sharon and J. Fineberg, Physical Review B **54**, 7128 (1996).
[26] E. Sharon and J. Fineberg, Nature **397**, 333 (1999).
[27] J. Fineberg, S. P. Gross, M. Marder, Physical Review B **45**, 5146 (1992).

[28] D. A. Kessler and H. Levine, Physical Review E **68**, 036118 (2003).

[29] K. Ravi-Chandar and W. G. Knauss, International Journal of Fracture **26**, 65 (1984).

[30] A. Livne, G. Cohen, and J. Fineberg, Physical Review Letters **94**, 224301 (2005).

[31] J. Fineberg, S. P. Gross, M. Marder, Physical Review Letters **67**, 457 (1991).

[32] E. Sharon, G. Cohen, and J. Fineberg, Physical Review Letters **88**, 085503 (2002).

[33] E. Sharon, S. P. Gross, and J. Fineberg, Physical Review Letters **74**, 5096 (1995).

[34] E. Sharon and J. Fineberg, Philosophical Magazine B**78**, 243 (1998).

[35] J. Astrom and J. Timomen, Physical Review B**54**, 9585 (1996).

[36] M. Adda-Bedia, Physical Review Letters **93** (2004).

[37] E. Bouchbinder, J. Mathiesen, and I. Procaccia, Physical Review E **71** (2005).

[38] S. P. Gross, J. Fineberg, M. Marder, Physical Review Letters **71**, 3162 (1993).

[39] J. F. Boudet, S. Ciliberto, and V. Steinberg, Europhysics Letters **30**, 337 (1995).

[40] O. Miller, L. B. Freund, and A. Needleman, Modelling and Simulation in Materials Science and Engineering **7**, 573 (1999).

[41] X. P. Xu and A. Needleman, Journal of the Mechanics and Physics of Solids **42**, 1397 (1994).

[42] J. Rottler, S. Barsky, and M. O. Robbins, Physical Review Letters **89**, 148304 (2002).

[43] J. A. Hauch and M. P. Marder, International Journal of Fracture **90**, 133 (1998).

[44] L. M. Sander and S. V. Ghaisas, Physical Review Letters **83**, 1994 (1999).

[45] A. Sagy, Z. Reches, and J. Fineberg, Nature **418**, 310 (2002).

[46] A. Sagy, J. Fineberg, and Z. Reches, Journal of Geophysical Research-Solid Earth **109** (2004).

[47] G. Perrin and J. R. Rice, Journal of the Mechanics and Physics of Solids **42**, 1047 (1994).

[48] J. R. Rice, Y. Benzion, and K. S. Kim, Journal of the Mechanics and Physics of Solids **42**, 813 (1994).

[49] Y. Ben Zion and J. Morrissey, Journal of the Mechanics and Physics of Solids **43**, 1363 (1995).

[50] J. W. Morrissey and J. R. Rice, Journal of the Mechanics and Physics of Solids **46**, 467 (1998).

[51] J. W. Morrissey and J. R. Rice, Journal of the Mechanics and Physics of Solids **48**, 1229 (2000).

[52] S. Ramanathan and D. S. Fisher, Physical Review Letters **79**, 877 (1997).

[53] J. R. Willis and A. B. Movchan, Journal of the Mechanics and Physics of Solids **43**, 319 (1995).

[54] J. R. Willis and A. B. Movchan, Journal of the Mechanics and Physics of Solids **45**, 591 (1997).

[55] E. Bouchaud, J. P. Bouchaud, D. S. Fisher, Journal of the Mechanics and Physics of Solids **50**, 1703 (2002).

[56] E. Sharon, G. Cohen, and J. Fineberg, Nature **410**, 68 (2001).

[57] J. Fineberg, E. Sharon, and G. Cohen, International Journal of Fracture **121**, 55 (2003).

[58] E. Sharon, G. Cohen, and J. Fineberg, Physical Review Letters **93**, 099601 (2004).

[59] D. Bonamy and K. Ravi-Chandar, Physical Review Letters **91**, 235502 (2003).

[60] D. Bonamy and K. Ravi-Chandar, Physical Review Letters **93**, 099602 (2004).

[61] A. Tsirk, in Advances in Ceramics:*Fractography of Glasses and Ceramics*, edited by J. Varner and V. Frechette (The AmericanCeramic Society, Westerville OH, USA, 1988), Vol. 22, p. 57–69.

[62] H. Wallner, Z. Physik **114**, 368 (1939).

[63] A. Sagy, G. Cohen, Z. Reches, Journal of Geophysic Research (to appear) .

[64] F. F. Abraham, D. Brodbeck, R. A. Rafey, Physical Review Letters **73**, 272 (1994).

[65] A. Omeltchenko, J. Yu, R. K. Kalia, Physical Review Letters **78**, 2148 (1997).

[66] P. Gumbsch, S. J. Zhou, and B. L. Holian, Physical Review B **55**, 3445 (1997).

[67] B. L. Holian and R. Ravelo, Physical Review B **51**, 11275 (1995).

[68] S. J. Zhou, P. S. Lomdahl, R. Thomson, Physical Review Letters **76**, 2318 (1996).

[69] E. Yoffe, Phil. Mag. **42**, 739 (1951).

[70] H. Gao, Journal of Mechanics and Physics of Solids **44**, 1453 (1996).

[71] H. Gao, Phil. Mag. Lett. **76**, 307 (1997).

[72] M. Adda-Bedia, M. Ben Amar, and Y. Pomeau, Physical Review E **54**, 5774 (1996).

[73] B. R. Baker, J. Applied Mech. **29**, 449 (1962).

[74] J. S. Langer and A. E. Lobkovsky, J. Mech. Solids **46**, 1521 (1998).

[75] E. S. C. Ching, J. S. Langer, and H. Nakanishi, Physical Review E **53**, 2864 (1996).

[76] E. S. C. Ching, J. S. Langer, and H. Nakanishi, Physical Review Letters **76**, 1087 (1996).

[77] I. S. Aranson, V. A. Kalatsky, and V. M. Vinokur, Physical Review Letters **85**, 118 (2000).

[78] A. Karma, D. A. Kessler, and H. Levine, Physical Review Letters **87**, 045501 (2001).

[79] A. Karma and A. E. Lobkovsky, Physical Review Letters **92**, 245510 (2004).

[80] H. Henry and H. Levine, Physical Review Letters **93**, 105504 (2004).

[81] M. Marder and X. M. Liu, Physical Review Letters **71**, 2417 (1993).

[82] M. Marder and S. Gross, Journal of the Mechanics and Physics of Solids **43**, 1 (1995).

[83] L. Slepyan, Sov. Phys. Dokl. **26**, 538 (1981).

[84] D. A. Kessler and H. Levine, Physical Review E **59**, 5154 (1999).

[85] D. A. Kessler, Physical Review E **61**, 2348 (2000).

[86] D. A. Kessler and H. Levine, Physical Review E **63**, 016118 (2001).

[87] J. Astrom and J. Timomen, Phys. Rev. **B54**, 9585 (1996).

[88] M. L. Falk, A. Needleman, and J. R. Rice, J. de Physique IV **11**, Pr5 43 (2001).

[89] M. Adda-Bedia, Journal of the Mechanics and Physics of Solids **52**, 1407 (2004).

[90] M. Adda-Bedia, Journal of the Mechanics and Physics of Solids **53**, 227 (2005).

[91] J. R. Rice, J. Applied Mech. **52**, 571 (1985).

[92] E. M. Dunham, P. Favreau, and J. M. Carlson, Science **299**, 1557 (2003).

[93] R. P. Kusy and D. T. Turner, Polymer **17**, 161 (1976).

[94] E. Bouchbinder and I. Procaccia, Physical Review E **72**, 055103 (2005).

Chapter 4

Optical Methods For Dynamic Fracture Mechanics

Hareesh V. Tippur

Dept of Mechanical Engineering
Auburn University, AL 36849
htippur@eng.auburn.edu

In this Chapter, prominent optical interferometers that have contributed to the advancement of dynamic fracture mechanics in the recent past are reviewed. Specifically, the methods of 2-D photoelasticty, Coherent Gradient Sensing (CGS) and moiré interferometry are discussed. Following a brief introduction of each method, sections on typical experimental set up, working principle and a few prominent examples of applications to dynamic fracture investigation are provided for each of these three methods.

1. Photoelasticity

Several milestones in the field of fracture mechanics in general and dynamic fracture mechanics in particular are attributed to photoelasticity. From the classical work of Post and Wells [1] that raised questions on crack tip K-dominance, Dally and coworkers' observations on $K_I - v$ relationship [2-4], Kobayashi and Ramulu [5]'s works on dynamic crack curving to the recent Shukla and Singh's [6] observation of Mach waves near dynamically growing interfacial cracks have all one thing in common. They all have used 2-D transmission photoelasticity as the investigative tool.

The method of transmission light photoelasticity uses transparent amorphous model materials which are optically isotropic in unstressed state and become anisotropic when stressed. This optical effect, first observed by Sir David Brewster (1816), is referred to as temporary optical birefringence and is similar to permanent double refraction observed in some naturally occurring crystals. The interference fringes, commonly referred to as isochromatics, observed by this method are proportional to the maximum in-plane shear stress at a point in the model when studied using a standard circular polariscope.

1.1 *Experimental set up*

A typical experimental set up used for dynamic photoelastic measurement is shown in Fig. 1(a). It consists of a pair of plane polarizers (P and A) and a pair of quarter-wave plates (Q_1 and Q_2) when monochromatic incoherent illumination is used. The light is polarized in, say, the vertical direction by the first polarizer P (polarization axis shown is by the solid arrow) before reaching the first quarter-wave plate Q_1. The 'fast' and 'slow' axes of Q_1 are oriented at $\pm 45^0$ relative to the polarizer axis such that the emergent light is circularly polarized before entering the planar photoelastic object. (Note that in Fig. 1, only the 'fast' axis is shown using a solid arrow and the 'slow' axis is perpendicular to the 'fast' axis and is not shown.) The state of stress at a generic point in the stressed object causes emerging light waves to be elliptically polarized. Upon transmission through the second quarter-wave plate Q_2, with its 'fast' and 'slow' axes reversed relative to Q_1, elliptically polarized light vector with an altered phase emerges. A component of the emergent light from the analyzer, with its polarization axis shown horizontal in this schematic, results in an intensity distribution as isochromatic patterns. An imaging device, typically a rotating mirror high-speed camera, is used in conjunction with a pulsed light source to record isochromatic patterns during a dynamic fracture event. It should be noted that when a high-intensity plane polarized pulsed laser is used instead of an extended light sources as shown in Fig. 1(a), the polarizer P can be entirely avoided.

In early dynamic fracture investigations, a spark-gap camera of the Cranz-Schardin type was relatively more common (see, Fig. 1(b)) when compared to the rotating mirror type camera (Fig. 1(a)). In this, an array

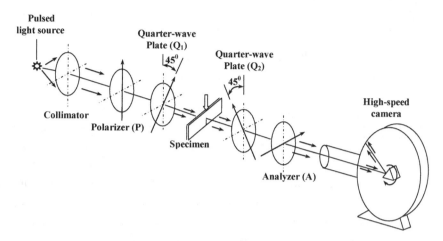

Figure 1(a). Schematic diagrams for 2-D dynamic photoelastic investigation of cracked bodies when a pulse laser is used as the strobe light source in conjunction with a rotating mirror high-speed camera.

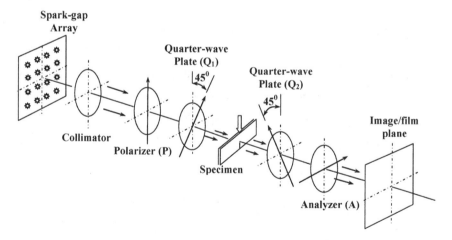

Figure 1(b). Schematic diagrams for 2-D dynamic photoelastic investigation of cracked bodies when a sequentially delayed spark-gap array is used as light sources in conjunction with a large format film plane.

of spark-gaps (commonly 4 x 4 array) acts as light sources which are switched 'on' and 'off' sequentially using a delay circuit. A large format film held stationary at the image plane is used to register fringe patterns at corresponding conjugate spatial locations.

1.2 Working principle

When an incident wave crosses into a stressed birefringent medium, double refraction occurs and two linear/plane polarized light beams, one called the 'ordinary beam' and the other an 'extraordinary beam' with polarization in two mutually perpendicular planes emerge from the stressed medium. Experiments show that these directions coincide with the directions of the two principal stresses at a generic point. These two refracted beams propagate at two distinct velocities in the medium depending upon the local stress state. Consequently a phase difference develops between them upon exiting the object. The two velocities can be interpreted in terms of two refractive indices for the medium. Hence, the phase difference α in radians can be expressed as,

$$\alpha = \frac{2\pi B}{\lambda}(n_{01} - n_{02}),   \tag{1.1}$$

where n_{01} and n_{02} are the two refractive indices, B is the sheet thickness and λ is the wavelength of light. For plane stress condition, it can be further shown that $(n_{01} - n_{02})$ is proportional to the principal stress difference $(\sigma_1 - \sigma_2)$ [7]. Hence, $(\sigma_1 - \sigma_2)$ can be expressed in terms of the phase difference to get,

$$(\sigma_1 - \sigma_2) = \frac{Nf_\sigma}{B}, \qquad N = \frac{\alpha}{2\pi} = 0, \pm 1, \pm 2...,   \tag{1.2}$$

where N is the fringe order, $f_\sigma = \left(\frac{\lambda}{D}\right)$ is called the photoelastic material constant where D is a constant of proportionality between the refractive index difference and the principal stress difference.

Formation of interference fringes can be understood by tracking variations to the complex light amplitude as light propagates through various optical elements of a standard circular polariscope shown schematically in Fig. 2. Here A_i (i = 1-14) represent complex amplitudes at various locations of the polariscope. The polarization axis of each element is shown by solid arrows in the figure and the 'fast axis' of each of the quarter-wave plates is marked 'F.A.'.

When an un-polarized light source is used, the light emerging from the Polarizer P is plane polarized along the polarization axis, as shown. The corresponding light vector can be expressed as, $A_1 = \hat{A}\cos\omega t$ where $\hat{A}$ is the amplitude and ω is the angular frequency of light. Upon incidence on the first quarter-wave plate Q_1, A_1 is resolved along at $\pm 45^0$ relative to the polarization axis as shown. If A_2 and A_3 denote the light vectors along $\pm 45^0$, then $A_2 = A_3 = \hat{B}\cos\omega t$ and $\hat{B} = \dfrac{1}{\sqrt{2}}\hat{A}$.

The quarter-wave plate Q_1 introduces a phase difference of $\pi/2$ (or, a path difference of $\lambda/4$) between A_2 and A_3 upon exiting the optical element. That is, the emergent complex amplitudes A_4 and A_5 can be expressed as,

$$A_4 = \hat{B}\cos\left(\omega t + \frac{\pi}{2}\right) = -\hat{B}\sin\omega t$$

$$A_5 = \hat{B}\cos\omega t$$

(1.3)

It should be noted that since A_4 and A_5 have identical amplitude $\hat{B}$, superposition of these two vectors represents a circularly polarized light. These light vectors A_4 and A_5 upon incidence on a birefringent stressed object at a generic point where principal directions are oriented at an arbitrary angle $\left(\dfrac{\pi}{4} + \beta\right)$ relative to the x- and the y-axes, are resolved along the principal directions as,

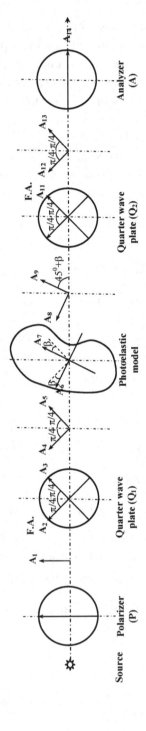

Figure 2. Schematic for describing polarization states of light as propagation occurs through various optical elements of a standard dark-field circular polariscope.

$$A_6 = A_4 \cos\beta - A_5 \sin\beta = -\hat{B}\sin(\omega t + \beta)$$

$$A_7 = A_4 \sin\beta + A_5 \cos\beta = \hat{B}\cos(\omega t + \beta)$$

$$(1.4)$$

In the above, β represents the angle relative to the polarization axes of the quarter wave plates. Upon exiting the stressed model, a phase difference α proportional to the principal stress difference at that point occurs between the two light vectors A_6 and A_7. Thus, light vector exiting the stressed model can be expressed as,

$$A_8 = -\hat{B}\sin(\omega t + \beta + \alpha)$$

$$A_9 = \hat{B}\cos(\omega t + \beta)$$

$$(1.5)$$

Light vectors A_8 and A_9 upon incidence on the second quarter-wave plate Q_2 are again resolved along the two polarization axes oriented at $\pm45^0$ relative to the horizontal and vertical directions as shown. If A_{10} and A_{11} are the components along the 'fast' and the 'slow' axes of Q_2,

$$A_{10} = A_8 \cos\beta + A_9 \sin\beta = -\hat{B}\sin(\omega t + \beta + \alpha)\cos\beta + \hat{B}\cos(\omega t + \beta)\sin\beta$$

$$A_{11} = -A_8 \sin\beta + A_9 \cos\beta = \hat{B}\sin(\omega t + \beta + \alpha)\sin\beta + \hat{B}\cos(\omega t + \beta)\cos\beta$$

$$(1.6)$$

Since the 'fast' and the 'slow' axes of Q_2 are reversed relative to those for Q_1, the phase difference of $\pi/2$ between the emerging light vectors A_{12} and A_{13} is introduced as follows:

$$A_{12} = A_{10}$$

$$A_{13} = \hat{B}\sin\left(\omega t + \beta + \alpha + \frac{\pi}{2}\right)\sin\beta + \hat{B}\cos\left(\omega t + \beta + \frac{\pi}{2}\right)\cos\beta \quad (1.7)$$

$$= \hat{B}\cos(\omega t + \beta + \alpha)\sin\beta - \hat{B}\sin(\omega t + \beta)\cos\beta$$

Next, when A_{12} and A_{13} are incident on the analyzer, a component of each of the two vectors along the polarization axis of the analyzer emerge and produce a light intensity distribution visualized by an observer or captured by an imaging device. If A_{14} is the amplitude distribution of the emerging light from the analyzer,

$$A_{14} = \hat{C} \left(2\cos\frac{1}{2}\{2\omega t + 4\beta + \alpha\}\sin\frac{\alpha}{2} \right) \qquad (1.8)$$

where $\hat{C} = \frac{1}{\sqrt{2}}\hat{B}$. Since the circular frequency of light is of the order of 10^{15} rad/sec, the cosine term in A_{14} can be ignored in optical metrology and experimental mechanics investigations. Thus the intensity observed at a generic point can be expressed as,

$$I = 4\hat{C}\sin^2\frac{\alpha}{2}.$$

Evidently, for the occurrence of destructive interference at a point, $\alpha = 2N\pi, \quad N = 0, \pm 1, \pm 2, \ldots$ where N represents the fringe order.

1.3 Applications

The demonstration of a unique relationship between instantaneous stress intensity factor ($K(t)$) and the velocity of a propagating crack (v) is among the major contributions of photoelasticity to dynamic fracture mechanics in the past two decades. This was researched in the late 70's by Dally and his coworkers [2-4] using epoxy and polyester compounds. An example [8] of the same is shown in Fig. 3 where a sequence of isochromatic fringe pattern in the vicinity of a dynamically propagating crack in Homalite-100 polyester sheet is shown. The images were recorded using a Cranz-Schardin type high-speed camera consisting of an array of sixteen spark gaps which were triggered sequentially by an event trigger, say, crack initiation. Each spark-gap acted as a broad-band point source which was subsequently collimated before illuminating the specimen in the circular polariscope. The images shown were recorded on a large format film with an observation window of approximately 275 μs. From such photoelastic fringes, instantaneous stress intensity

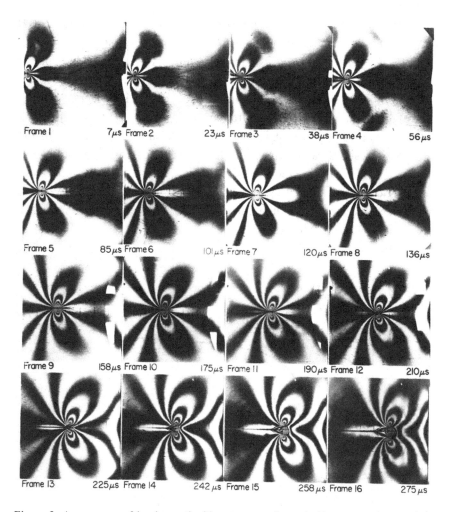

Figure 3. A sequence of isochromatic fringes near a dynamically propagating crack in Homalite-100 sheet. Successful crack branching of the main crack is evident at later time instants as crack reaches higher velocity. (Ref. 8, Courtesy: A. Shukla)

factors [9] and crack lengths and hence crack speed were extracted. Using separate experiments, the dynamic behaviors of a propagating crack from initiation to arrest and initiation to branching were characterized by relating instantaneous values of K and v. They were able to suggest K-v curve as a material property. A schematic representation of such a behavior is shown in Fig. 4. The values of K associated with $v = 0$ (denoted as K_{Im}) at crack arrest and $v = v_{max}$ (denoted as K_{Ib}) at successful branching were attached special significance. The arrest value of K was noted to have a slightly lower value compared to the crack initiation value K_{Ic}. Small increases in K above this minimum level resulted in very sharp increases in crack velocity followed by a transition zone to a saturation velocity. In the latter region, significant increases in K values are essential for producing even small increases to crack velocity. The fracture surface in the latter region shows very high roughness due to the formation of many small fractures in the vicinity of the main crack tip. The same is evident in the later frames (last row) in Fig. 3.

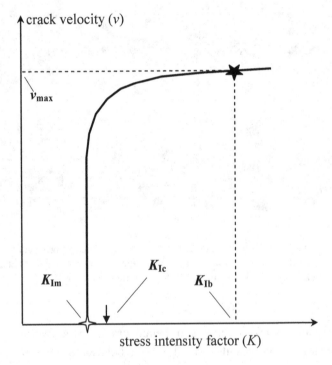

Figure 4. Schematic of variation of stress intensity factor with crack speed in Homalite-100.

Dynamic fracture mechanics of dissimilar material interfaces has drawn attention of the mechanics community in the past decade due to its relevance in a vast range of applications from microelectronic devices to aerospace structures. The photoelastic investigation by Singh and Shukla [6] of a dynamically growing crack along a coherent (bonded) interface of a polymer-metal bimaterial sheet is particularly worth noting in this context. The bimaterial system they studied was a Homalite-100 polyester sheet bonded to Al-6061 aluminum sheet of equal thickness (6.35 mm) along a square edge. The metallic half of the specimen was subjected to projectile impact (velocity 30 m/s) along the edge opposite to a pre-existing starter crack in the form of a disbond. The impact was achieved using a cylindrical steel projectile launched using a gas-gun. This loading scenario is schematically shown in Fig. 5. The impact being on the metallic half, the crack is subjected to dominant shear loading in the early stages after impact before wave reflections

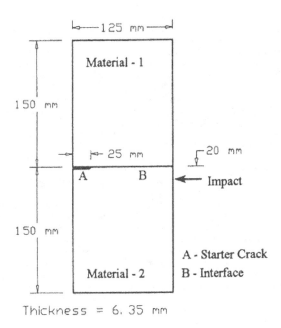

Figure 5. Specimen geometry and loading configuration used for interfacial crack growth studies in a polymer-metal bimaterial. (Ref. 6)

occur from the far-boundaries of the sample. This is because waves propagate much faster in steel compared to Homalite (longitudinal wave speed, shear wave speed and density of (Homalite, steel) are (2330, 5400) m/sec, (1330, 3195) m/sec at a strain rate of 1000 sec^{-1} and (1190, 7830) kg/m^3, respectively). Further, large mismatch in acoustic impedance between the two materials prevented transfer of stress waves across the interface during this period.   Figure 6 shows a set of isochromatic fringe patterns from this experiment where fringes are

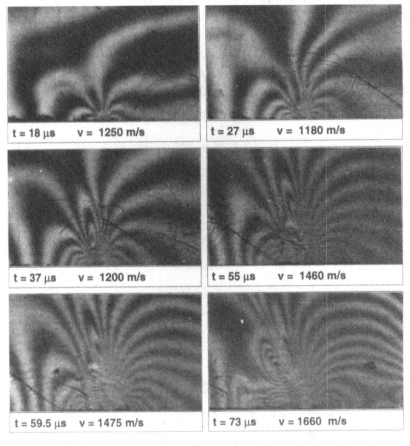

Figure 6. A select set of isochromatic fringe patterns obtained for dynamic crack growth along a Homalite-100/aluminum biomaterial interface subjected to impact loading. (Ref. 6, Courtesy R. P. Singh and A. Shukla)

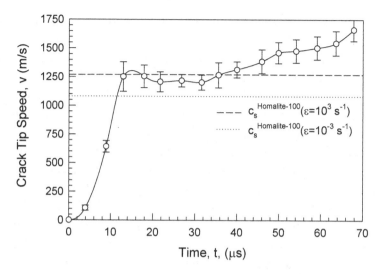

Figure 7. Crack-tip speed history for a dynamically growing crack along Homalite-100/aluminum bimaterial interface under impact loading. (Ref. 6, Courtesy R. P Singh and A. Shukla)

visible in the Homalite-100 half of the specimen. The crack speed history obtained from the images is plotted in Fig. 7. During the event, crack growth at sustained speeds in excess of shear wave speed of Homalite-100 is readily evident and hence crack growth is termed *intersonic* suggesting that crack speeds are bounded by the shear and longitudinal wave speeds of Homalite-100. Such an intersonic crack speed has a direct influence on the nature of crack tip shear stress field. At the beginning when the crack is propagating at subsonic speeds, fringes are smooth and continuous. When the crack speed is in excess of the shear wave speed for the material, fringes are squeezed in the propagation direction while elongated normal to the interface. Further, in the latter case the fringe pattern does not have a single focus along the interface. Instead, the rear and front fringe lobes intercept the interface over a finite length suggesting a large scale contact of the crack faces along the propagation direction. A second consequence of intersonic crack growth is that a discontinuity in the shear stress field is also evident implying the formation of a Mach wave in the stress field around the moving crack tip. The observed Mach wave matches the theoretical prediction of the same by Liu, et al. [10].

In the past few years, there has been an emerging interest investigating the mechanics of dynamic sliding between incoherent interfaces. Frictional sliding between geological layers causing earthquakes has motivated these investigations and initial results suggest many crack-like features integral to this problem and hence would be of great interest to the dynamic fracture mechanics community in the future. Among the photoelastic investigations in this area, the recent report by Coker et al., [11] is notable. In this work, a pair of identical Homalite-100 sheets (see, Fig. 8), initially pressed against each other along a common edge using far-field compression, was subjected to edge impact near the incoherent (or, unbonded) interface. The impact loading was imposed using a gas-gun fired cylindrical steel projectile traveling at 10-60 m/s. A steel buffer plate was bonded to the edge of the lower

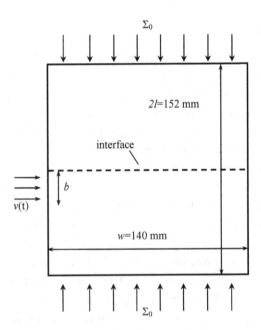

Figure 8. Specimen configuration to study frictional sliding behavior of two monolithic Homalite-100 sheets along an incoherent interface by projectile impact.

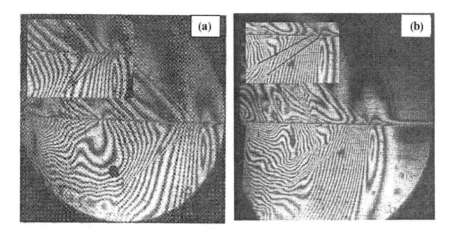

Figure 9. Isochromatic fringes near a sliding tip along frictionally held interfaces between equal thickness Homalite-100 sheets. The sliding tip is moving from the left to the right. (a) Static compressive stress 9.4 MPa and impact velocity 33 m/s, (b) Static compressive stress 9.4 MPa and impact velocity 42 m/s. Insets highlight Mach lines (solid line) associated with the sliding tip. (Ref.11, Courtesy, A. J. Rosakis).

sheet to prevent shattering of the polymer at high impact velocities. Initial impact induced an approximately planar wave and the measured loading profile was a trapezoid having a 10-20 µs rise time followed by a 40 µs steady velocity period. The stress fields in a 160 mm diameter field of view were recorded using photoelasticity and high-speed photography. Two representative results from these experiments for a compressive stress Σ_o of 9.4 MPa and impact velocities 32.7 and 42.2 m/s are shown in Fig. 9(a) and 9(b), respectively. In Fig. 9(a) isochromatic fringes for the lower velocity case are shown where the loading wave front (traveling at longitudinal wave speed C_L) arrives from the left. Behind this initial loading wave is the shear Mach cone seen as a sharp kink in an otherwise smooth fringe pattern highlighted by a solid line in the inset. This tip corresponding to the intersection of the Mach cone with the interface is the sliding tip traveling at intersonic speeds of approximately 1800 m/s. In the same isochromatic field there is also evidence of a Rayleigh wave due to the wave emanating from the projectile corners traveling at some distance behind the sliding tip as a second concentration of fringes. Other observations include (i) higher fringe density in the lower plate where impact occurs compared to the upper since the energy is not transferred easily across the interface and (ii) the fringe discontinuity at the interface due to sliding occurring in a crack-like mode. When the impact velocity is increased to 42 m/s, the propagation velocity of the sliding tip is about 1950 m/s. Although the general characteristics of Fig. 9(b) are same as the ones in Fig. 9(a), additional features are evident. In addition to the Mach line emanating from the sliding tip, a second Mach line of a shallower slope corresponding to a propagation speed of approximately 2600 m/s, which is supersonic with respect to the plane stress longitudinal wave of speed of Homalite-100, is visible. This second Mach wave eventually catches up to the first and only one Mach line continues to propagate in the material subsequently [11]. In each experiment, the speed of the leading sliding tip is found to remain constant and between $\sqrt{2}C_S$ and C_L.

2. Coherent Gradient Sensing (CGS)

Coherent Gradient Sensing is a relatively new addition to the family of optical methods used for dynamic fracture investigations. The technique was formulated and demonstrated in 1989 by the author and his colleagues [12] to study surface slopes and in-plane stress gradients near quasistatically loaded cracks. In view of its full-field capability coupled with a relatively simple optical arrangement for measuring field quantities which produce optical caustic near a crack tip, CGS offered an alternative method for investigating crack problems. Soon its potential for dynamic fracture investigations became quite evident when it was demonstrated successfully using high-speed photography. These early studies gave way to a period of widespread use of the method to study subsonic, intersonic and supersonic crack growths along dissimilar material interfaces, impact damage and fracture of fiber reinforced composites, dynamic fracture of homogeneous and graded particulate composites and foams, and dynamic crack-inclusion interactions, to name a few.

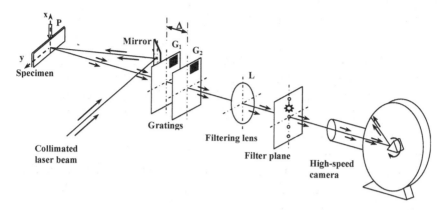

Figure 10. Schematic of the experimental set-up used for dynamic fracture studies using reflection-mode CGS.

2.1 *Experimental set-up*

Typical experimental set-up used for dynamic fracture investigation of opaque planar objects is shown schematically in Fig. 10. The set-up can be easily modified if a transparent object is to be studied instead. The object is illuminated by a collimated laser beam. The object waves reflected off of the object surface are transmitted through two parallel Ronchi gratings separated by a distance (Δ). The diffracted wave fronts are collected by a positive lens and the resulting diffraction spectrum is registered on its focal plane as shown. An aperture is used to block all but the necessary diffraction orders as depicted. The interferograms are recorded in real-time using a high-speed camera. It should be emphasized that the camera system comprising of the lens and the photo sensor array or film needs to be focused on the object surface during initial optical alignment. That is, the object and the image distances in the set-up obey the lens law.

The gratings used in CGS are typically chrome-on-glass line-pair depositions commonly referred to as Ronchi gratings. The choice of Ronchi grating plates is based on commercial availability or ease of fabrication but the method should function equally well with other grating profiles such as sinusoidal rulings. It should also be noted that anti-reflection coatings (typically, magnesium fluoride coatings) are essential for minimizing optical noise in the form of 'ghost fringes' which would otherwise occur due to multiple reflections in the cavity between the gratings. Other parameters such as grating pitch, grating separation distance and focal length of the imaging lens are chosen based on the need for (i) producing sufficient spatial shearing of the object wave front without unduly compromising the accuracy of derivative representation of the out-of-plane displacements, and (ii) resolving the diffraction spots sufficiently on the back focal plane of the imaging lens for easy filtering. The use of micro-positioning devices to achieve in-plane and out-of-plane parallelism between the two gratings limit the smallest grating separating distance to about 15 mm in most reported studied to date.

Specimen preparation of opaque planar objects for reflection-mode CGS study also needs attention. The test sample should be made optically flat and specularly reflective in the region of interest. If the material being studied is metallic, this is generally achieved by lapping the surface flat followed by a metallic (aluminum, gold) film deposition for high reflectivity. On the other hand, if the material being studied is a polymer or a metal matrix composite, an alternative method needs to be adopted. This involves machining the specimen surface flat followed by removing tool marks by sanding the surface using successively finer grade sandpapers. Next, an optical flat previously deposited with a thin layer of aluminum in a thermal evaporator (or a sputtering device) is transferred to the specimen surface. The transferring of metal film can be done using an suitable adhesive (say, low-viscosity epoxy) which preferentially bonds strongly with aluminum compared to the glass optical flat. This process is depicted in a self-explanatory schematic in Fig. 11.

2.2 *Optical alignment*

The shearing interferometer needs to be aligned to ensure (i) planarity of the incident laser beam used for interrogating the object surface and (ii) parallelism of the Ronchi gratings G_1 and G_2 as well as the grating

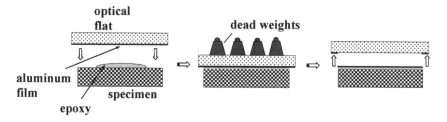

Figure 11. Schematic depicting aluminizing an opaque specimen surface for reflection-mode CGS study.

lines. This is accomplished by mounting gratings G_1 and G_2 on micro-positioning devices. One of the two gratings, say G_2, is mounted on a device with translational capability in the z-direction and rotational capability about the z-axis. Separate coarse rotational capabilities about the x- and the y-axes for G_1 are also preferable. First, parallelism between the gratings planes G_1 and G_2 can be ensured using the coarse positioning device for G_1 by rotating G_1 relative to G_2 about the x- and the y-axes. A high-degree of collimation of the laser beam can then be ensured using a front-coated mirror with a high-degree of flatness, say a $\lambda/8$-mirror, as the object. Assuming parallelism between grating lines of G_1 and G_2, when the incident beam is a spherical wave front $i(x,y) = \dfrac{x^2 + y^2}{2f}$, f being the focal length of the collimating lens, a fringe pattern as shown in Fig. 12(a) would result on the image plane when the principal direction of the Ronchi gratings is along the y-axis. Now, a planar wave-front can be obtained by adjusting the collimator to widen the fringe spacing until a uniform light fringe covers the entire field-of-view. For a planar wave front incident on the gratings, the rotational misalignment would result in equally spaced fringes along the principal direction of the gratings, as the one shown in Fig. 12(b). The formation of these fringes is due to coherent self-imaging phenomenon along the z-axis at the so-called Talbot intervals $Z_T = \dfrac{2p^2}{\lambda}$. These rotational fringes can be eliminated by rotating G_2 about the z-axis until the fringe spacing increases and a single light fringe covers the entire field-of-view. In reality, however, both these misalignments occur simultaneously resulting in equally spaced fringes at an arbitrary angle relative to the principal direction of the two gratings. Thus, optical alignment will have to be carried out iteratively until a uniform light fringe covers the entire field-of-view.

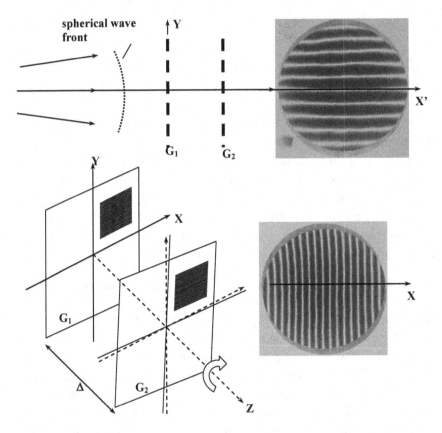

Figure 12. Optical alignment fringes when the principal direction of the gratings is along y-axis. (a) Fringes due to spherical wavefront, and (b) Fringes due to relative rotational misalignment between gratings.

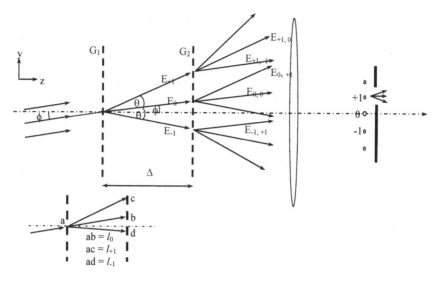

Figure 13. Schematic representation of diffraction and Fourier filtering of wave fronts in Coherent Gradient Sensing (CGS).

2.3 *Working principle*

Consider, an object wave making an angle ϕ with the optical axis in the y-z plane transmitted through a pair of Ronchi gratings G_1 and G_2 of pitch p with principal grating direction, say, along the y – axis (Fig. 13). The separation distance between the two gratings along the optical axis (the z-axis) be Δ. The diffracted light emerging from the first grating consists of a zero and several odd diffraction orders, each denoted by the corresponding complex amplitude distribution E. For simplicity only consider diffraction orders E_i ($i = 0, \pm1$) after the first grating G_1. These waves are propagating in discretely different directions according to the diffraction equation ($\theta = \sin^{-1}(\lambda/p) \approx (\lambda/p)$ for small angles). Each of these diffracted wave fronts diffract once more at the second grating G_2 plane and the corresponding wave fronts propagating in several discrete directions are denoted by $E_{(i, j)}$ ($i = 0, \pm1, j = 0, \pm1$) where the two subscripts correspond to diffraction order at the first and the second grating, respectively. The wave fronts $E_{\pm1, 0}$ and $E_{0, \pm1}$ contribute to the ±1 diffraction spot on the focal plane. And, diffracted wave fronts $E_{0, 0}$,

$E_{+1, -1}$ and $E_{-1, +1}$ contribute to the 0th order. By letting ± 1 diffraction order to pass through the filtering aperture, interference fringes resulting from the corresponding complex amplitudes can be evaluated. Let l_0 and $l_{\pm 1}$ denote the optical path lengths of E_0 and $E_{\pm 1}$, respectively, between the two gratings. Then, complex amplitudes E_0 and $E_{\pm 1}$ can be represented as $E_0 = A_0 \exp(ikl_0)$, $E_{\pm 1} = A_{\pm 1} \exp(ikl_{\pm 1})$, where A's denote amplitudes, k the wave number $(= 2\pi / \lambda)$, and $i = \sqrt{-1}$. Noting that no additional path difference occurs beyond the grating G_2, the intensity distribution on the image plane for a 1:1 magnification is proportional to,

$$I_{\pm 1} = (E_{0,\pm 1} + E_{\pm 1,0})(E_{0,\pm 1} + E_{\pm 1,0})^* = A_0^2 + A_{\pm 1}^2 + 2A_0 A_{\pm 1} \cos k(l_0 - l_{\pm 1}),$$

(2.1)

when ± 1 diffraction spot is let through the aperture at the spectrum plane. In the above, $(\bullet)^*$ denotes the complex conjugate. Based on geometrical considerations, the optical path lengths l_0 and $l_{\pm 1}$ can be expressed as,

$$l_0 = \frac{\Delta}{\cos\phi} = \frac{\Delta}{\left(1 - \dfrac{\phi^2}{2!} + \dfrac{\phi^4}{4!} - \cdots\right)}, \; l_{\pm 1} = \frac{\Delta}{\cos(\theta \pm \phi)} = \frac{\Delta}{\left(1 - \dfrac{(\theta \pm \phi)^2}{2!} + \dfrac{(\theta \pm \phi)^4}{4!} - \cdots\right)},$$

(2.2)

by expanding $\cos(\bullet)$ in the neighborhood of $(\bullet) = $ zero. For *small angles*, when the series is truncated by neglecting the terms of the order of $(\bullet)^3$ and beyond, Eq. (2.2) becomes,

$$l_0 \approx \frac{\Delta}{\left(1 - \dfrac{\phi^2}{2!}\right)} \approx \Delta\left(1 + \frac{\phi^2}{2}\right), \quad l_{\pm 1} \approx \frac{\Delta}{\left(1 - \dfrac{(\theta \pm \phi)^2}{2!}\right)} \approx \Delta\left(1 + \frac{(\theta \pm \phi)^2}{2}\right),$$

(2.3)

where binominal expansion

$$(1 \pm p)^q = \left[1 \pm \frac{q}{1!}p + \frac{q(q-1)}{2!}p^2 \pm \frac{q(q-1)(q-2)}{3!}p^3 + \cdots\right] \text{ for}$$

$(p^2 < 1, \; q \neq 0, 1, 2, \ldots)$ is used.

By calculating the path difference $(l_0 - l_{\pm1})$ from Eq. (2.3), the expression for intensity distribution Eq. (2.1) on the image plane when $\pm1^{st}$ diffraction order is allowed to pass through the filtering aperture is,

$$I_{\pm1} \approx A_0^2 + A_{\pm1}^2 + 2A_0 A_{\pm1} \cos\left[k\Delta\left(\frac{\theta^2}{2} \pm \theta\phi\right)\right]. \qquad (2.4)$$

The constructive interference occurs when the argument of the cosine term in the above equation is $2N_d\pi$ where $N_d = 0, \pm1, \pm2, \ldots$, is the fringe order. After substituting for the wave number we get, $\frac{\Delta}{\lambda}\theta\left(\frac{\theta}{2} \pm \phi\right) = N_d$. When the specimen is in the undeformed ($\phi = 0$) state, the fringe order of the uniform bright fringe of the undeformed object can be denoted by $N_u = \frac{\Delta}{p}\left(\frac{\theta}{2}\right) = $ constant ($N_u = 0, \pm1, \pm2, \ldots$), where $\theta \approx \lambda/p$ is utilized. By incorporating the expression for N_u, the magnitude of angular deflection of light at a point on the surface can be written as,

$$\phi = (N_d - N_u)\frac{p}{\Delta} = N\frac{p}{\Delta}, \qquad N = (N_d - N_u) = 0, \pm1, \pm2, \ldots$$

$$\qquad (2.5)$$

The fringe formation in the overlapping region can be schematically represented as shown in Fig. 14 where the solid line is used to represent the un-diffracted wave and the dotted line represents either +1 or -1 order wave front. Evidently, the sensitivity of the interferometer for measuring angular deflections of light rays depends on the ratio of the grating pitch p and the grating separation distance Δ. This offers experimental flexibility for controlling measurement sensitivity by suitably choosing p and Δ. Clearly this is an added experimental advantage over other methods used for dynamic fracture mechanics. Also, it should be noted that the interference fringes obtained by filtering all but +1 or −1 diffraction orders can be interpreted as forward and backward differences of the optical signal.

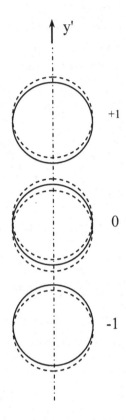

Figure 14. Overlapping but sheared wave fronts on the image plane (x'-y' plane) for filtering all but +1, 0, -1 diffraction spots. The solid and broken circles correspond to zero order and +1 or −1 diffraction orders passing though the aperture, respectively.

The angular deflections of light rays can be related to surface deformations of the object using a first order approximation. Let the propagation vector of the object wave front be $\underline{d} = \dfrac{\partial(\delta S)}{\partial x}\underline{e}_x +$ $\dfrac{\partial(\delta S)}{\partial y}\underline{e}_y + \underline{e}_z$ where δS denotes optical path difference and $\underline{e}_i$ is a unit normal in the i-th direction. For light rays in the y-z plane, $\dfrac{\partial(\delta S)}{\partial y} \approx \phi$ when small angles approximation is invoked.

When a transparent object is studied using transmission-mode CGS, the optical path S through the specimen can be expressed as $S = (n-1)B$ where n is the refractive index of the material and B is the thickness in the unstressed state. Upon deformation, the optical path length becomes $S + \delta S$ where $\delta S = C_\sigma B(\sigma_x + \sigma_y)$, $C_\sigma = \left(-\dfrac{v}{E} + C_o\right)(n-1)$ is the elasto-optic coefficient and C_o is a stress-optic constant of the material. It should be noted that C_σ accounts for stress induced refractive index changes as well as thickness changes due to Poisson effect. Thus, angular deflections of light rays in transmission mode CGS relate to the mechanical fields as,

$$\phi \approx \frac{\partial(\delta S)}{\partial y} = C_\sigma B \frac{\partial(\sigma_x + \sigma_y)}{\partial y} = N\frac{p}{\Delta}, \qquad N = 0, \pm 1, \pm 2, \ldots \quad (2.6)$$

When an optically opaque object is studied using reflection-mode CGS, the path difference is $\delta S = 2w$ where w is out-of-plane displacement along the z-direction. Then, angular deflections can be related to the mechanical fields (surface slopes) as,

$$\phi \approx \frac{\partial(\delta S)}{\partial y} = 2\left(\frac{\partial w}{\partial y}\right) = N\frac{p}{\Delta}, \qquad N = 0, \pm 1, \pm 2, \ldots, \quad (2.7)$$

where w is the out-of-plane deflection of the wafer in the z-direction.

2.4 Applications

The first demonstration of CGS for dynamic fracture studies was reported by Tippur, et al., [13] after the method was implemented to study deformations surrounding a growing crack in a nominally brittle solid. In this, transmission mode-CGS was used along with high-speed photography. A Plexiglas (PMMA) three-point bend specimen (dimensions 300 mm x 150 mm x 8.2 mm) having an edge notch (25 mm long) was subjected to low-velocity impact (impact velocity ~ 2 m/s) and dynamic crack growth was recorded using high-speed photography (framing rate - 100,000 fps and exposure time - 50 ns). A pair of Ronchi gratings was orientated along the initial crack direction such that the measured field quantity was proportional to $\dfrac{\partial(\sigma_x + \sigma_y)}{\partial x}$, x being the direction along the initial crack. Selected fringe patterns from that experiment are shown in Fig. 15. Highly discernable fringes in these early experiments clearly hinted at the potential of the method for dynamic fracture investigations. Also, by contrasting computed stress

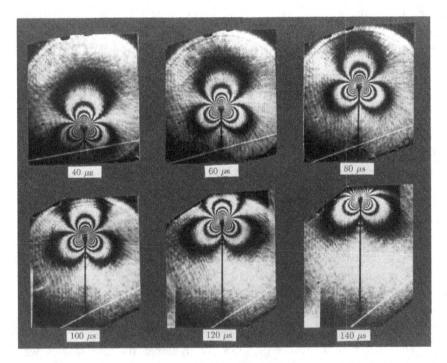

Figure 15. Selected transmission-mode CGS interferograms at different instants of dynamic growth in a PMMA sheet impact loaded in a drop-tower (impact velocity 2m/s) in 3-point bend configuration. Fringes are proportional to $\partial(\sigma_x + \sigma_y)/\partial x$ with x along the crack growth direction. (from Ref. 13)

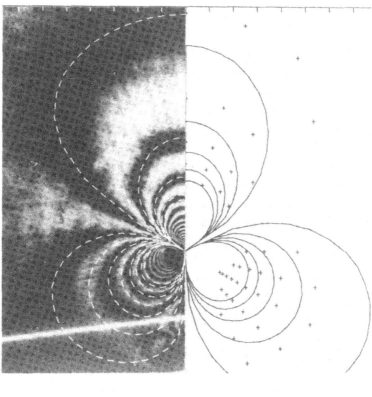

K_I^d-dominant analysis
Experimental data
Higher-order transient analysis

Figure 16. Synthetic fringe patterns reconstructed from on-term (solid lines) and higher-order (broken lines) transient analysis superposed on the corresponding transmission-mode CGS interferogram from a dynamic crack growth experiment on PMMA. (from Ref. 14).

intensity factors based on K-dominance assumption between quasi-static and dynamic experiments, they emphasized the significance of full-field transient effects neglected in earlier experiments using the method of optical caustics. An example of the same is shown in Fig. 16 where a fringe pattern from CGS technique is superposed with synthetic ones (dotted line in the left-half) obtained using over-deterministic analysis of optical data (shown as discrete data points seen in the right-half) using asymptotic expansion (six-terms in the expansion for $(\sigma_x + \sigma_y)$) for a steadily growing crack. The right-half of this figure shows a relatively poor correlation between the optical data and the synthetic field if K-dominance assumption were enforced. Shortly after this initial success of using transmission-mode CGS to study dynamic fracture problems, the method was extended to reflection-mode with an intention of studying opaque solids [14] using this method. The effectiveness of the method in the reflection-mode was demonstrated by capturing surface slope ($\partial w / \partial x$) fringes around a dynamically propagating crack in a AISI-4340 steel and a Plexiglas sample subjected to impact loading in three-point bend configuration. In case of the former, the sample surface had to be lapped and polished with diamond paste to obtain a flat mirror-like surface finish while in case of the latter the surface was simply deposited with a layer of aluminum using vacuum deposition. Another major contribution of CGS to dynamic fracture has been in the area of interfacial fracture mechanics. Until early 90's, there were no reports of optical investigations on dynamic crack growth along dissimilar material interfaces despite an overwhelming need for understanding failure behavior of layered materials and composite structures under mechanical shock and impact loading. In a ground-breaking optical investigation, Tippur and Rosakis [15] observed unusually high crack speeds during dynamic fracture of bimaterial specimens subjected to low-velocity impact. They studied a bimaterial beam made of equal thickness sheets of PMMA and aluminum bonded adhesively along a straight edge. A strong bond was created by roughening the aluminum half and using an adhesive based on methyl methacrylate monomer to minimize the possibility of introducing a distinctly different third material. The bimaterial was studied in a transmission-mode CGS apparatus using high-speed photography (framing rate 140,000 fps). The bimaterial specimen with a pre-cut edge notch along the interface was impact loaded on the aluminum half (velocity 2 m/s) close to the interface using a drop-tower in a three-point bend configuration. A record of angular deflections of light rays representing derivatives of stresses in the interfacial crack tip vicinity in the PMMA half obtained by them is shown in Fig. 17. In this sequence of interferograms, time instants shown correspond to times before (negative) and after (positive) crack initiation. The dynamic effects are clearly visible as sharp kinks in otherwise smooth fringes. Also, continuous rotation of fringe lobes before crack initiation suggests changing ratio of shear to normal stress at the crack tip. Initially the crack tip is experiencing dominant shear deformations due to higher wave speeds in the aluminum half relative to

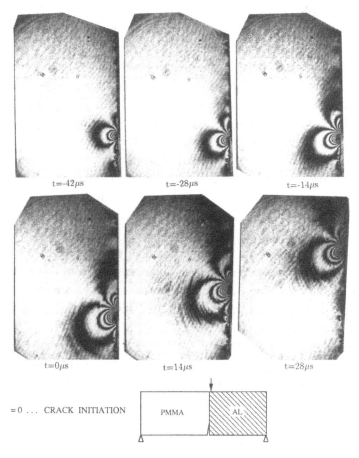

Figure 17. Transmission-mode CGS interferograms representing fringes proportional to to $\partial(\sigma_x + \sigma_y)/\partial x$ in the vicinity of a dynamically propagating interfacial crack in a PMMA/aluminum bimaterial. Fringes are visible only in the PMMA-half of the bimaterial. (from Ref. 15)

the PMMA half. With the passage of time stress waves returning from the far boundaries increase the tensile stress component at the crack tip up to crack initiation. In the post-initiation period, the crack tip fringes seem relatively self-similar. Interferograms were used to record crack growth and hence the crack speed history. A surprisingly high crack speed of about 80% of the Rayleigh wave speed (relative to the compliant constituent of the bimaterial) was recorded. At that juncture, due to the intrinsic difficulties associated with the dynamic problem, there were only a few theoretical studies on this topic [16-19] and explicit crack tip field equations to extract fracture parameters such as stress intensity factors, mode-mixity and energy release rate were absent. Additionally, there were disagreements between investigators on issues related to achievable/terminal interfacial crack speeds. The results of Willis [18] disagreed with that of Atkinson [19] regarding the terminal velocity in such a situation. The former claimed that the terminal velocity of an interfacial crack to be a little greater than the lower of the two Rayleigh wave speeds of the constituents while the latter suggested it to be equal to the lower of the two Rayleigh wave speeds.

Tippur and Rosakis' [15] experiments inspired several follow-up theoretical, numerical and experimental works on interfacial fracture of bimaterials. Yang, et al [20] reported field equations for a steadily propagating interfacial cracks which was later on enriched by a transient analysis of a dynamically propagating interfacial crack by Liu et al, [21]. An important conclusion of the new theoretical analyses was that as the crack tip approached the lower of the two Rayleigh wave speeds of a bimaterial a finite amount of energy was required for the crack tip to maintain propagation at Rayleigh wave speed with a non-zero complex stress intensity factor. This observation was unlike the ones for homogeneous counterparts requiring infinite amount of energy to maintain propagation at the Rayleigh wave speed [22, 23]. Thereby in a bimaterial case the restriction for the crack to extend at speeds in excess of the Rayleigh wave speed was absent. An experimental proof for this was reported by Lambros and Rosakis [24] who conducted optical investigations on PMMA-steel bimaterials using transmission-mode CGS apparatus. In their experiments a gas-gun was used instead of a drop-tower for impacting specimens. The interferograms were used to track crack length histories as well as map crack tip deformations. Their measurements revealed that interfacial crack speed exceeded not only the Rayleigh wave speed for PMMA but its shear wave speed as well. In a set of experiments when a blunt interfacial notch was used, they reported crack speeds approaching the longitudinal wave speed of PMMA demonstrating the possibility of interfacial crack propagation at intersonic (between shear and longitudinal wave speeds) speeds. They were also able to provide visual evidence of crack propagation in the intersonic regime relative to PMMA. By comparing CGS fringe pattern at early stages of propagation with the ones at intersonic speeds, they noted a squeezed appearance of main fringe lobes as well as 'shock-like' formations in interferograms. Crack propagation in the intersonic regime also pointed to special features about the crack tip. In the subsonic range the interfacial crack showed evidence of fringe lobes coming to focus at a single point along the interface which is the crack tip. When the crack attained intersonic speeds, however, the rear and front lobes showed interception of the interface over a finite length without evidence of a common focus suggesting large scale contact. This can be seen readily in a representative transmission-mode CGS interferogram (in the PMMA half) in Fig. 18 where the length between the front and rear lobes is marked by the letter 'd'. (Even though crack propagation occurred at speeds in excess of shear wave speeds, they could not visualize the shear stress discontinuity since transmission-mode CGS measures a quantity proportional to $(\sigma_x + \sigma_y)$. This necessitated the use of photoelasticity to accomplish the same by Rosakis and coworkers subsequently.) The availability of explicit crack tip fields also enabled estimation of relevant complex stress intensity factor histories (and hence energy release rate and mode-mixity) from interferograms using over-deterministic least-squares analysis of optical data. The energy release rate (G) history showed a highly unstable crack initiation event resulting in a steep drop in G values during the time period when crack accelerated to intersonic speeds after initiation. During the same time period, the mode-mixity continued to increase rapidly. They have noted that during interfacial crack growth crack tip opening angle remains constant.

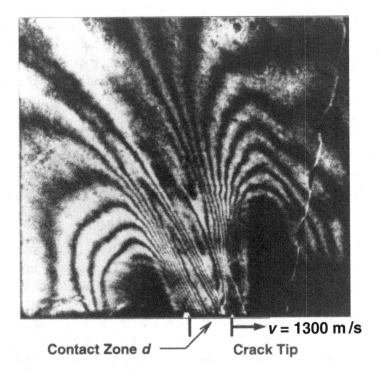

Figure 18. Enlarged view of transmission-mode CGS fringes in the intersonic crack growth regime showing occurrence of crack face contact. Here the crack is growing from left to right along a PMMA/steel interface. Fringes are visible only in the PMMA (upper) half of the bimaterial. (from Ref. 6, (Courtesy R. P. Singh and A. Shukla)

In the past decade there has been a growing interest in a family of new materials called Functionally Graded Materials (FGM). These biologically inspired materials are macroscopically isotropic and nonhomogeneous with continuously varying material properties along one or more spatial coordinates. FGM are proposed for mitigating deleterious effects of discrete interfaces encountered in conventional layered materials and engineering structures which experience extreme loading environments of thermal shock and/or mechanical impact. Material property variation is often achieved using multiphase materials in which volume fractions of the constituents are spatially tailored to suit an application. For example, in a metal matrix composite FGM [25], the variation could be from (100% metal, 0% ceramic) at one end to (0% metal, 100% ceramic) at the other in a continuous manner. Mechanical characterization in general and fracture behavior in particular is at the heart of understanding mechanical integrity of this relatively new and novel class of materials. Optical investigation of dynamic fracture behavior of FGM using the method of CGS has been part of this broader endeavor. For example, Tippur and coworkers [26-28] have examined the influence of continuous compositional gradients on dynamic fracture response of functionally graded composites. They have studied glass-filled epoxy particulate composite sheets having varying volume fraction of filler material. The FGM is made from solid A-glass microspheres (35 and 42 μm average diameter) or hollow glass microballoons (60 μm average diameter and ~0.5 μm wall thickness) as filler materials in epoxy matrix. The former is unidirectionally nonhomogeneous with a monotonic volume fraction variation of solid glass spheres from approximately 0% to 40% over a length of approximately 45 mm. In the latter, the same volume fraction variation is achieved using hollow microballoons resulting in a light-weight 'syntactic foam' material with a closed-cell microstructure. While solid glass spheres act as microscopic reinforcements increasing the stiffness of the mixture, hollow microballoons reduce the same. The resulting FGM is a microscopically heterogeneous material and is optically opaque. Hence it requires specimen preparation (as detailed earlier) to create an optically specular and flat surface for interrogation by a reflection-mode CGS apparatus. In Fig. 19, selected interferograms (framing rate 200,000 fps and 40 ns exposure time) from Kirugulige, et al., [28] are shown for crack growth occurring from an initial notch cut into the compliant edge of a syntactic foam FGM sheet towards the stiffer edge where impact occurs. In each of these interferograms, deformations associated with the impact point and the crack tip can be seen as two sets of fringe concentrations near the top and the bottom edges of the sheet, respectively. The three images are for different time instants with Fig. 19(a) just before crack initiation, Fig. 19(b) just after crack initiation and Fig. 19(c) after substantial dynamic crack growth. In Figs. 19(b) and (c), surface waves emanating from the crack tip can be seen as sharp kinks in otherwise smooth fringe patterns.

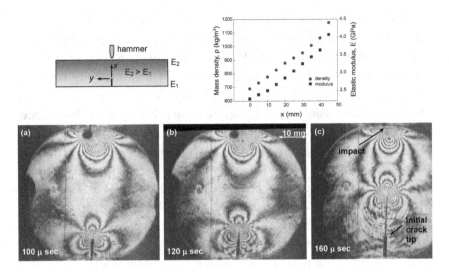

Figure 19. Selected reflection-mode CGS interferograms representing surface slopes (*dw/dx*) in crack growth direction in a functionally graded syntactic foam sheet subjected to symmetric impact relative to the crack plane. In this sheet volume fraction of hollow microballoons vary from ~40% at the bottom of the sheet to ~0% at the top. Time corresponds to instants after impact.

Similar experiments have also been reported when the direction of crack growth is reversed by situating an initial crack on the stiffer edge of the FGM and impacted on the compliant side. The stress intensity factor histories for the two cases of graded foam sheets were extracted using over-deterministic least-squares analysis and are shown in Fig. 20. (The inset in Fig. 20 is used to illustrate the least-squares analysis based fit obtained for the digitized optical data for the case of the graded foam with crack on compliant side. The experimental data (symbols) and the least-squares fit (solid line) shown in the figure are for 100 μs time instant after impact. The least-squares fit considering *K*-dominant solution shows good agreement with optical data.) In each case, crack tip stress intensification is evident only after ~50 μs after impact due to finite resolution of the optical technique. The stress intensity factors monotonically increase until crack initiation (indicated by vertical bands in the plot). The crack initiates earlier when situated on the compliant side of the sheet ($E_2 > E_1$) than on the stiffer side ($E_2 < E_1$). At crack

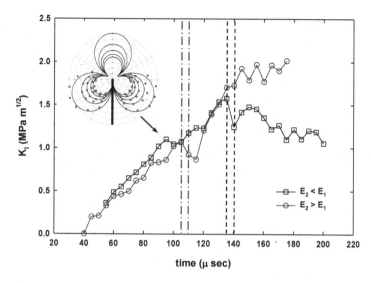

Figure 20. Mode-I dynamic stress intensity factor histories (Impact velocity=5.8 m/s) both for the crack on the compliant side ($E_2 > E_1$) and the crack on the stiffer side ($E_2 < E_1$) of functionally graded foam sheet. The crack initiation is indicated by the vertical bands, (dash-dash: $E_2 < E_1$, dash-dot: $E_2 > E_1$). Inset shows an example of the quality of least-squares fit of the digitized data for $E_2 > E_1$ at 100 μs. (Ref. 28)

initiation, a noticeable dip in the stress intensity factor history is seen suggesting a sudden release of stored energy from the initial notch. Subsequently, stress intensity factors for a growing crack increase for the case of a crack on the compliant side of the sheet while a noticeable reduction is seen when the crack is on the stiffer side. These can be attributed to the fact that fracture toughness of the foam decreases as volume fraction of microballoons increase in the foam.

Recently, Kirugulige and Tippur [29] have extended their study from dynamic mode-I fracture of FGM to mixed-mode crack growth under impact loading conditions using reflection-mode CGS and high-speed photography (framing rate of 200,000 fps and 40 ns exposure time). In this investigation they have used functionally graded particulate composite sheets made of solid glass microspheres (35 μm mean diameter) for matrix reinforcement. Samples with monotonic, unidirectional variation of volume fraction of filler with a crack on the compliant or the stiffer side are studied separately by one-point impact loading occurring eccentrically relative to the crack plane as shown in Fig. 21(a). The variations in the Young's modulus and mass density over the width of the sample are also shown (Fig. 21(b)). Representative optical results of this investigation are shown in Fig. 22. The images on the left and the right are for two separate situations of initial crack located close to the edge of the specimen with a lower and a higher volume fraction of reinforcement, respectively, compared to the edge where impact occurs. The crack paths in each of these images at time

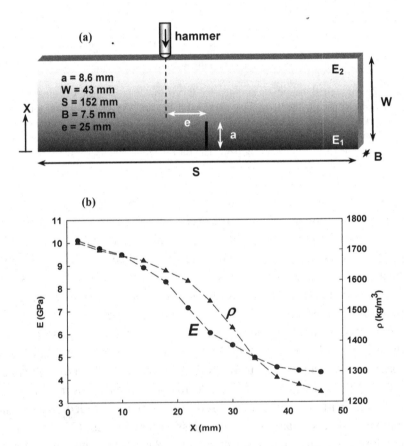

Figure 21. (a) Schematic of syntactic foam FGM specimens studied under one-point asymmetric impact. (b) Spatial variation of material properties in the FGM. (Ref. 29)

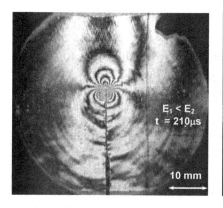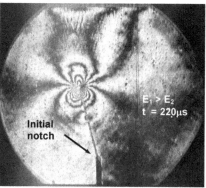

Figure 22. Fringe patterns representing surface slopes dw/dx (w being out-of-plane displacement) near dynamically propagating cracks in FGM subjected to impact loading as shown in Fig. 1. Distinctly different crack paths are obtained depending upon the functional gradient ($E_1/E_2 \sim 0.4$ in the left image and ~ 2.5 in the right image). (Ref. 29)

instants after significant crack growth from the initial crack tip are strikingly different. The crack path in the former ($E_1 < E_2$) is nearly self-similar even though the crack tip loading is a combined mode-I and –II type while in the latter ($E_1 > E_2$) the crack kinks from the initial tip and grows at a nearly constant angle relative to the pre-notch. Further, crack tip fringes in the FGM on the left show different asymmetries relative to the local crack orientation. The loading configuration and the sample geometries of the specimens studied being identical in these two cases, the differences in crack paths are attributed to compositional gradients (a combination of elastic and material strength gradients) present in the two FGM. A homogeneous sample of identical geometry and loading configuration is also studied for comparative purpose and the results are not shown here for brevity. Using over-deterministic least-squares analysis of interferograms corresponding to different time instants after impact, a history of stress intensity factors $K_I(t)$ and $K_{II}(t)$ were evaluated and hence mode-mixity angle histories $\psi(t) = \tan^{-1} \dfrac{K_{II}}{K_I}$ were estimated. Just prior to crack initiation $\psi(t)$ values approach zero suggesting initiation under predominantly mode-I conditions. In the post-initiation period mode-mixity histories show a small positive value of mixity in case of FGM with ($E_1 < E_2$) while it is a small negative value in case of FGM with ($E_1 > E_2$). Interestingly, the mixity history in the post initiation period for the homogeneous sample remains close to $\psi(t) = 0$ in the window of observation, consistent with dynamic crack growth in homogeneous brittle solids. These preliminary results raise the possibility of non-zero K_{II} during mixed-mode dynamic crack growth in brittle FGM if confirmed by further experimentation.

Reflection-mode CGS has also been used in dynamic fracture behavior studies on fiber reinforced composites. At the moment there are only a few reports of real-time measurement of crack tip fields in these materials. Lambros and Rosakis, [30] are the first ones to extend CGS technique to study fast-fracture in thick unidirectional graphite-epoxy composite plates (48 plies/6mm thick, fiber diameter 7.3 μm, fiber strength 231 GPa, fiber volume fraction 0.65) subjected to symmetric one-point impact loading. They observed that for low-impact velocities (4 m/s) a mode-I crack attained speeds of nearly 0.5 C_R where C_R is the Rayleigh wave speed traveling in the direction of fibers of a half-space. The consequence of material anisotropy is revealed optically as an elliptically shaped stress wave emanating from the initial notch (root radius 0.75 mm) upon crack initiation. They have analyzed fringes at time instants close to initiation to estimate crack initiation toughness ($K_{Ii} \sim 2.3$ MPa√m) of the composite. In a more recent study, Coker and Rosakis [31] have examined crack growth behaviors in unidirectional graphite-epoxy composites (mass density 1478 kg/m^3) also using CGS and high-speed photography but with an emphasis on shear dominated fracture behavior at higher impact velocities. The dynamic material characteristics for the coordinate shown in Fig. 23 are tabulated in the accompanying table. They have used an off-axis impact loading configuration (impact velocity 21 m/s) to generate dominant shear loading at the initial notch tip. Selected interferograms from their experiment is shown in Fig. 24. In this figure, time $t = 0$ corresponds to crack initiation. The nature of the crack tip fields suggests that the crack tip is primarily in mode-II condition prior to initiation with dominant side lobes tilted ahead of the crack tip. Soon after initiation,

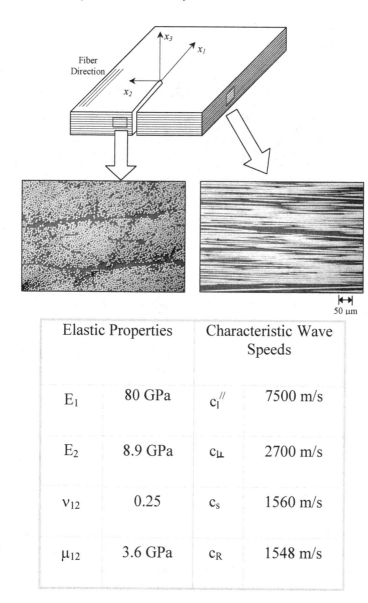

Elastic Properties		Characteristic Wave Speeds	
E_1	80 GPa	$c_l^{//}$	7500 m/s
E_2	8.9 GPa	c_L	2700 m/s
ν_{12}	0.25	c_s	1560 m/s
μ_{12}	3.6 GPa	c_R	1548 m/s

Figure 23. Cross-sectional view of fiber-reinforced unidirectional graphite-epoxy composite and definition of the coordinatesystem with respect to the material symmetry axes. The table shows plane strain elastic characteristics of the composite. (Ref. 31, Courtesy A. J. Rosakis).

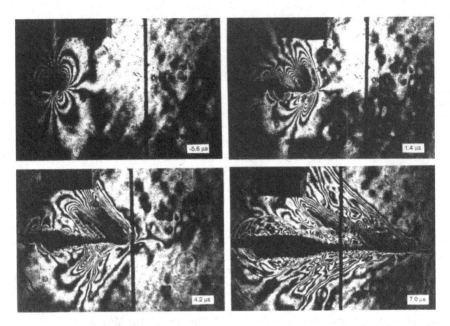

Figure 24. A sequence of reflection CGS interferograms for shear dominated crack tip loading in a unidirectional fiber-reinforced composite. The crack growth is along the fiber direction. The time $t = 0$ corresponds to crack initiation. The crack is propagating intersonically in the last two frames. (Courtesy - A. J. Rosakis)

however, the fringe lobes on the two sides of the crack tilt backwards, indicating high crack speeds and acceleration. During this early period soon after crack initiation, authors have estimated the crack speed to be ~2100 m/s, well above the shear wave speed for the material of 1560 m/s (see, Fig. 23), and an acceleration of 10^9 m/s^2. As the crack continues to propagate, the shape of the fringe lobes change even more drastically to a nearly triangular wedge shaped region with a line of highly concentrated fringes emanating from the crack tip at a distinct angle. This line represents a region of high stress gradients which eventually forms a discontinuity in stresses resulting in a shear shock wave. Based on their mode-I and mode-II crack growth observations, Rosakis and coworkers [30, 31] conclude that mode-I or opening mode cracks can propagate only subsonically along the fibers with an upper bound as the Rayleigh wave speed. However, mode-II or shear cracks can grow at intersonic speeds parallel to the fibers. They have observed crack acceleration from values close to shear wave speed to dilatational wave speed corresponding to the fiber direction.

3. Moiré interferometry

Ever since the development of a methodology for affixing high-frequency moiré gratings by a replication process by Post and his colleagues [32, 33], the technique of moiré interferometry has found widespread utilization in a variety of quasi-static elastic [34] and elasto-plastic [35] fracture investigations. Yet, its utility in the area of dynamic fracture mechanics has remained relatively sparse. Nevertheless the potential of the method to measure displacement components on object surfaces at extremely high optical resolutions cannot be overlooked. This feature is particularly relevant as dynamic fracture research proceeds towards microscales placing a severe demand on resolution. An exception in this regard has been the works of Kobayashi and his co-workers [36] who have extended the method to dynamic fracture mechanics applications in recent years.

3.1 *Experimental method*

The first step in moiré interferometry is the creation of a high-density grating pattern capable of following surface deformations. This is done by producing a high-frequency master grating (grating density up to 4000 lines/mm has been achieved) on a photosensitive glass plate. A standing wave on the emulsion side of a photographic plate is created by two uniform intersecting coherent laser beams. Processing of the photographic plate produces high-density undulations in the emulsion. The processed photographic plate is next deposited (using vacuum deposition, sputtering, etc.) with a metallic (typically aluminum or gold) film. This undulating grating pattern on the photographic plate is subsequently transferred to the specimen surface using a process [33] similar to the one depicted in Fig. 11. Generally, the thickness of the adhesive layer used in the grating transferring process ranges between 15-25 μm. If this is too thick a layer to faithfully follow deformations in the underlying substrate, Ifju and Post [37] recommend using the so-called 'zero-thickness' gratings. This is made of highly adherent reflective film deposited directly on the specimen surface and coated with photoresist. The photoresist is then exposed to intersecting laser beams and chemically etched to create a grating pattern on the substrate/specimen directly.

A typical experimental set-up for dynamic fracture studies is shown in Fig. 25. Due to short exposure times needed for freezing transient moiré fringe patterns, a high-power laser (an argon-ion laser) is commonly used as the light source. It also provides the necessary coherent illumination. The laser beam before expanding is modulated using an acousto-optical device (either internal or external to the laser cavity) to accomplish shuttering when used in conjunction with a continuous access high-speed camera such as a rotating mirror camera. The light beam is collimated using a collimating lens and subsequently split into two symmetric coherent beams using a front coated optical mirror before illuminating the cracked specimen surface. The optical arrangement is such that the specimen, printed with a high-frequency grating of pitch p, is illuminated by the two beams at an angle $\pm\alpha$ and the equation $\dfrac{1}{p} = \dfrac{2}{\lambda \sin \alpha}$ is obeyed. The diffracted light beams bouncing off of the specimen surface create an interference pattern representative of the surface displacements in the direction of the principal grating direction.

In situations where complete determination of surface strain field is needed, two orthogonal displacement fields are to be recorded simultaneously during a transient event. Kokaly, et al., [34] describe an optical set-up where they use two pairs of collimated laser beams in two mutually perpendicular planes. Resulting dual pairs of interfering object waves carry surface displacement information but are polarized in orthogonal planes when the laser beam used in the experiments is initially plane polarized. They recommend using a dichoric cube beam splitter to separate the two displacement fields before recording using a single high-speed camera or using two different ones if available.

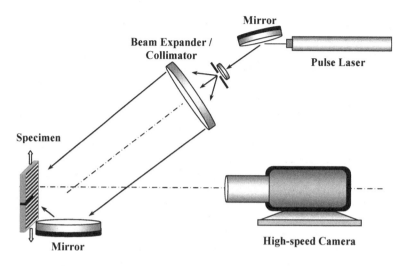

Figure 25. Schematic of moiré interferometry set-up used for dynamic fracture investigations.

3.2 *Working principle*

For simplicity of analysis, let the transmission function of gratings affixed to the specimen be a simple sinusoidal function of the form,

$$t(y) = A + 2B\cos\frac{2\pi y}{p} \qquad (3.1)$$

The above represents linear gratings of pitch p with the y-direction as the principal direction and A and B are constants. Now, consider a plane wave illuminating the gratings at an angle α to the z-axis as shown in Fig. 26. The complex amplitude distribution E_i on the specimen plane can be represented in this case as $E_i\big|_{z=0} = C\exp(iky\sin\alpha)$ where C is the amplitude and k is the wave number of light. The resulting complex amplitudes corresponding to the diffracted light beams $E_{r,0}$, $E_{r,+1}$ and $E_{r,-1}$ can be obtained (in case of an opaque specimen) as,

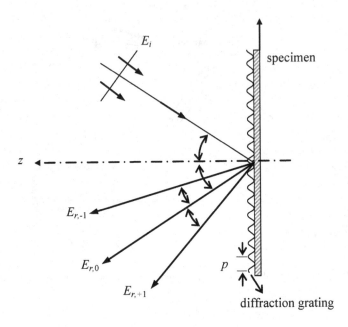

Figure 26. Diffraction of light waves when a linear sinusoidal grating on a specimen surface is illuminated by a plane wave.

$$E_r\big|_{z=0} = C\exp\left(iky\sin\alpha\right)\left[A + B\exp\left(i\frac{2\pi y}{p}\right) + B\exp\left(-i\frac{2\pi y}{p}\right)\right]$$

$$= AC\exp\left(i\frac{2\pi}{\lambda}y\sin\alpha\right) + BC\exp\left[i2\pi y\left(\frac{\sin\alpha}{\lambda} + \frac{1}{p}\right)\right]$$

$$+ BC\exp\left[i2\pi y\left(\frac{\sin\alpha}{\lambda} - \frac{1}{p}\right)\right] = E_{r,0} + E_{r,+1} + E_{r,-1} \quad (3.2)$$

In the above, subscripts 0, ±1 are used to represent various diffraction orders and diffraction angle θ is related to the wavelength of light and the grating pitch as $\dfrac{1}{p} = \dfrac{\sin\theta}{\lambda}$. Then,

$$E_r\big|_{z=0} = AC \exp\left(i\frac{2\pi y}{\lambda}\sin\alpha\right) + BC \exp\left[i\frac{2\pi y}{\lambda}(\sin\alpha + \sin\theta)\right]$$

$$+ BC \exp\left[i\frac{2\pi y}{\lambda}(\sin\alpha - \sin\theta)\right] \tag{3.3}$$

Each term in the above represents a plane wave propagating in a distinctly different direction. If the illumination angle is adjusted such that = , then

$$E_r\big|_{z=0} = AC \exp\left(i\frac{2\pi y}{\lambda}\sin\theta\right) + BC \exp\left[i\frac{2\pi y}{\lambda}(2\sin\theta)\right] + BC \tag{3.4}$$

where

$$AC \exp\left(i\frac{2\pi y}{\lambda}\sin\theta\right) = E_{r,0}, \ BC \exp\left[i\frac{2\pi y}{\lambda}(2\sin\theta)\right] = E_{r,+1}, \ BC = E_{r,-1}$$

In the above $E_{r,-1}$ is a plane wave propagating in the z-direction. When the object undergoes deformation, the grating pitch p changes locally to say $p' = p$ p, p $p(x, y)$ and $\dfrac{1}{p'} = \dfrac{\sin\theta'}{\lambda}$ where $\theta' \equiv \theta'(x, y)$.

Then

$$E_r'\big|_{z=0} = AC \exp\left(i\frac{2\pi y}{\lambda}\sin\alpha\right) + B'C \exp\left[i\frac{2\pi y}{\lambda}(\sin\alpha + \sin\theta')\right]$$

$$+ B'C \exp\left[i\frac{2\pi y}{\lambda}(\sin\alpha - \sin\theta')\right] \tag{3.5}$$

where $\ AC \exp\left(i\frac{2\pi y}{\lambda}\sin\alpha\right) = E_{r,0}'$, $\ B'C \exp\left[i\frac{2\pi y}{\lambda}(\sin\alpha + \sin\theta')\right] = E_{r,+1}'$,

and $\ B'C \exp\left[i\frac{2\pi y}{\lambda}(\sin\alpha - \sin\theta')\right] = E_{r,-1}'$ and the prime notation is used to denote the parameters after deformation.

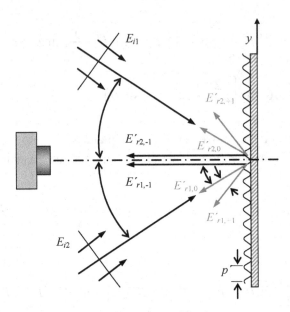

Figure 27. Schematic of diffracted waves due to symmetric dual illumination of a specimen grating by two plane waves.

Now consider a situation when the deformed object is illuminated by two coherent collimated light beams at angles + and − as shown in Fig. 27. The two incident plane waves are,

$$E_{i1} = C \exp(iky \sin\alpha)\big|_{z=0} \,, \; E_{i2} = C \exp(-iky\sin\alpha)\big|_{z=0} \quad (3.6)$$

and the diffracted light beams for each incident beam can be written as,

$$E'_{r2} = AC\exp(-iky\sin\alpha) + B'C\exp(-iky(\sin\alpha+\sin\theta'))$$
$$+ B'C\exp(-iky(\sin\alpha-\sin\theta')) = E'_{r2,0} + E'_{r2,+1} + E'_{r2,-1} \quad (3.7)$$

and

$$E'_{r1} = AC\exp(iky\sin\alpha) + B'C(iky(\sin\alpha+\sin\theta'))$$
$$+ B'C(iky(\sin\alpha-\sin\theta')) = E'_{r1,0} + E'_{r1,+1} + E'_{r1,-1} \quad (3.8)$$

Then $E'_{r1,-1}$ and $E'_{r2,-1}$ are propagating towards the imaging device. The intensity distribution recorded will be proportional to

$$I = \left(E'_{r1,-1} + E'_{r2,-1} \right)\left(E'_{r1,-1} + E'_{r2,-1} \right)^{*}$$

$$= 2\left(B'C\right)^{2}\left[1 + \cos\left(2ky\left(\sin\alpha - \sin\theta'\right)\right)\right]$$

$$= 2\left(B'C\right)^{2}\left[1 + \cos 2\pi\, y\left(2\left(\frac{1}{p} - \frac{1}{p'}\right)\right)\right] \qquad (3.9)$$

where $(\bullet)^{*}$ represents the complex conjugate. Recognizing that   and   are equal, the argument of the cosine term represents a low frequency intensity variation in the form of interference fringes. Note that the argument of the cosine term contains the factor '2' resulting in doubling of the fringe sensitivity compared to traditional moiré methods. A constructive interference would occur when,

$$2\pi\, y\left[2\left(\frac{1}{p} - \frac{1}{p'}\right)\right] = 2n\pi \quad n = 0, \pm 1, \pm 2,... \qquad (3.10)$$

The similarity of the above expression with the ones for geometric moiré fringe formation [38] when two linear gratings of slightly different pitch p and p' are superimposed on each other is obvious. The only difference between the two being the factor '2' in moiré interferometry, the governing equation relating optical interference with surface deformations v is,

$$v = N\frac{p}{2}, \quad (N = 0, \pm 1, \pm 2,...) \qquad (3.11)$$

3.3 *Applications*

Among the few instances when moiré interferometry has been used to study dynamic fracture behavior of materials, the works of Kobayashi and his co-workers [36] are notable. They have used moiré interferometry in conjunction with high-speed photography to measure transient displacement fields around a running crack in 7075-T6 aluminum alloy. They have employed a single edge notch (SEN) specimen geometry with either a fatigue crack or a blunt notch to produce low or high crack velocities, respectively. They have utilized a relatively low-spatial frequency of 40 lines/mm on a mirror finished specimen surface due to the presence of large-scale plastic yielding of the material in the crack tip vicinity. A IMACON-790 high-speed camera has been used to record four to eight moiré fringe patterns corresponding to either vertical or horizontal displacements at framing rates of 10,000 or 100,000 frames per second for fatigue cracked or blunt notched specimens, respectively. Multiple recordings of identically loaded SEN specimens at different delay timings have been used to record the entire fracture event lasting approximately 1.2 ms. They have been able to successfully record either the horizontal (u-field) or vertical (v-field) displacement field only at a given time instant due the lack of two high-speed cameras and experimental difficulties associated with image shifting method. A composite of u- and v- displacement fields in

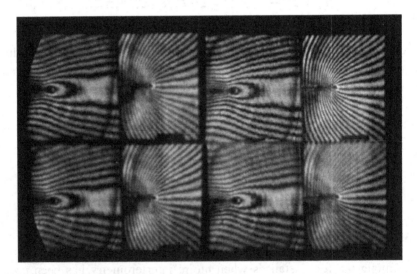

Figure 28. Dynamic moiré interferometry fringes at the onset of crack propagation in 7075-T6 aluminum alloy. A composite image of the two displacement fields was created from different experiments.

the crack tip vicinity at the onset of crack propagation is shown in Fig. 28. In their experiments crack velocities have varied between 35 and 300 m/sec depending upon a fatigued pre-crack or a blunt notch were used, respectively. Based on the measurements, they note that crack tip opening angle (CTOA) depends on crack velocity. A slowly propagating crack propagates with a CTOA of around 4° while the one propagating at the higher velocity has a CTOA of approximately 7.5°. Interestingly, a blunt notch transitioning from higher velocity to the lower velocity shows transitioning of CTOA values between the two levels.

References

1. Wells, A. A. and Post, D. 'The dynamic stress distribution surrounding a running crack – A photoelastic analysis', Proceedings of SESA, 16(1), pp 69-92, 1958.

2. Dally, J. W. 'Dynamic photoelastic studies of fracture', Experimental Mechanics, xx (3), pp 349-361, 1979.

3. Kobayashi, T. and Dally, J. W., 'The relation between crack velocity and stress intensity factor in birefringent polymers', ASTM-STP 627, pp 257-273, 1977.

4. Irwin, G. R., Dally, J. W., Kobayashi, T., Fourney, W. L., Etheridge, M. J., and Rossmanith, H. P., 'On the determination of a-K relationship for birefringent polymers', Experimental Mechanics, 19(4), pp 121-128, 1979.

5. Ramulu, M., and Kobayashi, A. S., 'Dynamic crack curving – A photoelastic investigation', Experimental Mechanics, xx(1), pp 1-9, 1983,

6. Singh, R. P., and Shukla, A., 'Subsonic and intersonic crack growth along a bimaterial interface', Journal of Applied Mechanics, 63, pp 919-924, 1996.

7. Dally, J. W., and Riley, W. F., Experimental Stress Analysis, 3rd edition, McGraw Hill, 1991.

8. Shukla, A., Practical Fracture Mechanics in Design, 2nd edition, Mercel and Dekker, 2005.

9. Chona, R., 'Extraction of fracture mechanics parameters from steady state crack tip fields', Optics and Lasers in Engineering, 19, pp 171-199, 1993.

10. Liu, C., Lambros, J., and Rosakis, A. J., 'Shear dominated transonic interfacial crack growth in a bimaterial – II: Asymptotic fields and favorable velocity regimes, Journal of the Mechanics and Physics of Solids, 43, pp 189-206, 1995.

11. Coker, D., Lykotrafitis, G., Needleman, A., Rosakis, A. J., 'Fractional sliding modes along an interface between identical elastic plates subject to shear impact loading', Journal of the Mechanics and Physics of Solids, 53, pp 884-922, 2005.

12. Tippur, H. V., Krishnaswamy, S., and Rosakis, A. J., 'A coherent gradient sensor for crack tip measurements: analysis and experimental results', International Journal of Fracture, 48, pp 193-204, 1991.

13. Tippur, H. V., Krishnaswamy, S., and Rosakis, A. J., 'Optical mapping of crack tip deformations using the methods of transmission and reflection coherent gradient sensing: A study of crack tip K-dominance', International Journal of Fracture, 52, pp 91-117, 1991.

14. Krishnaswamy, S., Tippur, H. V., and Rosakis, A. J., 'Measurement of transient crack tip deformation fields using the method of coherent gradient sensing', Journal of the Mechanics and Physics of Solids, 40(2), pp 339-372, 1992.

15. Tippur, H. V., and Rosakis, A. J., 'Quasi-static and dynamic crack growth along bimaterial interfaces: A note on crack tip field measurements using coherent gradient sensing', Experimental Mechanics, 31, pp 243-251, 1991.

16. Goldshtein, R. V., 'On surfaces in joined elastic materials and their relation to crack propagation along the junction', Applied Mathematics and Mechanics, 31, pp 496-502, 1967.

17. Brock, L. M., and Achenbach, J. D., 'Extension of an interface flaw under the influence of transient waves', International Journal of Solids and Structures, 9, pp 53-67, 1973.

18. Willis, J. R., 'Self similar problems in elastodynamics', Philosophical Transaction of the Royal Society of London', 274, pp 435-491, 1973.

19. Atkinson, C., 'Dynamic problems in dissimilar media', in Mechanics of Fracture – 4: Elasto-Dynamic Crack Problems', (Sih, G. C., Editor), pp 213-248, Noorhoff, Leyden, 1977.

20. Yang, W., Suo, Z. and Shih, C. F., 'Mechanics of elastodynamic debonding', Proceedings of the Royal Society of London, A433, pp 679-697, 1991.

21. Liu, C., Lambros, J. and Rosakis, A. J., 'Highly transient elastodynamic crack growth in a bimaterial interface: Higher order asymptotic analysis and experiments', Journal of the Mechanics and Physics of Solids, 41, pp 1887-1954, 1993.

22. Broberg, K. B., 'The propagation of a Griffith crack', Ark. Fys. 18, pp 159-192, 1960.

23. Freund, L. B., 'Dynamic Fracture Mechanics', Cambridge University Press, Cambridge, 1990.

24. Lambros, J., 'Dynamic decohesion of bimaterials: Experimental observations and failure criteria', International Journal of Solids and Structures', 32(4), pp 2677-2702, 1995.

25. Koizumi, M., 'Overview of FGM research in Japan', MRS Bulletin, 20, pp 19-21, 1995.

26. Marur, P. R., and Tippur, H. V., 'Dynamic response of bimaterial and graded interface cracks under impact loading', International Journal of Fracture, 103, pp 95-109, 2003.

27. Rousseau, C.-E., and Tippur, H. V., 'Dynamic fracture of functionally graded materials with cracks along the elastic gradient: Experiments and analysis', Mechanics of Materials, 33, pp 403-421, 2001.

28. Kirugulige, M. S., and Kitey, R. A., and Tippur, H. V., 'Dynamic fracture behavior of model sandwich structures with functionally graded core: A feasibility study', Composites Science and Technology', 65, pp 1052-1068, 2005.

29. Kirugulige, M. S., and Tippur, H. V., 'Dynamic mixed-mode crack growth in functionally graded glass-filled epoxy', Experimental Mechanics, to appear, 2006.

30. Lambros, J. and Rosakis, A. J., 'Dynamic crack initiation and growth in thick unidirectional graphite/epoxy plates', Composites Science & Technology, 57, 55-65, 1997.

31. Coker D. and Rosakis, A. J., 'Experimental observation of intersonic crack growth in asymmetrically loaded unidirectional composite plates', Philosophical Maganize A, 81(3), 571-595, 2001.

32. Post, D., 'Development of moiré interferometry', Optical Engineering, 21(3), pp 458-467, 1982.

33. Post, D., Han, B., and Ifju, P., High Sensitivity Moire, Springer-Verlag, 1994.

34. Smith, C. W., 'Experiments in 3-D fracture problems', Experimental Mechanics, 33(4), pp 249-262, 1993.

35. Chiang, F.-P., 'Moiré and speckle methods applied to elastic-plastic fracture mechanics', in Experimental Techniques in Fracture, Epstein, J. S., (Editor), VCH Publishers, pp 291-325, 1993.

36. Kokaly, M. T., Lee, J., and Kobayashi, A. S., 'Moire interferometry for dynamic fracture study', Optics and Lasers in Engineering, 40, pp 231-247, 2003.

37. Ifju, P., and Post, D., 'Zero-thickness specimen grating for moiré interferometry', Optical Engineering, 29(7), pp 564-569, 1988.

38. Chiang, F.-P., 'Moire methods of strain analysis', (Chapter 6) in Manual of Experimental Stress Analysis, 4th Edition, (Kobayashi, A. S., Editor), Society of Experimental Mechanics, 1983.

Chapter 5

On The Use of Strain Gages In Dynamic Fracture

Venkitanarayanan Parameswaran

Department of Mechanical Engineering
Indian Institute of Technology Kanpur, Kanpur 208 016, India
venkit@iitk.ac.in

Arun Shukla

Department of Mechanical Engineering and Applied Mechanics
University of Rhode Island, Kingston RI 02881, USA
shuklaa@egr.uri.edu

Strain gages are the most commonly used type of strain sensors because of the fact that they are easy to use and relatively inexpensive. The application of strain gages to investigate propagating cracks in homogeneous materials, composites and bi-material interfaces is discussed in this chapter. A brief review on strain gage instrumentation and data acquisition is provided at the outset. The importance of using appropriate gage location and gage orientation relative to the crack path in obtaining reliable measurement of the fracture parameters is highlighted. The theory and analysis procedure for obtaining the crack speed and stress intensity factor are discussed for each type of material. Typical results, available from existing literature, are also presented.

1. Introduction

There are very few techniques available for real-time investigation of fast running cracks in materials. Photoelasticity, Moiré interferometry, Caustics and Coherent gradient Sensing are the commonly used optical techniques. Coupled with high speed imaging these techniques gather detailed information about the mechanical fields near the tip of a propagating crack in real time. While the data obtained from optical techniques is full field in nature and therefore more comprehensive, these methods are not simple, call for an imaging system with very high framing rates and not easy to use outside laboratory environments.

Electrical resistance strain gages on the other hand are the industry standard for stain measurement, relatively noninvasive, easy to install, less expensive and can be used with ease in most environments. However, measurements can be made only at selected locations and the measured strains are in fact local averages. The last one and half decade has seen considerable progress on research aimed at the application of strain gages to measure fracture parameters for both stationary and propagating cracks. Along with the availability of extremely small strain gages, high bandwidth strain conditioners and high speed recording instruments, this technique has matured into a reliable and robust method for investigating dynamic fracture in a wide variety of materials. The contents of this chapter are structured primarily based on the works of Berger *et. al* [1], Sanford *et. al* [2] and Shukla *et. al* [3-7].

For the sake of completeness, a brief overview of strain gage instrumentation and signal conditioning is included. Subsequently, the application of strain gages to obtain fracture parameters for cracks propagating in isotropic materials, orthotropic materials and cracks propagating along the interface of dissimilar materials is discussed in detail.

2. Strain gage instrumentation

Electrical resistance strain gages have been in use for over five decades. A typical strain gage installation for investigating dynamic fracture comprises of foil type strain gages, strain conditioners and the recording system. In this section, a brief description of these systems is provided to apprise the reader about the desired system characteristics for obtaining reliable information.

The foil type strain gage is bonded to the material and forms a part of it. When the material deforms, the gage is subjected to the same deformation which changes the resistance of the gage material. This resistance change is measured in terms of voltage change using a Wheatstone bridge circuit. The voltage differential from the bridge circuit is typically on the order of 5-10 $\mu V/\mu\varepsilon$ (micro volts per micro strain) and needs to be amplified.

Typically the crack speeds encountered in dynamic fracture are in the range of 200 m/sec to over 1200 m/sec depending on the loading and material type. The strain signals during dynamic fracture are of short duration (microseconds) and not repetitive in nature. Therefore adequate bandwidth for both the amplifier and the data acquisition system along with high sampling rates are required to ensure that the strain signals are recorded without distortion. A recording instrument capable of sampling rates of 1 Mega samples/sec or greater will provide adequate temporal resolution. This could be a digital oscilloscope or PC based data acquisition. The foil strain gages themselves have a fast response time of less than a micro second [8] and recent investigations indicate that their bandwidth can exceed 300 kHz [9]. Strain conditioners of bandwidth 200 kHz have been used successfully in the investigations listed earlier. A high bandwidth differential probe coupled with an oscilloscope can also be used.

In dynamic fracture, the total time of data collection is typically few milliseconds and therefore thermal drift of the system and fluctuations in excitation are not a serious concern. Any offset caused due to these can be corrected during post processing of the data. However, averaging errors are not negligible and can be minimized by selecting gages of smaller grid size and by judicious selection of the gage location and orientation as will be explained later. Further information on strain gage instrumentation is available in Dally [8].

3. Dynamic fracture in isotropic and homogeneous media

It is a well established fact that the stress field near the tip of stationary and propagating cracks in homogeneous isotropic materials exhibits the classical inverse square root singular nature. Thus the stress and strain will have steep gradients as one approaches the crack tip. Assuming small scale yielding, linear elastic fracture mechanics postulates the existence of a singularity dominant zone near the crack tip, in which the stress field can be characterized by a single parameter. For fast moving cracks, the stress field close to the crack tip is characterized by the single parameter called the dynamic stress intensity factor (DSIF). The determination of the instantaneous crack speed and the DSIF for the duration of crack propagation constitutes the crux of any dynamic fracture investigation. This would require real time recording of an appropriate mechanical field such as stress, strain or displacement and analyzing this information using the relevant elastodynamic field equations. In the case of strain gages, the field equations would imply the representation of the strain field surrounding the propagating crack. These field equations are developed in the next section.

3.1 Dynamic strain field equations

The strain field equations are derived from the available solution for the stress field around a crack tip moving with constant velocity (c) under steady state conditions [10]. For a two dimensional case, the dynamic stress field for a moving crack is written in a series form as

$$\sigma_{xx} = D\left\{(1+2\lambda_1^2 - \lambda_2^2)\,\mathrm{Re}\,\Gamma_1 - \frac{4\lambda_1\lambda_2}{(1+\lambda_2^2)}\mathrm{Re}\,\Gamma_2 + (1+2\lambda_1^2 - \lambda_2^2)\,\mathrm{Re}\,Y_1 - (1+\lambda_2^2)\,\mathrm{Re}\,Y_2\right\}$$

$$\sigma_{yy} = D\left\{-(1+\lambda_2^2)\,\mathrm{Re}\,\Gamma_1 + \frac{4\lambda_1\lambda_2}{(1+\lambda_2^2)}\mathrm{Re}\,\Gamma_2 + (1+\lambda_2^2)(\mathrm{Re}\,Y_2 - \mathrm{Re}\,Y_1)\right\}$$

$$\sigma_{xy} = D\left\{2\lambda_1(\mathrm{Im}\,\Gamma_2 - \mathrm{Im}\,\Gamma_1) - 2\alpha_l\,\mathrm{Im}\,Y_1 + \frac{(1+\lambda_2^2)^2}{2\lambda_2}\,\mathrm{Im}\,Y_2\right\} \qquad (3.1)$$

where $D = \dfrac{1+\lambda_2^2}{4\lambda_1\lambda_2 - (1+\lambda_2^2)^2}$, $\lambda_1 = 1 - \dfrac{\rho c^2}{c_l^2}$, $\lambda_2 = 1 - \dfrac{\rho c^2}{c_s^2}$

The complex functions Γ_1, Γ_2, Y_1 and Y_2 are defined as a series as follows

$$\Gamma_1(z_1) = \sum_{n=0}^{N} C_n z_1^{n-1/2} \text{ and } \Gamma_2(z_2) = \sum_{n=0}^{N} C_n z_2^{n-1/2}$$

$$Y_1(z_1) = \sum_{m=0}^{M} D_m z_1^{m} \text{ and } Y_2(z_2) = \sum_{m=0}^{M} D_m z_2^{m} \qquad (3.2)$$

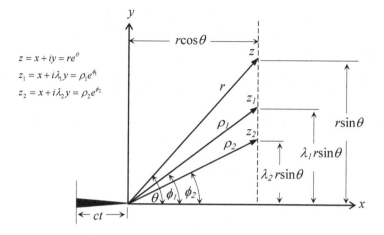

Figure 3.1. Crack tip coordinate reference.

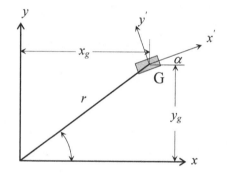

Figure 3.2. Location and orientation of strain gage.

The complex variables z_1 and z_2 are defined in figure 3.1, c_l is the dilatational wave speed and c_s is the shear wave speed in the material. A judicious location and orientation of the strain gage with respect to the crack path has to be established such that the strain signals recorded will have clear signatures from which the instantaneous position of the crack tip relative to the gage location can be determined unambiguously. In order to establish this, a strain gage oriented at an angle α to the direction of crack propagation is considered as shown in figure 3.2. The rotated x'-y' axis is fixed to the strain gage and orients with it as shown in figure

3.2. Using Hooke's law and the strain transformation relations, the strain at the center of the gage, point G, can be obtained in terms of the inplane stress components as

$$2\varepsilon_{x'x'} = \frac{1}{E}\left\{(1-v)(\sigma_{xx}+\sigma_{yy})-(1+v)(\sigma_{yy}-\sigma_{xx})\cos 2\alpha + 2(1+v)\sigma_{xy}\sin 2\alpha\right\} \quad (3.3)$$

where E is the elastic modulus of the material and v is the Poisson's ratio. Substituting for the stresses from equation 3.1, we get

$$2\mu\varepsilon_{x'x'} = D\Big\{\kappa(\lambda_1^2 - \lambda_2^2)[\mathrm{Re}\,\Gamma_1(z_1)+\mathrm{Re}\,Y_1(z_1)]+(1+\lambda_1^2)[\mathrm{Re}\,\Gamma_1(z_1)+\mathrm{Re}\,Y_1(z_1)]\cos 2\alpha$$
$$-\frac{4\lambda_1\lambda_2}{1+\lambda_2^2}\mathrm{Re}\,\Gamma_2(z_2)\cos 2\alpha - (1+\lambda_1^2)\mathrm{Re}\,Y_2(z_2)\cos 2\alpha$$
$$+\left(2\lambda_1[-\mathrm{Im}\,\Gamma_1(z_1)+\mathrm{Im}\,\Gamma_2(z_2)-\mathrm{Im}\,Y_2(z_2)]+\frac{(1+\lambda_2^2)^2}{4\lambda_1\lambda_2}\mathrm{Im}\,Y_2(z_2)\right)\sin 2\alpha\Big\}$$

$$(3.4)$$

where $\kappa = \dfrac{1-v}{1+v}$ and μ is the shear modulus. Taking $N=1$ and $M=0$ in equation 3.2, a three parameter representation of the strain can be obtained as

$$2\mu\varepsilon_{x'x'} = C_0 f_0 + C_1 f_1 + D_0 g_0 ;$$
$$f_0 = D\Big\{\rho_1^{-1/2}\Big[\kappa(\lambda_1^2 - \lambda_2^2)\cos\frac{\phi_1}{2}+(1+\lambda_1^2)\cos 2\alpha\cos\frac{\phi_1}{2}+2\lambda_1\sin 2\alpha\sin\frac{\phi_1}{2}\Big]$$
$$-\rho_2^{-1/2}\Big[\frac{4\lambda_1\lambda_2}{1+\lambda_2^2}\cos 2\alpha\cos\frac{\phi_2}{2}+2\lambda_1\sin 2\alpha\sin\frac{\phi_2}{2}\Big]\Big\}$$
$$f_1 = D\Big\{\rho_1^{1/2}\Big[\kappa(\lambda_1^2 - \lambda_2^2)\cos\frac{\phi_1}{2}+(1+\lambda_1^2)\cos 2\alpha\cos\frac{\phi_1}{2}-2\lambda_1\sin 2\alpha\sin\frac{\phi_1}{2}\Big]$$
$$-\rho_2^{1/2}\Big[\frac{4\lambda_1\lambda_2}{1+\lambda_2^2}\cos 2\alpha\cos\frac{\phi_2}{2}-2\lambda_1\sin 2\alpha\sin\frac{\phi_2}{2}\Big]\Big\}$$
$$g_0 = D(\lambda_1^2 - \lambda_2^2)(\kappa + \cos 2\alpha)$$

$$(3.5)$$

In equation 3.5, there are three unknown constants, C_0, C_1 and D_0. Of these three, the constant C_0 is related to the DSIF, K_{ID}, as $K_{ID}=C_0\sqrt{(2\pi)}$ and D_0 is related to the constant non-singular stress acting parallel to the crack. If the strain gage is located in the singularity dominant zone, the first term of equation 3.5 corresponding to C_0 alone is sufficient to provide accurate representation of the strain field. The value of C_1 is very close to zero in specimen geometries like the single edge notch [11]. One can notice that g_0, the term corresponding to D_0 will vanish if the gage orientation is such that

$$\cos 2\alpha = -\kappa \tag{3.6}$$

Equation 3.6, provides two values for angle α for a given material, one will be acute angle (α_1) and the other obtuse angle (α_2). Either choice will cancel the effect of the term corresponding to D_0 in the strain field and neglecting the term corresponding to C_1 the gage strain can be related to the DSIF directly as

$$2\mu\varepsilon_{x'x'} = D\frac{K_{ID}}{\sqrt{2\pi}}\left\{\rho_1^{-1/2}\left[\kappa(\lambda_1^2 - \lambda_2^2)\cos\frac{\phi_1}{2} + (1+\lambda_1^2)\cos 2\alpha\cos\frac{\phi_1}{2} + 2\lambda_1\sin 2\alpha\sin\frac{\phi_1}{2}\right]\right.$$
$$\left. - \rho_2^{-1/2}\left[\frac{4\lambda_1\lambda_2}{1+\lambda_2^2}\cos 2\alpha\cos\frac{\phi_2}{2} + 2\lambda_1\sin 2\alpha\sin\frac{\phi_2}{2}\right]\right\} \tag{3.7}$$

3.2 Synthetic strain profiles

For cracks following a straight path, the vertical location of the gage from the crack path (y_g in figure 3.2) remains constant but the coordinate x_g keeps changing. The strain recorded by the gage will vary as the crack approaches and passes past the gage location. The variation of the strain as a function of the distance of the gage relative to the crack tip has to be first characterized for different crack speeds keeping the DSIF constant to understand the type of strain signals one can expect. The strain variation for both acute and obtuse angle orientation of the gage generated using equation 3.7, is shown in figure 3.3. In generating this figure the dilatational and shear wave speeds were assumed as 2220 m/sec and 1270 m/sec respectively. For the assumed Poisson's ratio

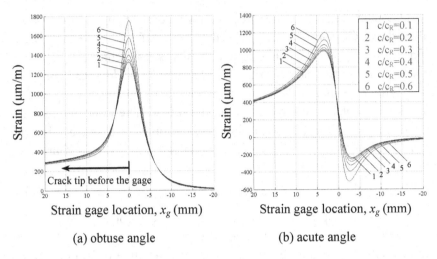

Figure 3.3. Strain as a function of gage location relative to the crack tip for different crack velocities (c_R = 1186 m/s).

of 0.35, the gage orientations are α_1 =59.4° and α_2 =120.6° respectively. The strain profiles for different crack speeds were generated assuming a DSIF of 1 MPa√m.

Figure 3.3 indicates that for both gage orientations, the strain initially increases and then decreases as the crack traverses the gage location. The abscissa in figure 3.3 is the gage location, x_g, relative to the instantaneous crack tip. In the case of the obtuse angle, the strain attains peak value when the gage is directly below the crack tip for all crack speeds, thereby providing a definite signature for determining the crack tip location directly from the strain profile. For acute angle orientation the strain attains peak value just before the crack reaches the gage location. More importantly in this case the nature of the strain changes from positive to negative (zero crossing) once the crack has passed the gage location. The location of the crack tip relative to the gage when peak strain is recorded is relatively insensitive to the crack speed. However, this is not the case for zero crossing. Another important aspect is that for the same value of DSIF, the obtuse angle gage records peak strains which are 30% larger than that for the acute angle and hence provides better sensitivity.

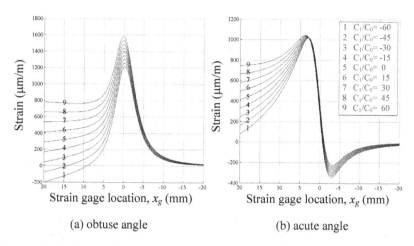

Figure 3.4. Strain as a function of gage location relative to the crack tip for different C_1/C_0 ratios ($c/c_R = 0.3$).

If the strain gage is not within the singularity dominant zone and C_1 values are not negligible, equation 3.7 can give erroneous results. In such cases, the effect of the term corresponding to C_1 in equation 3.5 on the strain profile has to be considered. To characterize this, strain profiles corresponding to different ratios of C_1/C_0 are generated using equation 3.5 keeping the rest of the parameters same as that used in generating figure 3.3.

Figure 3.4 indicates that for obtuse angle, the peak strain magnitude is highly sensitive to C_1/C_0 ratio where as the peak strain is recorded when the crack tip is very close to the gage location (within 0.3 mm). The peak strain magnitude however varies to the extent of ±15 % for the case shown in figure 3.4. This would imply that one can still locate the crack tip with sufficient accuracy, however; to obtain accurate values of DSIF, equation 3.7 is not sufficient. In the case of acute angle, it is interesting to observe from figure 3.4 that the magnitude of the peak strain is almost insensitive to C_1/C_0 where as the relative location of the crack tip when peak strain is recorded is influenced by C_1/C_0. The analysis procedure employed in each case, acute and obtuse, is discussed in the next section.

3.3 *Analysis of strain gage data*

An array of strain gages have to be used to obtain the crack speed and DSIF history. These gages can be placed at regular spacing and the data from each recorded on a separate channel of the data acquisition system. The gage could be oriented in either angle and typically placed at a vertical distance in the range of $h/2 \leq y_g \leq h$, where h is the specimen thickness.

3.3.1 *Acute angle gage configuration*

In the acute angle configuration, the velocity of the crack can be determined with sufficient accuracy using the time corresponding to either the peak strain or the zero crossing. Knowing the gage spacing and the time from each strain signal corresponding to peak or zero crossing, a crack tip location versus time graph is constructed from which the crack speed can be obtained. The subsequent procedure is to locate the crack tip relative to each of the gages. This is done by using the zero crossing feature of the strain profile along with a triangulation and iteration scheme [1]. For this, equation 3.5 is written as

$$\frac{2\mu\varepsilon_{x'x'}}{C_0} = f_1 + \frac{C_1}{C_0}f_2 \tag{3.8}$$

In equation 3.8, f_1 and f_2 are functions of crack velocity, c, and the coordinate, x_g, which keeps changing as the crack propagates. For a pair of adjacent gages a and b, we will consider the instant of time t defined as the time corresponding to zero crossing for gage a. At this time, the crack has advanced past gage a yet not reached the location of gage b. It is assumed that the crack speed and the ratio C_1/C_0 remains constant as the crack propagates from gage a to gage b. With both y_g and c known, a value of C_1/C_0 is assumed and the value of x_{ga} is determined through iteration, such that equation 3.8 gives zero strains for gage a. In order to ensure the correctness of the assumed C_1/C_0, and hence the calculated x_{ga}, a second check is necessary. This is done by using the ratio $(\varepsilon_a/\varepsilon_b)$ of the strains recorded from gages a and b. Having determined x_{ga}, the position of gage b, i.e. x_{gb} is easily calculated from the gage spacing.

First the ratio ($\varepsilon_a/\varepsilon_b$) is calculated theoretically using equation 3.8 for the assumed value of C_1/C_0 at a time $t-\Delta t$. This ratio is compared with the ratio obtained from the experimental strain profiles corresponding to the same time. If the comparison is not satisfactory, then C_1/C_0 is revised and the procedure repeated until the difference between the two ratios is negligibly small. The DSIF corresponding to time t is then calculated using the converged values of C_1/C_0 and x_{gb} and the experimentally measured $\varepsilon_b(t)$, in equation 3.8. This procedure is repeated for each pair of gages in succession.

Berger *et. al* [1] used the acute angle gage orientation to investigate dynamic fracture in 12.7 mm thick 4340 alloy steel plate using the compact tension specimen shown in figure 3.5. Shallow side grooves of depth 5% of the specimen thickness were made on each side of the specimen to guide the crack path. Six strain gages were installed at a distance of $y_g=10.4$ mm and at an angle of 61.3°. The specimen was loaded to failure and the strain signals were recorded through an oscilloscope. The strain profiles obtained from the experiment are shown in figure 3.6. It can be noticed that these are very much similar to the theoretically generated ones shown in figure 3.3 (b). The peaks and zero crossings are very distinct. The crack propagated with a constant velocity of 650 m/sec. The DSIF and C_0/C_1 ratio for each gage pair calculated through the triangulation procedure described earlier are given in table 3.1. The results indicate a decreasing DSIF and increasing C_0/C_1 ratio.

Table 3.1. Fracture parameters for propagating crack in 4340 steel [1]

Gage pair	DSIF (MPa√m)	C_0/C_1
1-2	131.4	-26.2
2-3	123.8	-27.7
3-4	122.4	-27.5
5-6	120.9	-30.9

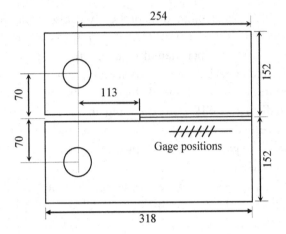

Figure 3.5. Compact tension specimen for studying dynamic fracture in 4340 steel [1]. (all dimensions in mm)

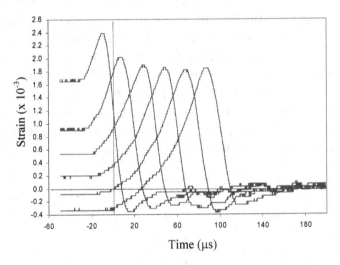

Figure 3.6. Strain records for a propagating crack in 4340 steel [1].

3.3.2 *Obtuse angle gage configuration*

Figure 3.3 indicates that the obtuse angle configuration provides a well defined peak in the strain profiles when the crack tip is directly below the strain gage. In an experiment, the strain is recorded as a function of time and the time corresponding to the peak strain value will provide the time at which the crack tip is directly under the gage. By recording the time corresponding to the strain peaks from each of the gages, one can locate the crack tip as a function of time from which the crack speed can be obtained. If the gage is placed within the singularity dominant zone, then the DSIF can be calculated from the magnitude of the peak strain using equation 3.7. When the crack is directly below the gage, $x_g=0$ resulting in $\theta=\phi_1=\phi_2=\pi/2$, $r=y_g$ and neglecting the C_1 term, the DSIF will be given by the expression

$$K_{ID} = \frac{4\mu}{D}(\varepsilon_g)_{peak}\sqrt{\pi y_g}\left\{\frac{1}{\lambda_1}\left[\kappa(\lambda_1^2 - \lambda_2^2)+(1+\lambda_1^2)\cos 2\alpha + 2\lambda_1\sin 2\alpha\right]\right.$$
$$\left.-\frac{1}{\lambda_2}\left[\frac{4\lambda_1\lambda_2}{1+\lambda_2^2}\cos 2\alpha + 2\lambda_1\sin 2\alpha\right]\right\}^{-1} \qquad (3.9)$$

where $(\varepsilon_g)_{peak}$ is the peak strain value obtained from the strain profile of each gage.

In certain specimen geometries the effect of C_1 cannot be ignored and in such cases, equation 3.9 does not provide accurate estimate of DSIF. In such cases, a revised procedure similar to that employed by Sanford *et. al* [2] for the acute angle configuration can be used. It can be noticed from figure 3.4 (a) that, even when the effect of C_1 is considered, the peak strain occurs when the crack is directly under the gage. The crack speed can therefore be calculated accurately following the procedure explained earlier. The term C_1 influences the magnitude of peak strain and also the spread of the strain signal. For example, the range of x_g (or the equivalent time duration) for which the strain is greater than 75% of $(\varepsilon_g)_{peak}$ is influenced considerably by the value of C_1. A master curve which gives this 75% width can be generated theoretically for different ratios of C_1/C_0 and crack speeds. From the knowledge of the crack speed and the 75% width obtained from the experimental strain signal one can easily determine the appropriate value of C_1/C_0 using the master curve. This value of C_1 can be used in equation 3.5 to calculate the DSIF when the crack is directly below the gage.

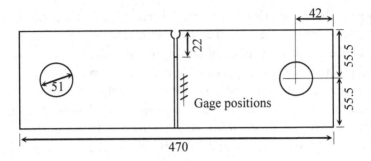

Figure 3.7. Single edge notched specimen for studying dynamic fracture in 7075-T6 Aluminum [3]. (all dimensions in mm)

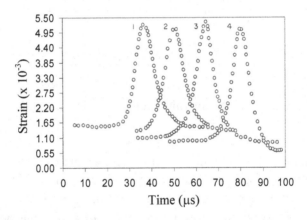

Figure 3.8. Strain records for a propagating crack in Aluminum [3].

Shukla *et. al*, [3] investigated dynamic fracture in 7075-T6 Aluminum, using the obtuse angle configuration. They used the single edge notch specimen shown in figure 3.7 with four strain gages of grid size, 0.79 x 0.81 mm^2. The specimen thickness was 6.25 mm. Side grooves were provided for guiding the crack path. The experimental strain profiles shown in figure 3.8 closely resemble those generated theoretically. The crack speed in this study was 1075 m/sec. The technique of dynamic photoelasticity in reflection was also employed to investigate the dynamic fracture simultaneously. Figure 3.9 shows DSIF as a function of crack length obtained from both measurements. There is

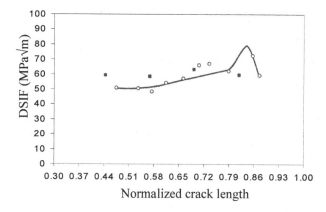

Figure 3.9. DSIF for Aluminum as a function of crack length [3].

clearly a good agreement between the values of DSIF obtained from the strain gage and the photoelastic full field data establishing the feasibility of using strain gages for studying propagating cracks in isotropic homogeneous materials.

3.4 *Discussion*

By properly choosing the location of the gage and its orientation, both crack speed and the DSIF can be determined with sufficient accuracy. The two gage orientations, acute and obtuse each have their own characteristic merits and limitations. In both cases, it is possible to accurately determine the crack velocity. In the case of obtuse angle gage, locating the crack tip is relatively easier and DSIF can be directly obtained provided the gage is located within the singularity zone.

When investigating low yield strength materials, the gages might have to be placed further away from the crack path to ensure that it is outside the plastic zone. In such cases a single parameter representation of the strain field is not acceptable and the acute angle configuration is more appropriate. However when investigating hard and highly brittle materials such as ceramics, the expected strain levels will be very small and the obtuse angle configuration offers increased sensitivity.

4. Dynamic fracture in orthotropic composites

With the advent of composite materials, there is considerable interest in their failure patterns. These materials due to their opaque nature are not amenable to optical techniques in transmission mode. Due to the presence of the fibers, which are considerably stiff compared to the matrix, and the associated orthotropic behavior the dynamic fracture of fiber composites is more complex compared to isotropic materials.

4.1 Dynamic strain field equations for orthotropic composites

Using the dynamic stress fields in the vicinity of a fast propagating crack tip in an orthotropic medium [12], Khanna and Shukla [5] obtained the expression for strain at any point along a particular direction with respect to the crack path assuming steady state crack growth. The crack propagation direction (x axis in figure 3.1) was assumed to coincide with one of the elastic symmetry axis. Only the final expression for the strain is presented here and further details are available elsewhere [5, 12]. For a strain gage oriented at an angle of ϕ with the x axis (similar to figure 3.2), following the procedure similar to that in section 3.1, the strain for a crack propagating with constant velocity (c) can be written as follows

$$
\begin{aligned}
\varepsilon_{x'x'} = AC_{66}C_0 & \left\{ -\frac{1}{\sqrt{r_1}}\cos(\frac{\phi_1}{2}) \left[\frac{l_1}{\alpha}(S_{11}\cos^2\phi + S_{12}\sin^2\phi) - p^2 l_3(S_{12}\cos^2\phi + S_{22}\sin^2\phi) \right] \right. \\
& \left. + \frac{1}{\sqrt{r_2}}\cos(\frac{\phi_2}{2}) \left[\frac{pl_2 l_5}{\alpha q l_6}(S_{11}\cos^2\phi + S_{12}\sin^2\phi) + \frac{pq l_4 l_5}{l_6}(S_{12}\cos^2\phi + S_{22}\sin^2\phi) \right] \right\} \\
& + Apl_5 C_0 \left\{ \frac{1}{\sqrt{r_1}}\sin(-\frac{\phi_1}{2}) - \frac{1}{\sqrt{r_2}}\sin(-\frac{\phi_2}{2}) \right\} \sin\phi\cos\phi \\
& + AC_{66}B_0 \left[\frac{p^2 l_2 l_3}{\alpha q^2 l_4} - \frac{l_1}{\alpha} \right] \left\{ S_{11}\cos^2\phi + S_{12}\sin^2\phi \right\}
\end{aligned}
$$

$$\tag{4.1}$$

$$AC_{66} = \frac{l_6}{pql_4l_5 - p^2l_3l_6};$$

$$l_1 = \frac{2\beta q^2}{(\alpha - p^2)(1 - M_1^2)} + (2\beta - \alpha); \quad l_2 = \frac{2\beta q^2}{(\alpha - q^2)(1 - M_1^2)} + (2\beta - \alpha);$$

$$l_3 = (1 - M_2^2) - \frac{2\beta}{\alpha - p^2}; \quad l_4 = (1 - M_2^2) - \frac{2\beta}{\alpha - q^2}; \quad l_5 = -(l_3 + M_2^2); \quad l_6 = -(l_4 + M_2^2)$$

$$p = \left[a_1 - \sqrt{(a_1^2 - a_2)} \right]^{1/2}; \quad q = \left[a_1 + \sqrt{(a_1^2 - a_2)} \right]^{1/2}; \quad 2a_1 = \alpha + \alpha_1 - 4\beta\beta_1; \quad a_2 = \alpha\alpha_1$$

$$2\beta = \frac{C_{12} + C_{66}}{C_{11}(1 - M_1^2)}; \quad 2\beta_1 = \frac{C_{12} + C_{66}}{C_{66}(1 - M_2^2)}; \quad \alpha = \frac{C_{66}}{C_{11}(1 - M_1^2)}; \quad \alpha_1 = \frac{C_{22}}{C_{66}(1 - M_2^2)}$$

$$M_1 = \frac{c}{c_1}; \quad M_2 = \frac{c}{c_2}; \quad c_1 = \sqrt{\frac{C_{11}}{\rho}}; \quad c_2 = \sqrt{\frac{C_{66}}{\rho}}$$

$$r_1 = \left\{ x^2 + \frac{y^2}{p^2} \right\}^{1/2}; \quad r_2 = \left\{ x^2 + \frac{y^2}{q^2} \right\}^{1/2}; \quad \phi_1 = \tan^{-1}(\frac{y}{px}); \quad \phi_2 = \tan^{-1}(\frac{y}{qx})$$

where C_{ij} and S_{ij} are the material constants [5, 12] and ρ the mass density. The coefficient C_0 is related to the DSIF, K_{ID} as $K_{ID} = C_0\sqrt{(2\pi)}$ and B_0 is related to the constant non-singular stress acting parallel to the crack. The contribution of this term, B_0, can be eliminated by satisfying the following condition

$$\tan^2 \phi = -\frac{S_{11}}{S_{12}} = \frac{1}{v_{LT}} \tag{4.2}$$

Equation (4.2) gives us two gage orientations, one acute and the other obtuse, and these angles depend only on the Poisson's ratio v_{LT} of the material. The rest of the procedure is more or less similar to what was followed in section 3.2. Synthetic strain profiles are to be generated for different crack speeds for both gage orientation to establish the appropriate signatures in the strain profile that will enable the determination of the crack speed and crack tip location. The study by Khanna and Shukla [5] indicate that the strain profiles are similar to that for homogeneous materials given in figure 3.3. The crack tip is exactly below the gage when peak strain is recorded in the case of the obtuse angle. In the case of acute angle, the crack tip is ahead of the gage when peak strain is recorded, but there is a definite zero crossing. They also observed that the strain recorded by an obtuse angle gage is more sensitive to the crack speed than the acute angle gage (see figure 3.3). Also for the same crack speed and DSIF the obtuse angle gage has more sensitivity. They also point out that for the same crack speed and DSIF, the magnitude of the peak strain is a strong function of the degree of anisotropy (E_L/E_T). With increasing degree of anisotropy, the peak strain reduces considerably for both gage orientations and one should therefore choose the instrumentation carefully to sense the small strain. In the case of obtuse angle, the DSIF can be obtained directly from the peak strain using equation 4.1. Further details are available in Khanna and Shukla [5].

5. Dynamic interfacial fracture of isotropic bi-materials

Multi phase materials like fiber reinforced composites, layered armor systems, particle and whisker reinforced composites all have interfaces between the different phases involved. The overall response of these systems is intrinsically linked to the behavior of these interfaces, which are often the weakest link. Rapid failure of these interfaces leads to physical separation of the phases causing loss of system integrity. Dynamic interfacial failure is an area of considerable interest to the fracture mechanics community [13-19].

In a bi-material system involving two dissimilar materials joined together, the dynamic behavior of an interfacial crack is influenced by the combination of the materials constituting the interface in addition to the loading and geometry. Due to the elastic mismatch the fracture is inherently mixed mode. Crack speeds can be in the subsonic or intersonic regime depending on the elastic mismatch. The application of strain gages to evaluate the fracture parameters in such systems is not as straightforward as that for homogeneous materials. The contents of this section are based on the work by Ricci et. al [6].

5.1 *Dynamic strain field equations in isotropic bi-materials*

A bi-material system consists of a stiff component (Material 2) and a compliant component (Material 1), joined together as shown in figure 5.1. Across the interface there is a discontinuity in the material characteristics. Crack propagation is subsonic if crack speed is less than the shear wave speed, c_s, of the more compliant material i.e. Material 1. The stress field for subsonic interfacial crack growth given by Yang *et. al* [13] is a coupled oscillatory field scaled by the complex stress intensity factor as given in equation 5.1.

$$\sigma_{ij} = \frac{\text{Re}\{K^d r^{(i\varepsilon v)}\}}{\sqrt{2\pi r}} \tilde{\sigma}_{ij}^1(\theta,v) + \frac{\text{Im}\{K^d r^{(i\varepsilon v)}\}}{\sqrt{2\pi r}} \tilde{\sigma}_{ij}^2(\theta,v) \qquad (5.1)$$

where r and θ are polar coordinates in the translating crack tip reference, v is the crack speed, $K^d = K_1^d + iK_2^d$ is the complex stress intensity factor and $\tilde{\sigma}_{ij}^1$ and $\tilde{\sigma}_{ij}^2$ are real dimensionless angular functions [15]. The oscillatory index ε in equation 5.1 is given as

$$\varepsilon = \frac{1}{2\pi} \ln \frac{1-\beta}{1+\beta} \qquad (5.2)$$

where β, which is a function of the elastic constants of materials 1 and 2 and crack speed, is given by Deng [15]. Using Hooke's law with

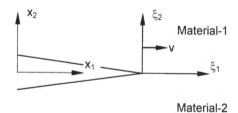

x₁, x₂ Fixed coordinates.
ξ₁, ξ₂ Coordinates moving with crack tip

Figure 5.1. Coordinate system for a crack propagating along a bi-material interface. (interface along the x_1 axis)

equation 5.1, the strain at a point (r, θ) in Material 1, along a direction θ_g with the x_1 axis can be written as

$$\varepsilon = C_1 K_1^d - C_2 K_2^d$$

$$C_1 = \frac{1}{E_1\sqrt{2\pi r}} \left\{ \cos(\varepsilon \ln r)\left[(\tilde{\sigma}_{11}^1 - v_1\tilde{\sigma}_{22}^1)\cos^2\theta_g + (\tilde{\sigma}_{22}^1 - v_1\tilde{\sigma}_{11}^1)\sin^2\theta_g + 2\tilde{\sigma}_{12}^1\cos\theta_g\sin\theta_g\right] \right.$$

$$\left. + \sin(\varepsilon \ln r)\left[(\tilde{\sigma}_{11}^2 - v_1\tilde{\sigma}_{22}^2)\cos^2\theta_g + (\tilde{\sigma}_{22}^2 - v_1\tilde{\sigma}_{11}^2)\sin^2\theta_g + 2\tilde{\sigma}_{12}^2\cos\theta_g\sin\theta_g\right] \right\}$$

$$C_1 = \frac{1}{E_1\sqrt{2\pi r}} \left\{ \sin(\varepsilon \ln r)\left[(\tilde{\sigma}_{11}^1 - v_1\tilde{\sigma}_{22}^1)\cos^2\theta_g + (\tilde{\sigma}_{22}^1 - v_1\tilde{\sigma}_{11}^1)\sin^2\theta_g + 2\tilde{\sigma}_{12}^1\cos\theta_g\sin\theta_g\right] \right.$$

$$\left. - \cos(\varepsilon \ln r)\left[(\tilde{\sigma}_{11}^2 - v_1\tilde{\sigma}_{22}^2)\cos^2\theta_g + (\tilde{\sigma}_{22}^2 - v_1\tilde{\sigma}_{11}^2)\sin^2\theta_g + 2\tilde{\sigma}_{12}^2\cos\theta_g\sin\theta_g\right] \right\}$$

$$(5.3)$$

where E_1 and v_1 are the elastic modulus and Poisson's ratio of the compliant material (Material 1). In equation 5.3, there are three unknowns, the crack speed, v, K_1^d and K_2^d. If the crack speed is known, one needs strains from two different locations measured at the same time instant for evaluating K_1^d and K_2^d using equation 5.3. The details of the scheme for measuring the crack speed and K_1^d and K_2^d are discussed in the following sections.

5.2 *Selection of strain gage location and orientation*

The strain gage should be positioned and oriented such that the strain signals will have distinctive features that enable the location of the instantaneous crack tip. Since only the singular term is considered in equation 5.3, the strain gage should be located within the singularity dominant zone. This can be ensured by placing the gage at a vertical distance of 0.5B from the interface, where B is the thickness of the material.

The strain gage orientation should be such that the crack tip can be located from the strain peaks. This particular orientation can be identified by investigating iso-strain contours around the crack tip for gage orientation, θ_g, varying from 0 to 179 while keeping K^d constant. A typical case of Aluminum-PSM-1 interface, for which the elastic properties are given in table 5.1, is considered for illustration.

Table 5.1. Properties of Aluminum-PSM-1 bi-material system

	PSM-1	Aluminum 6061-T6
Elastic modulus E (GPa)	2.76	71
Poisson's ratio	0.38	0.33
Density (kg/m^3)	1200	2770

For this particular case, the strain contours for a gage angle of θ_g=114° are shown in figure 5.2 for two crack speeds, 0.3c_s and 0.6c_s (c_s is the shear wave speed in PSM-1). Figure 5.2 indicates that the apogee of all the contours fall on the vertical axis passing through the crack tip. This suggests that a gage oriented at this angle would record peak strain when the crack tip is in the vicinity directly below the gage. For other gage orientations the strain lobes would tilt with respect to the interface hence, locating the crack tip from strain records will be difficult. Subsequently, the strain profiles are generated for a given K^d and crack speed by changing the gage orientation slightly from the chosen angle (114°). The location of the crack tip relative to the gage is further inspected for this set of angles as shown in figure 5.3. It can be seen from figure 5.3 that the location of the crack tip relative to the gage datum is considerably influenced by the gage orientation and figure 5.3 reiterates the fact that for θ_g=114° the crack tip is almost directly below the gage when the peak strain is recorded.

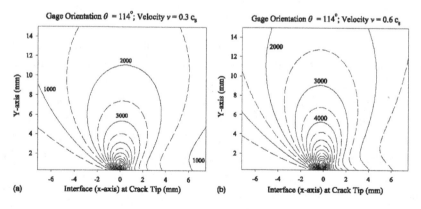

Figure 5.2. Iso-strain contours for a gage orientation of 114°C in an Aluminum-PSM-1 bi-material system [6].

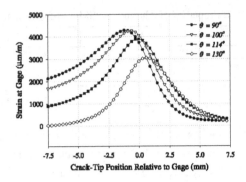

Figure 5.3. Effect of gage orientation on the crack-tip location relative to the gage at the time of peak strain ($v=0.3c_s$) [6].

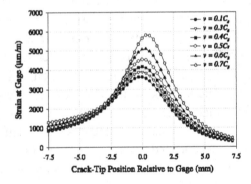

Figure 5.4. Strain as a function of crack tip location relative to the gage datum for different crack speeds for a gage orientation of $114°$ [6].

Once the optimum angle is identified as explained earlier, one has to investigate the effect of crack speed on the strain profiles especially the exact location of the crack tip relative to the gage datum when the peak strain is recorded. This can be done by calculating the gage strain for the chosen gage orientation for different positions of the crack tip relative to the gage datum while keeping a constant value of K^d. As shown in figure 5.4, it is clear that as crack speed is changed the position of the crack tip when peak strain is recorded is not exactly at the gage datum. However, the shift in crack tip position is very small and for the range of crack speeds considered in figure 5.4, one can conclude that the crack tip is very close to the gage datum when the peak strain occurs.

Another factor to be considered in fracture of bi-material systems is the effect of mode mixity on the strain profile, especially the peak strain and location of the crack tip when peak strain is recorded. Mode mixity, ϕ, gives the relative strength of K_2^d to K_1^d and is defined as arctan of (K_2^d / K_1^d). A plot of the strain profiles for different mixity in the range of $15°$ to $60°$ is shown in figure 5.5 for a crack speed of $0.3c_s$. This figure indicates that for a gage oriented at $114°$, the crack tip is either directly below the gage or very near to the gage datum for the range of mixity considered.

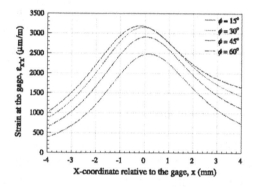

Figure 5.5. Effect of mixity on the location of the crack tip corresponding to peak strain for a gage orientation of $114°$ and crack speed of $0.3c_s$ [6].

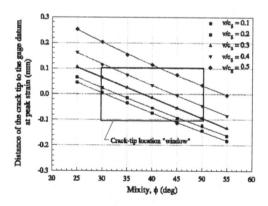

Figure 5.6. Crack tip location relative to gage datum for a gage orientation of $114°$ for varying mixity and crack speed [6].

By generating a comprehensive plot of the crack tip location relative to the gage datum for different mixity and crack speed, as shown in figure 5.6, one can establish a crack tip location window which specifies the possible crack tip locations when peak strain is recorded. If one were to assume that the crack tip is directly below the gage when the peak strain is recorded by the gage, the maximum error induced by this assumption is indicated by this window for an anticipated range of crack speeds and mixity values.

For the particular bi-material system considered, the maximum error is ±0.1mm, which is in fact negligibly small. By this way the optimum location and orientation of the strain gage can be determined for any bi-material system and for an expected range of mixity and crack speed one can also establish the bounds of uncertainty in locating the crack tip from the peak strain.

5.3 *Analysis of strain gage data*

An array of regularly placed strain gages located at a constant height from the interface (crack path) is needed for obtaining the fracture parameters. Once the strain-time records have been obtained, an initial estimate of the crack speed can be made by assuming that the crack is directly below the gage at the time instant corresponding to the peak strain. From the strains recorded by two adjacent gages, *1* and *2*, at an instant of time at which the crack tip is in the vicinity of the gage datum of gage *1*, K_1^d and K_2^d can be calculated using the following relations

$$K_1^d = \frac{C_2^2 \varepsilon^1 - C_2^1 \varepsilon^2}{C_2^2 C_1^1 - C_2^1 C_1^2}$$

$$K_2^d = \frac{C_1^2 \varepsilon^1 - C_1^1 \varepsilon^2}{C_2^2 C_1^1 - C_2^1 C_1^2}$$

(5.4)

where superscripts indicate the gage label, ε^1 and ε^2 the strain recorded by adjacent gages *1* and *2* corresponding to the time instant when the crack tip is below gage *1*. C_1 and C_2 defined in equation 5.3 have to be calculated using the coordinates of each gage relative to the instantaneous crack tip. From this estimate of K_1^d and K_2^d the mixity can be calculated and using this mixity and the crack speed one can correct for the error in crack tip location from figure 5.6. With the corrected crack tip location, K_1^d and K_2^d can be recalculated using equation 5.4 and the procedure can be repeated until the mixity converges.

Another scheme of analysis is to use the strain from three adjacent gages *1*, *2*, and *3* to simultaneously determine the location of the crack tip and K_1^d and K_2^d. The crack speed is estimated first from the time corresponding to the peak strain of each of the three gages and the gage spacing. From this information, the instantaneous crack speed when the crack tip is near gage *2* can be obtained. Now for this crack speed, K_1^d and K_2^d are extracted using equation 5.4 from the strain records of gages *1* and *2* by shifting continuously the crack tip location around datum of gage *2*. This provides us with a record of K_1^d and K_2^d for different crack tip locations centered on gage *2*. Through a similar approach K_1^d and K_2^d are calculated from the strain records of gages *2* and *3* for different crack tip positions. These two sets of K_1^d and K_2^d obtained from gage pairs (*1*, *2*) and (*2*, *3*) are then compared as shown in figure 5.7. The intersection of the two plots gives the exact location of the crack tip and the corresponding values of K_1^d and K_2^d.

Ricci *et al.* [6] investigated interfacial crack propagation in Aluminum-PSM-1 bi-material given in table 4.1 using strain gages. Simultaneously the crack propagation was investigated using dynamic

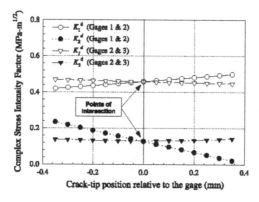

Figure 5.7. Simultaneous determination of crack tip location and K_1^d and K_2^d from three gages. Point of intersection gives the exact crack tip location and K_1^d and K_2^d [6].

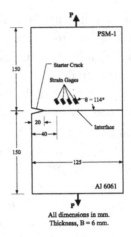

All dimensions in mm.
Thickness, B = 6 mm.

Figure 5.8. Specimen geometry and gage configuration used in dynamic interfacial fracture study [6].

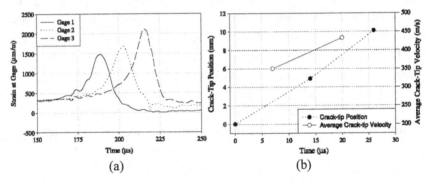

(a)   (b)

Figure 5.9. (a) Strain records for a propagating interfacial crack in Aluminum-PSM-1 bi-material. (b) Crack tip position and crack velocity. (assuming peak strain occurs when crack tip is directly below the gage) [6].

photoelasticity and high-speed photography. The specimen geometry shown in figure 5.8 was loaded in tension until crack propagation. Four strain gages, each having a gage length 0.38 mm and grid width 0.51 mm, installed on the PSM-1 half at an angle of $114°$ as indicated in figure 5.8 were used. The gages were positioned at a height of 5mm above the interface and had a spacing of 5mm between them. PSM-1 being the compliant material among the two will give larger strain magnitudes for the same stress intensity factor. Figure 5.9 shows the experimentally recorded strain profiles, the crack tip location history and crack speed history calculated from the time of peak strain for each gage.

The magnitude of the dynamic complex stress intensity factor, $\left|K^d\right|$, obtained by the two gage method of analysis and from the three gage analysis is shown in figure 5.10 as a function of time along with the values from photoelastic data. Ricci *et al.* [6] also provide a master plot which can be used to select the optimum gage orientation for any given mismatch parameter as shown in figure 5.11. This figure is generated for propagating interfacial cracks considering a range of mixity values and subsonic crack speeds such that the crack tip location window (see figure 5.6) is centered on the gage datum.

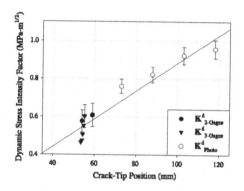

Figure 5.10. Magnitude of complex stress intensity factor as a function of time for interface crack propagation in Aluminum-PSM-1 bi-material [6].

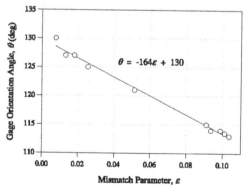

Figure 5.11. Optimum strain gage orientation for studying subsonic interfacial fracture as function of mismatch parameter [6].

The static mismatch parameter ε used in plotting figure 5.11 is defined as

$$\varepsilon = \frac{1}{2\pi} \ln \left[\frac{\dfrac{\kappa_1}{\mu_1} + \dfrac{1}{\mu_2}}{\dfrac{\kappa_2}{\mu_2} + \dfrac{1}{\mu_1}} \right] \qquad (5.5)$$

where $\kappa_{1,2}=(3-\nu_{1,2})/(1+\nu_{1,2})$ for plane stress $\kappa_{1,2}=(3-4\nu_{1,2})$ for plane strain and $\mu_{1,2}$ are the shear modulus and $\nu_{1,2}$ are the Poisson's ratio for materials 1 and 2 respectively.

5.4 Discussion

The method described in this section for obtaining the fracture parameters, for propagating interfacial cracks in bi-materials, using strain gages considers only the singular term $(r^{-1/2})$ in the expansion of the stress/strain field. However, the stress field around the crack tip is represented as a series and the higher order terms in fact have an effect on the strain field. In particular the term corresponding to (r^0) which is referred to as the non-singular stress (σ_{0x}) or the T-stress does have a physical significance as well.

In the case of homogeneous materials (section 3), the effect of this term (D_0) has been nullified by proper selection of the gage orientation. In the case of bi-materials determining a gage orientation which will nullify the effect of the non-singular stress and still give strain signals with clear signatures indicating the crack tip location is difficult due to the complexity of the stress field equations. The gage orientation is therefore determined to satisfy the primary requirement that the crack tip is very close to the gage datum when the peak strain is recorded and changes in crack speed and mixity do not alter this requirement significantly.

The non-singular stress can change in magnitude and sign as the crack propagates. When calculating K_1^d and K_2^d without considering the effect of the non-singular stress, K_1^d and K_2^d will be forced to compensate for the effect of non-singular stress. This causes the discrepancy between

K^d_{photo} and K^d_{gage} in figure 5.10 as explained by Ricci *et al* [6]. However by using strain gages of small grid size and placing them close to the crack path, such errors can be minimized to a level acceptable for practical applications. If one decides to install the gages in the stiffer half i.e. Material 2, then the optimum gage orientation has to be determined following the steps outlined earlier.

6. Dynamic fracture of interface between isotropic and orthotropic materials

The use of composite materials is on the rise in recent years and many times hybrid systems consisting of composite and metallic parts joined together are used. Invariably this joining has to be through a bond which is essentially an interface across which there is a sudden change in the elastic properties along with a change in the material nature. Though metals are isotropic, composites being orthotropic will bring in additional complexities depending on the orientation of the material axes relative to the interface. Ricci *et al* [7] have effectively used strain gages to study the behavior of cracks propagating along the interface between an isotropic and orthotropic material.

6.1 *Strain field equations*

Crack propagation is considered subsonic for crack-tip velocities, v, below the shear wave velocity, c_s, of the more compliant material. Figure 6.1 illustrates the moving coordinate system at the tip of a crack propagating subsonically along an interface between two elastic materials: Material-1, the more compliant material and Material-2, the stiffer orthotropic material. Considering the orthotropic constitutive relations for Material 2 along with the stress field equations given by Deng [15] and Lee [20], the stress field for the present system was obtained by Lee [21]. The stress field is coupled oscillatory in the mismatch parameter similar to that given in equation 5.1; however, the mismatch parameter will further depend on the orientation of the material axis with respect to the interface. Using this stress field and the appropriate stress-strain relation one can obtain the singular strain field

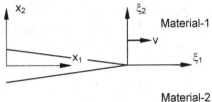

x₁, x₂ Fixed coordinates.
ξ₁, ξ₂ Coordinates moving with crack tip

Figure 6.1. Coordinate system for a crack propagating along a bi-material interface.

for either material. For the sake of brevity, only the final expressions are given and the reader is referred to [7, 21] for the definition of the terms and constants in the following equations. For material 1, the in plane strain components are

$$
\begin{aligned}
(\varepsilon_x)_1 = {} & \frac{K_1^d}{2\sqrt{2\pi r}\,\mu_1 D_1 \cosh(\varepsilon\pi)} \left\{ \left[-e^{\varepsilon(\pi-\theta_l)}\overline{A_1}\cos\left(\varepsilon\ln\frac{r_l}{a}-\frac{\theta_l}{2}\right) + e^{-\varepsilon(\pi-\theta_l)}A_1\cos\left(\varepsilon\ln\frac{r_l}{a}+\frac{\theta_l}{2}\right) \right] f_l(\theta) \right. \\
& \left. - \beta_s \left[e^{\varepsilon(\pi-\theta_s)}\overline{B_1}\cos\left(\varepsilon\ln\frac{r_s}{a}-\frac{\theta_s}{2}\right) + e^{-\varepsilon(\pi-\theta_s)}B_1\cos\left(\varepsilon\ln\frac{r_s}{a}+\frac{\theta_s}{2}\right) \right] f_s(\theta) \right\} \\
+ {} & \frac{K_2^d}{2\sqrt{2\pi r}\,\mu_1 D_1 \cosh(\varepsilon\pi)} \left\{ \left[e^{\varepsilon(\pi-\theta_l)}\overline{A_1}\sin\left(\varepsilon\ln\frac{r_l}{a}-\frac{\theta_l}{2}\right) - e^{-\varepsilon(\pi-\theta_l)}A_1\sin\left(\varepsilon\ln\frac{r_l}{a}+\frac{\theta_l}{2}\right) \right] f_l(\theta) \right. \\
& \left. + \beta_s \left[e^{\varepsilon(\pi-\theta_s)}\overline{B_1}\sin\left(\varepsilon\ln\frac{r_s}{a}-\frac{\theta_s}{2}\right) + e^{-\varepsilon(\pi-\theta_s)}B_1\sin\left(\varepsilon\ln\frac{r_s}{a}+\frac{\theta_s}{2}\right) \right] f_s(\theta) \right\}
\end{aligned}
$$

$$
\begin{aligned}
(\varepsilon_y)_1 = {} & \frac{K_1^d}{2\sqrt{2\pi r}\,\mu_1 D_1 \cosh(\varepsilon\pi)} \left\{ \beta_l^2 \left[e^{\varepsilon(\pi-\theta_l)}\overline{A_1}\cos\left(\varepsilon\ln\frac{r_l}{a}+\frac{\theta_l}{2}\right) - e^{-\varepsilon(\pi-\theta_l)}A_1\cos\left(\varepsilon\ln\frac{r_l}{a}-\frac{\theta_l}{2}\right) \right] f_l(\theta) \right. \\
& \left. + \beta_s \left[e^{\varepsilon(\pi-\theta_s)}\overline{B_1}\cos\left(\varepsilon\ln\frac{r_s}{a}+\frac{\theta_s}{2}\right) + e^{-\varepsilon(\pi-\theta_s)}B_1\cos\left(\varepsilon\ln\frac{r_s}{a}-\frac{\theta_s}{2}\right) \right] f_s(\theta) \right\} \\
+ {} & \frac{K_2^d}{2\sqrt{2\pi r}\,\mu_1 D_1 \cosh(\varepsilon\pi)} \left\{ -\beta_l^2 \left[e^{\varepsilon(\pi-\theta_l)}\overline{A_1}\sin\left(\varepsilon\ln\frac{r_l}{a}+\frac{\theta_l}{2}\right) - e^{-\varepsilon(\pi-\theta_l)}A_1\sin\left(\varepsilon\ln\frac{r_l}{a}-\frac{\theta_l}{2}\right) \right] f_l(\theta) \right. \\
& \left. - \beta_s \left[e^{\varepsilon(\pi-\theta_s)}\overline{B_1}\sin\left(\varepsilon\ln\frac{r_s}{a}+\frac{\theta_s}{2}\right) + e^{-\varepsilon(\pi-\theta_s)}B_1\sin\left(\varepsilon\ln\frac{r_s}{a}-\frac{\theta_s}{2}\right) \right] f_s(\theta) \right\}
\end{aligned}
$$

$$\left(\gamma_{xy}\right)_1 = \frac{K_1^d}{4\sqrt{2\pi r}\,\mu_1 D_1 \cosh(\varepsilon\pi)}\left\{2\beta_l\left[e^{\varepsilon(\pi-\theta_l)}\overline{A_1}\sin\left(\varepsilon\ln\frac{r_l}{a}-\frac{\theta_l}{2}\right)+e^{-\varepsilon(\pi-\theta_l)}A_1\sin\left(\varepsilon\ln\frac{r_l}{a}+\frac{\theta_l}{2}\right)\right]f_l(\theta)\right.$$

$$+\left(1+\beta_s^2\right)\left[e^{\varepsilon(\pi-\theta_s)}\overline{B_1}\sin\left(\varepsilon\ln\frac{r_s}{a}-\frac{\theta_s}{2}\right)-e^{-\varepsilon(\pi-\theta_s)}B_1\sin\left(\varepsilon\ln\frac{r_s}{a}+\frac{\theta_s}{2}\right)\right]f_s(\theta)\right\}$$

$$+\frac{K_2^d}{4\sqrt{2\pi r}\,\mu_1 D_1 \cosh(\varepsilon\pi)}\left\{2\beta_l\left[e^{\varepsilon(\pi-\theta_l)}\overline{A_1}\cos\left(\varepsilon\ln\frac{r_l}{a}-\frac{\theta_l}{2}\right)+e^{-\varepsilon(\pi-\theta_l)}A_1\cos\left(\varepsilon\ln\frac{r_l}{a}+\frac{\theta_l}{2}\right)\right]f_l(\theta)\right.$$

$$+\left(1+\beta_s^2\right)\left[e^{\varepsilon(\pi-\theta_s)}\overline{B_1}\cos\left(\varepsilon\ln\frac{r_s}{a}-\frac{\theta_s}{2}\right)-e^{-\varepsilon(\pi-\theta_s)}B_1\cos\left(\varepsilon\ln\frac{r_s}{a}+\frac{\theta_s}{2}\right)\right]f_s(\theta)\right\}$$

$$(6.1)$$

For material 2, the strain components are

$$\left(\varepsilon_x\right)_2 = \frac{K_1^d}{2\sqrt{2\pi r}\,D_2 \cosh(\varepsilon\pi)}\left\{p_l\left[e^{-\varepsilon(\pi+\theta_l)}\overline{A_2}\cos\left(\varepsilon\ln\frac{r_l}{a}-\frac{\theta_l}{2}\right)+e^{\varepsilon(\pi+\theta_l)}A_2\cos\left(\varepsilon\ln\frac{r_l}{a}+\frac{\theta_l}{2}\right)\right]f_l(\theta)\right.$$

$$-p_s\left[e^{-\varepsilon(\pi+\theta_s)}\overline{B_2}\cos\left(\varepsilon\ln\frac{r_s}{a}-\frac{\theta_s}{2}\right)+e^{\varepsilon(\pi+\theta_s)}B_2\cos\left(\varepsilon\ln\frac{r_s}{a}+\frac{\theta_s}{2}\right)\right]f_s(\theta)\right\}$$

$$+\frac{K_2^d}{2\sqrt{2\pi r}\,D_2 \cosh(\varepsilon\pi)}\left\{-p_l\left[e^{\varepsilon(\pi+\theta_l)}\overline{A_2}\sin\left(\varepsilon\ln\frac{r_l}{a}-\frac{\theta_l}{2}\right)+A_2\sin\left(\varepsilon\ln\frac{r_l}{a}+\frac{\theta_l}{2}\right)\right]f_l(\theta)\right.$$

$$+p_s\left[e^{-\varepsilon(\pi+\theta_s)}\overline{B_2}\sin\left(\varepsilon\ln\frac{r_s}{a}-\frac{\theta_s}{2}\right)+B_2\sin\left(\varepsilon\ln\frac{r_s}{a}+\frac{\theta_s}{2}\right)\right]f_s(\theta)\right\}$$

$$\left(\varepsilon_y\right)_2 = \frac{K_1^d}{2\sqrt{2\pi r}\,D_2 \cosh(\varepsilon\pi)}\left\{pq_l\left[e^{-\varepsilon(\pi+\theta_l)}\overline{A_2}\cos\left(\varepsilon\ln\frac{r_l}{a}-\frac{\theta_l}{2}\right)+e^{\varepsilon(\pi+\theta_l)}A_2\cos\left(\varepsilon\ln\frac{r_l}{a}+\frac{\theta_l}{2}\right)\right]f_l(\theta)\right.$$

$$-qq_s\left[e^{-\varepsilon(\pi+\theta_s)}\overline{B_2}\cos\left(\varepsilon\ln\frac{r_s}{a}-\frac{\theta_s}{2}\right)+e^{\varepsilon(\pi+\theta_s)}B_2\cos\left(\varepsilon\ln\frac{r_s}{a}+\frac{\theta_s}{2}\right)\right]f_s(\theta)\right\}$$

$$+\frac{K_2^d}{2\sqrt{2\pi r}\,D_2 \cosh(\varepsilon\pi)}\left\{-pq_l\left[e^{-\varepsilon(\pi+\theta_l)}\overline{A_2}\sin\left(\varepsilon\ln\frac{r_l}{a}-\frac{\theta_l}{2}\right)+A_2\sin\left(\varepsilon\ln\frac{r_l}{a}+\frac{\theta_l}{2}\right)\right]f_l(\theta)\right.$$

$$+qq_s\left[e^{-\varepsilon(\pi+\theta_s)}\overline{B_2}\sin\left(\varepsilon\ln\frac{r_s}{a}-\frac{\theta_s}{2}\right)+B_2\sin\left(\varepsilon\ln\frac{r_s}{a}+\frac{\theta_s}{2}\right)\right]f_s(\theta)\right\}$$

$$
\begin{aligned}
\left(\gamma_{xy}\right)_2 =\ & \frac{K_1^d a_{66}}{4\sqrt{2\pi}\, D_2 \cosh(\varepsilon\pi)}\left\{\alpha_l\left[e^{-\varepsilon(\pi+\theta_l)}\overline{A_2}\sin\left(\varepsilon\ln\frac{r_l}{a}-\frac{\theta_l}{2}\right)-e^{\varepsilon(\pi+\theta_l)}A_2\sin\left(\varepsilon\ln\frac{r_l}{a}+\frac{\theta_l}{2}\right)\right]f_l(\theta)\right. \\
& -\alpha_s\left[e^{-\varepsilon(\pi+\theta_s)}\overline{B_2}\sin\left(\varepsilon\ln\frac{r_s}{a}-\frac{\theta_s}{2}\right)-e^{\varepsilon(\pi+\theta_s)}B_2\sin\left(\varepsilon\ln\frac{r_s}{a}+\frac{\theta_s}{2}\right)\right]f_s(\theta)\bigg\} \\[2mm]
& +\frac{K_2^d a_{66}}{4\sqrt{2\pi}\, D_2 \cosh(\varepsilon\pi)}\left\{\alpha_l\left[e^{-\varepsilon(\pi+\theta_l)}\overline{A_2}\cos\left(\varepsilon\ln\frac{r_l}{a}-\frac{\theta_l}{2}\right)-e^{\varepsilon(\pi+\theta_l)}A_2\cos\left(\varepsilon\ln\frac{r_l}{a}+\frac{\theta_l}{2}\right)\right]f_l(\theta)\right. \\
& -\alpha_s\left[e^{-\varepsilon(\pi+\theta_s)}\overline{B_2}\cos\left(\varepsilon\ln\frac{r_s}{a}-\frac{\theta_s}{2}\right)-e^{\varepsilon(\pi+\theta_s)}B_2\cos\left(\varepsilon\ln\frac{r_s}{a}+\frac{\theta_s}{2}\right)\right]f_s(\theta)\bigg\}
\end{aligned}
$$

$$\text{(6.2)}$$

For a strain gage oriented at an angle θ_g with the interface (x_l) axis, the strain can be obtained through appropriate strain transformation relations. The final expressions can be written as follows

$$\varepsilon = C_1(\varepsilon,r,\theta,\theta_g,v)K_1^d - C_2(\varepsilon,r,\theta,\theta_g,v)K_2^d \qquad (6.3)$$

where $K^d = K_1^d + iK_2^d$ is the complex stress intensity factor.

One can easily appreciate the fact that the strain field equations have a form similar to equation 5.3 for the isotropic case; however, the angular functions, mismatch parameter and the other constants used to obtain C_1 and C_2 are defined differently. The overall methodology for obtaining the preferred gage orientation therefore is very much similar to that discussed in section 5.4; however, the orientation of the material axis of the orthotropic material with respect to the interface (denoted as α subsequently) will be an additional governing parameter.

6.2. *Optimum gage orientation and location*

To ensure that the singular strain field describes the gage strain accurately, the gage has to be located at a vertical distance of 0.5B from the interface, B being the thickness of the material. For explanation, a bi-material consisting of PSM-1 as the isotropic half and Soctchply 1002, a unidirectional glass fiber composite in epoxy matrix as the orthotropic half will be considered with the strain gage installed in the orthotropic half (Material 2). The material properties of this system are provided in table 6.1. The two cases of $\alpha = 0°$, for which the fibers are oriented parallel to the interface i.e. along the x_1 direction and $\alpha = 90°$, for which the fibers are perpendicular to the interface will be analyzed.

Table 6.1. Properties of Scotchply1002 -PSM-1 bi-material system

	PSM-1[*]	Scotchply® 1002[**]
Young's Modulus, E (GPa)	2.76	$E_L = 30$ $E_T = 7$
Poisson's Ratio, ν	0.38	$\nu_{LT} = 0.25$
Density, ρ (kg/m³)	1200	1860

[*] Manufactured by Measurements Group, Raleigh, NC, USA
[**] 3M® unidirectional, glass-fiber-reinforced epoxy

The first step is to arrive at a gage orientation which will record maximum strain when the crack tip approaches the gage datum for the range of anticipated crack speeds. This can be determined by plotting the iso-strain contours around the crack tip for a given value of K^d for the range $0 < \theta_g < 180$ and identifying the angle for which the contour apogee falls on the vertical axis through the crack tip. For the material combination used, this exercise gives angle of $\theta_g=55°$ for $\alpha = 0°$ and $\theta_g=100°$ for $\alpha = 90°$. Subsequently theoretical strain profiles are generated for a given K^d for different gage orientations which are close to the values chosen from the earlier exercise to reconfirm optimality.

The distance of the crack tip from the gage datum at peak strain is then inspected for a range of crack speeds and mixity. Figure 6.2 shows the location of crack tip when gage records peak strain for both fiber orientations ($\alpha=0$ and 90). For the case of $\theta_g=55°$ and $\alpha = 0°$, shown in figure 6.2 (a), the crack tip is very close to the gage datum when the gage records peak strain and change in velocity or mixity does not shift the location of crack tip considerably.

For $\theta_g=100°$ which corresponds to $\alpha = 90°$, figure 6.2 (b) indicates that the crack tip is behind the gage datum when peak strain is recorded by the gage. However, the relative shift in the crack tip location with

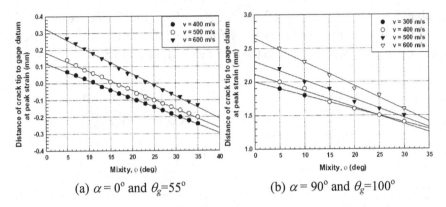

(a) $\alpha = 0°$ and $\theta_g = 55°$ (b) $\alpha = 90°$ and $\theta_g = 100°$

Figure 6.2. Crack-tip location relative to the gage datum for varying mixity and velocity [7].

change in velocity and mixity is still considerably small. This leads to the conclusion that if one assumes that the crack tip is at the gage datum when peak strain is recorded by the gage, the error in the velocity estimate due to this assumption is still negligible. However the same is not true for in the case of K_1^d and K_2^d as the absolute location of the crack tip relative to the gages is required in this case.

6.3 *Analysis of strain gage data*

The analysis procedure is identical to that given in section 5.3 for the isotropic bi-materials. The crack speed is first estimated from the time records of the peak strains from each gage and the gage spacing. Subsequently, a two gage analysis or three gage analysis described earlier can be used to determine K_1^d and K_2^d. Ricci *et al* [7] investigated crack propagation along the interface between PSM-1 and Scotchply 1002 for two different fiber orientations, $\alpha = 0$ and 90 using the specimen geometry and loading similar to that shown in figure 5.8. The fracture phenomena was simultaneously studied using dynamic photoelasticity. Typical strain pulses for the case of $\alpha=0$ are shown in figure 6.3. The estimated crack speed from the strain peaks was 600 m/sec and this was in close agreement with that estimated from the

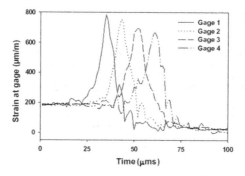

Figure 6.3. Typical strain records for a propagating interfacial crack in a isotropic-orthotropic bi-material for fibers parallel to the interface, $\alpha = 0°$ [7].

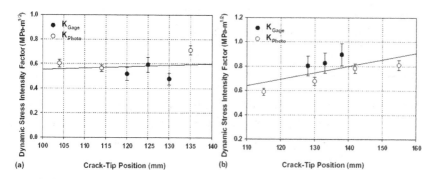

Figure 6.4. Magnitude of Complex Stress intensity factor for (a) $\alpha = 0°$, (b) $\alpha = 90°$.

high speed photographs. The magnitude of the complex stress intensity factor as a function of the instantaneous crack length obtained from the strain signals as well as from the analysis of photoelastic fringes is shown in figure 6.4 for both fiber orientations. One could see that there is close agreement between the two.

7. Closure

Experimental observation of dynamic failure of materials resulting from fast crack propagation is highly challenging due to the rapidness of the phenomena and a comprehensive thought process should precede the design of the experiment and choice of the technique. This chapter and the papers cited [1-7] demonstrate the applicability of the strain gage technique, which is simple, elegant and robust, to obtain reliable information on the behavior of propagating cracks in a variety of situations involving systems of different nature. Appropriate choice of the gage size, its location and the orientation is mandatory for the success of this approach. This is achieved through an extensive parametric investigation of the strain profiles before the experiment is undertaken.

Finally, we would like alert the reader that the cases discussed in this chapter are mostly steady state crack growth. The crack velocity is essentially constant and there are no drastic changes in stress and strain field with time. For situations that are different from this, one might have to use gages that are more closely spaced and reevaluate the whole procedure for the particular problem at hand.

Acknowledgements

Arun Shukla would like to acknowledge the support of the National Science Foundation, Air Force Office Of Scientific Research and the Office of Naval Research over the years for his research on dynamic failure.

References

1. Berger, J. R., Dally, J. W. and Sanford, R. J., "Determining the dynamic stress intensity factor with strain gages using a crack tip locating algorithm", *Engineering Fracture Mechanics*, Vol. 36, No. 1, pp. 145-156 (1990).
2. Sanford, R. J., Dally, J. W. and Berger, J. R., "An Improved Strain gage method for measuring K_{ID} for a propagating crack", *Journal of Strain Analysis*, Vol.25, No.3, pp. 177-183 (1990).
3. Shukla, A., Agarwal, R. K. and Nigam, H., "Dynamic fracture studies of 7075-T6 aluminum and 4340 steel using strain gages and photoelastic coatings", *Engineering Fracture Mechanics*, Vol.31, No.3, pp.501-515 (1988).
4. Shukla, A. and Khanna, S. K., "On the use of strain gages in dynamic fracture mechanics", *Engineering Fracture Mechanics*, Vol. 51, No.6, pp. 933-948 (1995).
5. Khanna, S. K. and Shukla, A., "Development of stress field equations and determination of stress intensity factor during dynamic fracture of orthotropic composite materials", *Engineering Fracture Mechanics*, Vol. 47, No.3, pp. 345-359 (1994).

6. Ricci, V., Shukla, A. and Kavaturu, M., "Using strain gages to investigate subsonic interfacial fracture in an isotropic-isotropic biomaterial", *Engineering Fracture Mechanics*, Vol. 70, pp. 1303-1321 (2003).
7. Ricci, V., Shukla, A., Chalivendra, V. B. and Lee, K. H., "Subsonic interfacial fracture using strain gages in an isotropic-orthotropic biomaterial" *Theoretical and Applied Fracture Mechanics*, Vol. 39, pp. 143-161 (2003).
8. Dally, J. W. and Riley, W. F., "Experimental Stress Analysis", Third Edition, College House Enterprises LLC, Tennessee, USA
9. Ueda, K. and Umeda, A., "Dynamic Response of Strain Gages up to 300 kHz", Experimental Mechanics, Vol. 38, No. 2, pp. 93-98 (1998).
10. Irwin, G. R., "Series representation of the stress field around constant speed cracks", University of Maryland Lecture Notes (1980).
11. Shukla, A. and Chona, R., "Dynamic crack tip stress fields for propagating cracks", *Advances in Fracture Research*, Vol. 5, pp. 3167-3176 (1984).
12. Piva, A. and Viola, E., "Crack propagation in an orthotropic medium", *Engineering Fracture Mechanics*, 29(5), pp. 535-548, (1988).
13. Yang, W., Suo, Z. and Shih, C. H., "Mechanics of dynamic debonding", *Proceedings of the Royal Society (London)*, A433, 679-697, (1991).
14. Wu, K. C., "Explicit crack-tip fields of an extending interface crack in an anisotropic material", *International Journal of Solids and Structures*, 27(4), 455-466, (1991).
15. Deng, X., "Complete complex series expansions of near-tip fields for steadily growing interface cracks in dissimilar isotropic materials", *Engineering Fracture Mechanics*, 42(2), 237-242, (1992).
16. Liu, C., Lambros, J. and Rosakis, A. J., "Highly transient elastodynamic crack growth in a bimaterial interface: higher order asymptotic analysis and optical experiments", *Journal of the Mechanics and Physics of Solids*, 41(12), 1857-1954, (1993).
17. Singh, R. P. and Shukla, A., "Characterization of isochromatic fringe patterns for a dynamically propagating interface crack", *International Journal of Fracture*, 76, 293-310, (1996).
18. Singh, R. P. and Shukla, A.,"Subsonic and intersonic crack growth along a bimaterial interface", *Journal of Applied Mechancis*, 63, 919-924, (1996).
19. Singh, R. P., Kavaturu, M. and Shukla, A., "Initiation, propagation and arrest of an interface crack subjected to controlled stress wave loading", *International Journal of Fracture*, 83, 291-304, (1997).
20. Lee, K.H., Hawong, J. S. and Choi, S. H., "Dynamic stress intensity factors K_I, K_{II} and dynamic crack propagation characteristics of orthotropic material", *Engineering Fracture Mechanics*, 53, 119–140, (1997).
21. Lee, K.H., "Stress and displacement fields for propagation crack along interface of isotropic–orthotropic materials under dynamic mode II and II load", *Transactions of the KSME*, A23, 1463–1475, (1999).

Chapter 6

Dynamic And Crack Arrest Fracture Toughness

Richard E. Link

*Mechanical Engineering Department, United States Naval Academy,
590 Holloway Road, Annapolis, MD 21402, USA*

Ravinder Chona

*United States Air Force Research Laboratory,
Wright-Patterson Air Force Base, OH 45433, USA*
Ravi.Chona@wpafb.af.mil

This chapter is primarily concerned with the dynamic propagation behavior of a crack in a brittle material and the arrest event that occurs when self-sustaining, rapid crack propagation ceases. The basic fracture mechanics concepts of dynamic fracture and crack arrest are reviewed. Experimental procedures for measuring the dynamic and crack arrest fracture toughness are presented and the limitations of the procedures are discussed.

1. Basic Concepts

In order for a crack to initiate and propagate in a structure, the stresses and strains in the vicinity of the crack tip must be high enough to separate the material and there must be sufficient energy available to form the new surfaces and drive the crack forward. Under linear elastic conditions, the stress intensity factor, K, provides a single parameter that describes the stress and strain fields in the vicinity of the crack tip, outside the crack tip plastic zone and is related to the energy release rate,

G, through the relationship: $G = K^2/E$. A crack will initiate when the applied stress intensity equals the critical value for the material, termed the fracture toughness, K_c.

Once initiated at the onset of a fast fracture event, a rapidly propagating crack experiences a sudden increase in crack velocity. The crack will continue to propagate as long as the crack driving force, expressed in terms of G or K, exceeds the resistance to fracture of the material. The dynamic fracture toughness of a rapidly propagating crack can be considerably less than the initiation toughness and, quite often, crack initiation is followed by complete and catastrophic failure of the structure. Whether or not a crack will propagate completely through a structure is a function of the evolution of the crack driving force as the crack extends and the material dynamic fracture toughness of the material.

The rate of crack tip advance or, crack speed, is a function of the material's dynamic fracture toughness, $\dot{a} = \dot{a}(K_{ID}, T)$, where $\dot{a}$ is the crack speed, K_{ID} is the dynamic fracture toughness of the propagating crack and T is the temperature of the material. This relationship is shown schematically in Figure 1. At low stress intensities, the crack speed increases rapidly with small changes in the stress intensity factor.

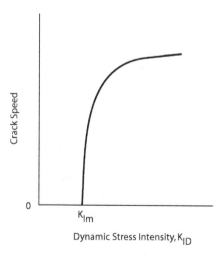

Fig. 1. Schematic representation of the characteristic relationship between crack speed and the dynamic stress intensity, K_{ID}.

At high stress intensities, the crack speed approaches a terminal velocity and is relatively insensitive to changes in the stress intensity. As the crack approaches the terminal velocity, it may begin to branch into two or more active cracks. The value of the stress intensity factor that corresponds to zero velocity is the minimum stress intensity required to maintain crack growth, K_{Im}. The crack will arrest if the stress intensity falls below this value and K_{Im} represents the true crack arrest fracture toughness of the material. The steep decrease in crack velocity as the dynamic stress intensity approaches K_{Im} leads to an abrupt arrest event.

Analytical techniques for predicting the dynamic fracture behavior of a structure require characterization of the dynamic fracture toughness (as a function of crack speed) and the crack arrest toughness of the relevant structural material. Several approaches for characterizing the dynamic and crack arrest fracture toughness have evolved over the last 50 years. This chapter reviews several aspects of dynamic fracture toughness and crack arrest including:

- experimental characterization of the dynamic fracture toughness of various materials;
- initial attempts at characterizing the crack arrest performance of structural steel alloys;
- a summary of the research leading to the development of a standard method for measurement of the crack arrest toughness;
- large-scale, dynamic fracture experiments used to validate small specimen test procedures and analytical procedures for predicting structural behavior; and
- recent advances in characterizing the crack arrest behavior of steel alloys.

2. Measurement of dynamic fracture toughness

Direct measurements of the dynamic fracture toughness of a rapidly propagating crack and the corresponding crack speed relationship of engineering structural materials are relatively rare due to the difficulty in making such measurements. The experimental techniques for measuring the dynamic fracture toughness of running cracks include optical methods such as photoelasticity[1] and the method of caustics[2,3] and strain gage-based methods[4]. Indirect methods couple experimental measurements of key boundary conditions with numerical analysis techniques, such as finite elements, to infer the fracture toughness during dynamic crack propagation[5].

Photoelasticity was used to make some of the earliest experimental observations of the stress field surrounding a running crack[6]. High-speed flash photography was used to record the isochromatic fringe patterns associated with a running crack in a brittle polymer. Irwin[7] demonstrated that the stress intensity factor could be determined from the size and shape of the isochromatic fringe loops surrounding the crack tip.

Dynamic photoelasticity was subsequently used for many model studies of the dynamic fracture phenomenon. Kobayashi and Dally[1] used the technique to measure the relationship between crack speed and the stress intensity factor for a birefringent polymer, Homalite 100, as shown in Figure 2. Their work showed that there was a unique relationship

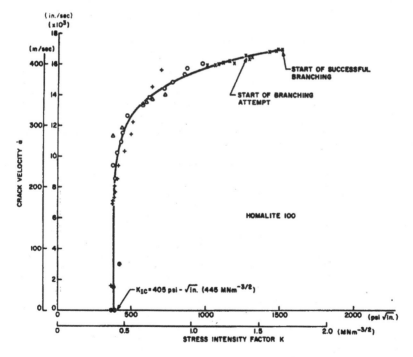

Fig. 2. Crack speed vs. stress intensity relationship for a brittle polymer, Homalite 100[1]

between the instantaneous crack velocity and the dynamic stress intensity factor for each material, independent of the specimen geometry. They observed that the $\dot{a}\,(K_{ID})$ relationship was bounded by a terminal velocity at high values of stress intensity and a minimum stress intensity for continued crack propagation, K_{Im}.   Crack branching was noted at sufficiently high values of the stress intensity once the terminal velocity had been achieved.

The dynamic photoelasticity technique was not limited to transparent materials and strictly model studies.   Kobayashi and Dally[8] were successful in using a photoelastic coating applied to an AISI 4340 alloy steel compact specimen to determine the $\dot{a}\,(K_{ID})$ relationship for that steel.   The $\dot{a}\,(K_{ID})$ relationship for the 4340 steel specimens, shown in Figure 3, exhibited the same general characteristics as the curves

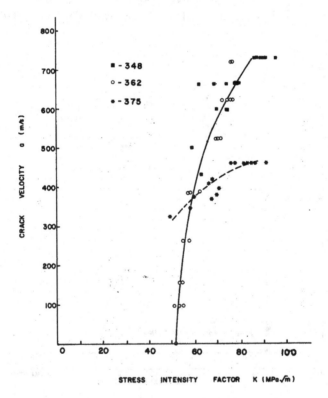

Fig. 3. Crack speed vs. stress intensity relationship for 4340 alloy steel [8].

developed for the photoelastic polymers. Two of the three specimens produced consistent $\dot{a}$ (K_{ID}) relationship. The third specimen had a much lower terminal velocity than the first two and a much smoother fracture surface. They attributed this difference to a variation in the heat treatment of that specimen.

The method of caustics coupled with high-speed photography has been used by several investigators to measure the dynamic stress intensity factor in transparent polymers and in high strength structural alloys[2,9,10]. The high stresses surrounding the crack tip produce local out-of-plane contraction of the material surrounding the crack tip. This deformation acts as a diverging lens and scatters the light rays passing through a transparent specimen. The resulting shadow is called a caustic and the size of the spot can be related to the stress intensity factor through the asymptotic crack tip stress fields and the laws of optics for the particular optical arrangement used in the experiment. For opaque materials such as structural metals, the technique can be applied in a reflection mode by observing the highly polished surface of the specimen.

Kalthoff, *et al.*[2,9] used the technique to quantify dynamic effects in typical fracture specimen geometries during dynamic crack propagation and arrest. Their work demonstrated significant difference between the true dynamic stress intensity factor compared with that determined from a static analysis. Immediately after initiation, the dynamic stress intensity fell below the statically calculated value as stored elastic energy was converted to kinetic energy in the specimen. At longer times, the dynamic stress intensity exceeded the static value as kinetic energy in the specimen was converted back into strain energy. The dynamic effects were most severe for rectangular double-cantilever beam specimens and decreased for tapered DCB specimens and were the smallest for a compact specimen geometry. They observed significant oscillations in the stress intensity factor after crack arrest which damped out with time.

Strain gage techniques have long been used for observations of dynamic fracture events. Brittle fracture experiments by Videon and coworkers[11] employed strain gages for estimating the crack speed in large plate fracture experiments. Eftis and Kraft[12] analyzed the strain gage results reported by Videon *et al*[11]., to demonstrate a relationship between the crack speed and the stress intensity factor. It should be noted that Eftis and Krafft[12] assumed a one-parameter, static representation of the crack tip stress field which leads to some errors in the actual results. Nevertheless, their results qualitatively showed the typical crack speed vs. K relationship reported by others using optical methods.

Berger, *et al.*[13] developed a strain gage-based procedure for determining both the crack tip position and the stress intensity factor in dynamic fracture experiments. They employed a three-parameter characterization of the dynamic crack tip stress field to relate the strains measured near the crack path to the crack tip position and the instantaneous driving force. The technique was used to determine the dynamic stress intensity factor and the crack speed history in a wide-plate crack arrest test. Since the wide-plate specimen had a thermal gradient along the crack path, the resulting $\dot{a}$ (K_{ID}, T) relationship was actually a trajectory of a path on the crack speed - stress intensity - temperature surface for the plate material as shown in Figure 4.

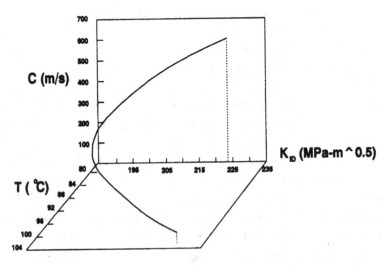

Fig. 4. Crack speed vs. stress intensity and temperature for A533B steel plate from a wide plate experiment[4].

3. Characterization of crack arrest toughness

3.1 *Temperature-based approach to crack arrest*

The earliest studies of crack arrest were directed at characterizing the conditions of stress and temperature that would prevent catastrophic failure of steel plates in which a fast-running crack was initiated. Robertson[14] devised a test to determine the ductile arrest temperature of steel plates. The test piece, shown in Figure 5, was welded to pull tabs and subjected to a remote tensile force. A temperature gradient was imposed across the specimen width by cooling the notched region and heating the opposite end of the specimen. A brittle crack was initiated by impacting the rounded end of the specimen with a projectile fired from a gun. The brittle crack propagated across the specimen into progressively warmer, and therefore tougher, material until it arrested. A series of tests were conducted at various stress levels to determine the relationship between the arrest temperature and the stress level. A typical result from Robertson's experiments is shown in Figure 6. At the lower temperatures, there is a relationship between the arrest temperature and the applied stress. At a certain temperature, which Robertson called the ductile arrest temperature, the crack always arrested independent of the

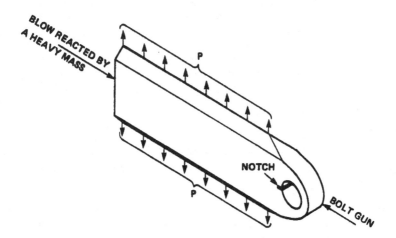

Fig. 5. Schematic drawing of the test specimen in the Robertson crack arrest test. The test specimen was welded to pull tabs and subjected to a remote tensile force.

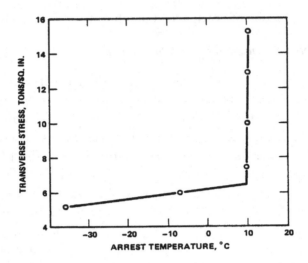

Fig. 6. Typical result from a series of Robertson crack arrest tests. The ductile arrest temperature in this example is 10°C.

applied stress level. The ductile arrest temperature was shown to be a function of specimen thickness.

Feely and co-workers[15] at Esso Research and Engineering Co. devised a similar test, the Esso test, that was used to determine the "Esso brittle temperature." In the Esso test, a large single-edge notch specimen (16 in. wide x 60 in. long) is subjected to a remote tensile stress of 18 ksi. and cooled to a uniform temperature. A brittle crack is initiated by firing a wedge from a gun into a short transverse notch on one edge of the specimen. This is a "go or no-go" test in which the crack either propagates entirely across the specimen or arrests after a short distance. The lowest temperature at which the crack arrests is termed the "Esso brittle temperature".

The explosion bulge test was developed by Pellini and co-workers[16] at the Naval Research Laboratory to determine the upper temperature limit for crack propagation under purely elastic loads. This temperature was designated the nil-ductility transition temperature (NDTT). In the explosion bulge test, the test piece has a brittle weld placed in the center of one surface and the weld bead is notched. The specimen is clamped in a rigid circular die and a crack is initiated by subjecting the specimen to

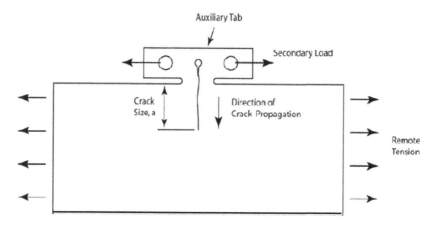

Fig. 7. Double tension crack arrest specimen geometry.

an explosive load. The NDTT is the lowest temperature at which the crack arrests before propagating through the specimen. Subsequently, a smaller specimen, the drop-weight specimen was developed and standardized in ASTM E208, to determine the NDTT.

Another test used in early crack arrest studies is the double tension test developed by Yoshiki and Kanazawa[17]. The double tension crack arrest specimen is shown in Figure 7. Similar to the Esso test, this is a "go or no-go" test when performed as an isothermal test. The specimen is subjected to a remote tension and the crack is initiated in this test by means of an auxiliary loading tab on the notched edge of the specimen. The notch is embrittled to promote initiation of a brittle fracture and the crack either arrests when it enters the main test section or propagates all the way across the specimen. This test is used to determine a crack arrest temperature that is a function of the applied stress level and specimen thickness.

3.2 *Fracture mechanics approach to crack arrest*

The temperature-based approach to crack arrest had several significant drawbacks. The crack arrest temperatures were specific to a particular stress level and material thickness. Results could not be easily

transferred to other loading conditions and the tests required large
specimens and were expensive to perform. A fracture mechanics based
approach to crack arrest could potentially avoid some of the short-
comings of the temperature-based approach and could lead to a
predictive methodology that the temperature-based approach could not.
Irwin and Wells[18] proposed that crack arrest could be viewed as a simple
time reversal of the crack initiation event. Viewed in this manner, it was
suggested that there might exist a material property, the crack arrest
toughness, K_{IA}, analogous to the fracture initiation toughness, K_{Ic}.

3.3 Measurement of the Crack Arrest Fracture Toughness

Some of the earliest reported measurements of the crack arrest fracture
toughness were reported by Crosley and Ripling[19,20]. They used tapered,
double cantilever beam (TDCB) specimens to investigate the dynamic
crack initiation and arrest fracture toughness of pressure vessel steels.
The tapered specimen design, shown in Figure 8, had the advantage of
having a constant ratio of stress intensity, K, to force, P, over a wide
range of crack lengths. Only the initiation and arrest loads were needed
to determine the fracture toughness. Under fixed displacement
conditions, the stress intensity factor in the TDCB specimen decreases

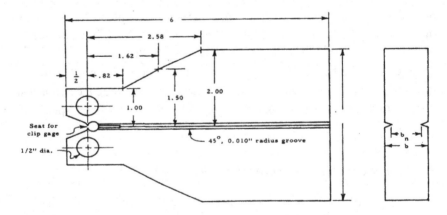

Fig. 8. Tapered double cantilever beam specimen employed in initial crack arrest tests of
Crosley and Ripling[19].

with increasing crack length so that the crack would tend to arrest as it propagated across the ligament. They calculated the crack arrest toughness based on a static analysis of the specimen and the load measured immediately after the crack had arrested. They designated the crack arrest toughness with the symbol, K_{Ia}. Their results are presented in Figure 9.

They observed that the crack arrest toughness, K_{Ia}, was insensitive to the amount of unstable crack extension and exhibited less scatter than the initiation toughness. The crack arrest data established a lower bound to the initiation toughness that was not as sensitive to specimen thickness as the initiation toughness, K_{Ic}. The crack arrest data for the 25.4 mm (1 inch) thick specimens tested at low loading rates showed an increase in K_{Ia}. It was suggested that this increase was due to low crack speeds in these tests resulting from non-plane-strain conditions in the thinner specimens.

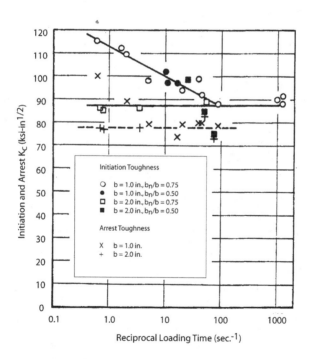

Fig. 9. Crack initiation as a function of loading rate and crack arrest toughness from DCB specimen tests of Crosley and Ripling[9].

Soon after Crosley and Ripling published their initial results, it was argued by Hahn and co-workers[21] that a static approach was unsuitable for crack arrest determinations and that a dynamic approach was necessary. Kanninen et al.[22] performed a dynamic analysis of the DCB specimen using a beam on elastic foundation model of the specimen and a finite difference approach to solve the equations of motion for the specimen. They computed the dynamic energy release rate, G, from the relationship:

$$G = \frac{1}{B}\left\{\frac{dW}{da} - \frac{dU}{da} - \frac{dT}{da}\right\} \qquad (1)$$

where

B = specimen thickness,
W = work done by external forces,
U = elastic strain energy,
T = kinetic energy,
a = crack length

The dynamic analysis showed that inertial effects were important to properly describe the dynamic crack propagation, arrest and possible reinitiation events in specimens of finite geometry. Their calculations predicted that the loading system compliance could have a great influence on the crack propagation and arrest behavior in a DCB specimen. Figure 10 compares calculations of the crack extension in a contoured DCB specimen loaded by a wedge, which results in a relatively stiff system, and by a more compliant, machine-loaded system. The wedge-loaded specimen exhibits a single run-arrest event while the machine-loaded system produces several bursts of crack initiation and arrest followed by reinitiation.

The calculations also predicted that the dynamic stress intensity factor would exhibit significant post-arrest oscillations about the static arrest value as shown in Figure 11. These oscillations could lead to reinitiation of the arrested crack and further supported the notion that a static analysis was unsuitable for the analysis of crack arrest behavior. The analysis further indicated that K_{Ia} was sensitive to the amount of

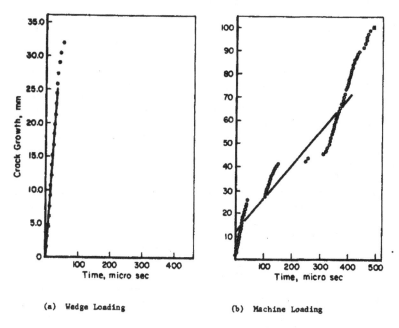

(a) Wedge Loading   (b) Machine Loading

Fig. 10. Predicted dynamic crack growth vs. time for a specimen with a stiff loading system (a), and a complint loading system (b), showing dynamic behavior of system influences the crack growth and arrest behavior [22].

crack jump and the initiation value, K_Q, during the test. The static approach was therefore concluded to only be valid for very small crack jumps and low crack velocities.

Experimental measurements of K_{ID} and crack velocity made by Kalthoff and Winkler[2] lent support to the idea that a dynamic analysis was necessary to accurately determine the crack arrest fracture toughness. They used high speed photography and the method of caustics to make direct observations of the local deformations surrounding a rapidly propagating crack tip in brittle epoxies. The local contraction of the material surrounding the crack tip acts as a diverging lens and produces a shadow spot pattern at the crack tip. The size of the shadow spot is related to the instantaneous stress intensity factor. An example of their results for the wedge loaded DCB specimen is presented in Figure 12. The figure shows that K_{ID} falls below the statically

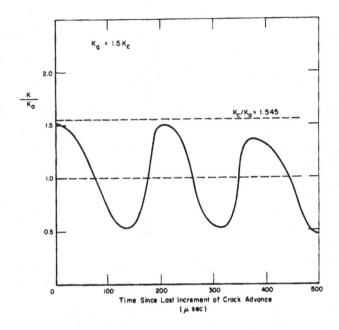

Fig. 11. Post crack arrest oscillations in the stress intensity factor from dynamic analysis of a run-arrest event in a DCB specimen[22].

calculated value of K just after initiation and then increases above the static value as the crack propagates through the specimen. At arrest, the dynamically calculated value is considerably higher than the value calculated based on a static analysis - up to 50% for specimens that achieved the highest crack velocities and the longest crack jumps. For very small crack jumps and low crack velocities, the difference between the static and dynamic values was shown to be small. The experiments by Kalthoff and co-workers also demonstrated that the dynamic stress intensity factor exhibited large oscillations after arrest in the DCB specimen as shown in Figure 13. This confirmed the behavior predicted from the computational models of Kanninen, et al.[22].

An alternative crack arrest test procedure, based on a dynamic analysis of the crack propagation and arrest process, was developed by researchers at Battelle Columbus Laboratories[23]. They utilized a duplex DCB specimen that had a high strength AISI 4340 steel crack starter

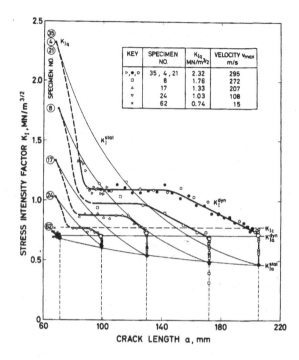

Fig. 12. Experimental measurements of the dynamic stress intensity vs. crack length compared with the statically calculated value in a series of tests on DCB specimens[2].

section that was electron beam welded to the test material. The specimen had a blunt notch to permit a large amount of elastic energy to be stored in the specimen prior to crack initiation. The high strength starter section limited the amount of plasticity at the blunt notch.

The initiation load was used to determine the stress intensity at crack initiation, K_Q, and the amount of crack extension to the point of arrest, Δa, was measured. The dynamic stress intensity during the test, K_{ID}, and the crack velocity, were determined from a calculated relationship between the ratio K_{ID}/K_Q, Δa, and crack velocity shown in Figure 14. In this manner, each crack arrest test yielded an estimate of K_{ID} and crack velocity for a specimen. The K_{ID} vs. crack velocity results from a series of tests were fit to a straight line to determine the value of K_{ID} corresponding to zero velocity that represented an estimate of K_{Im}.

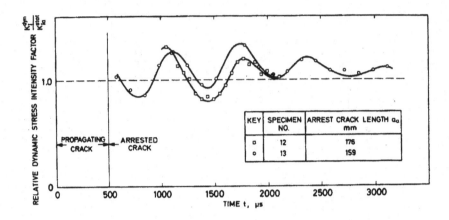

Fig. 13. Experimental measurements of the post-arrest oscillations in the stress intensity factor [2].

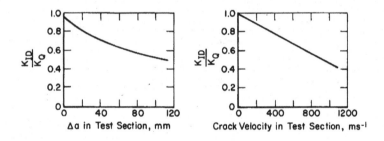

Fig. 14. Calibration curves for determining K_{ID} and crack speed in a duplex crack arrest specimen[23].

The American Society for Testing and Materials (ASTM) initiated the Cooperative Test Program to examine the two procedures for measuring the crack arrest toughness with the aim of developing a standard test method[24]. A wedge-loaded compact specimen design, the compact crack arrest (CCA) specimen, was utilized in the test program. The specimen, shown in Figure 15, is loaded by a wedge thru a split pin inserted along the crack line. This produces a stiff loading system that approaches fixed displacement conditions during the run-arrest event. A notched, brittle weld bead serves as a crack starter at the root of the

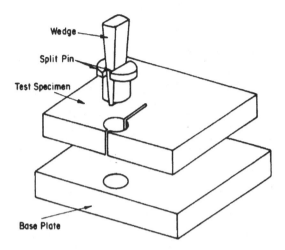

Fig. 15. Compact crack arrest specimen and split-pin wedge loading arrangement used in the cooperative test program.

machined notch. An alternative design employed a duplex specimen that consisted of a hardened AISI 4340 steel crack starter section that was electron beam welded to the test material. The specimens had sharp grooves machined along the intended crack path to help control the crack path and suppress the tendency of the crack to branch. The test is conducted by heating or cooling the specimen to the desired test temperature and then slowly driving the wedge into the split pins until a brittle crack initiates and arrests. The static approach calculated K_{Ia} based on the arrested crack length and the crack opening displacement at crack arrest measured on the notched edge of the specimen a short time after arrest. The dynamic approach used the procedure described in the preceding paragraphs to determine K_{ID} based on the initiation stress intensity, K_Q, and the crack jump, Δa.

The results of the Cooperative Test Program supported the static approach to calculating K_{Ia}[25]. The crack arrest values calculated using the static approach were insensitive to the initiation value, K_Q, as shown in Figure 16. The K_{ID} values increased as a function of K_Q as shown in Figure 17 and this behavior would be expected if the crack velocity was

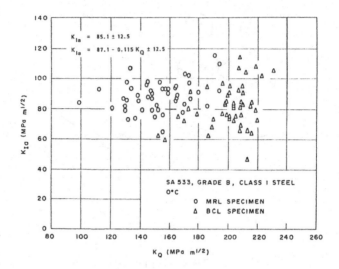

Fig. 16. Summary of K_{Ia} as a function of initiation stress intensity from tests conducted in the cooperative test program[25].

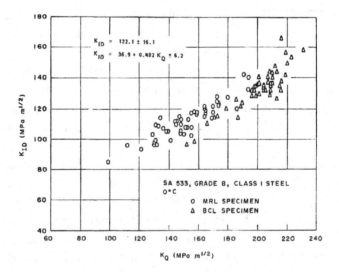

Fig. 17. Summary of K_{ID} as a function of initiation stress intensity, K_Q from tests conducted in the cooperative test program. Note that the minimum K_{ID} values approach the K_{Ia} values plotted in Figure 16[25].

a function of K_{ID}. Higher K_Q values provide more stored elastic energy and should drive the crack at a higher velocity. The lower bound of the K_{ID} data are in agreement with the scatter band in the K_{Ia} data, showing that the dynamic and static approaches yielded similar values of the crack arrest toughness, $K_{Im} \approx K_{Ia}$.

The similarity between K_{Im} using the dynamic approach and K_{Ia} from the static approach was explained by further experimental measurements of K_{ID} and crack velocity in steel specimens using the compact specimen geometry. Kalthoff, *et al.*[9] again used high speed photography and the method of caustics in a reflection mode to determine K_{ID} and the crack velocity in steel specimens of various geometries including the DCB and compact specimen. The results presented in Figure 18 and 19 showed that the transverse-wedge loaded compact specimen was not as sensitive to dynamic effects as the DCB specimen and that the statically calculated K_{Ia} was very close to the experimentally measured value. The DCB specimens had a much greater dynamic effect which was attributed to the flexibility of the loading arms of the specimen.

Based on the results of the Cooperative Test Program, ASTM conducted a follow-on round-robin to validate a standardized procedure for measuring the crack arrest fracture toughness using the compact crack arrest specimen geometry[26]. The procedure used in the round-robin eventually became the Standard Method for the Determination of Plane-Strain Crack-Arrest Fracture Toughness, ASTM E1221. The test method incorporates several validity requirements to ensure that the measured crack arrest fracture toughness is insensitive to specimen size and that the static approach employed in the analysis is suitable. For example, the crack must propagate beyond the initial plastic zone that develops at the notch tip and the arrested crack tip must be sufficiently removed from the specimen edge so that the crack tip plastic zone at arrest is completely surrounded by elastic material. There are additional requirements on crack front straightness, specimen thickness and the extent of unbroken ligaments behind the crack tip.

It can be difficult to obtain strictly valid test results using the test method depending upon the temperature selected for conducting the crack arrest test. In the upper transition region it can be difficult to

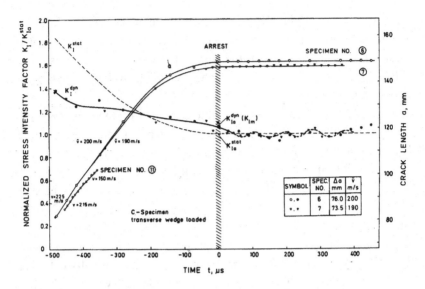

Fig. 18. Experimentally measured dynamic stress intensity, K_{ID}, vs. time in a compact specimen compared with the statically calculated stress intensity factor[9].

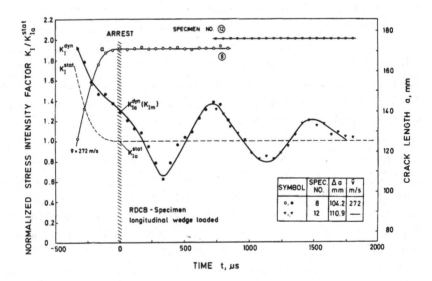

Fig. 19. Experimentally measured dynamic stress intensity factor, K_{ID} vs. time for a DCB specimen compared with statically calculated value[9].

initiate a brittle crack reliably. Once a crack does initiate, it may propagate too far, branch or curve from the center plane of the specimen. The test method is also unsuitable for relatively tough materials in thinner sections where the specimen has insufficient thickness to meet the validity requirements. The success rate for the tests conducted as part of the ASTM round-robin was less than 50%[26]. Other researchers have also reported a low success rate with the test procedure when using it to evaluate weldments and tough steel alloys used for modern ship hulls[27,28]. Notwithstanding these limitations, the test method is still a useful tool.

Bonenberger and Dally[29] suggested several modifications to the test method in an attempt to increase the success rate and extract additional information from the test. The most notable suggestion is the use of a chevron-style sidegroove. The crack front is very narrow at the initial crack tip to increase constraint and promote crack initiation at a low load. As the crack propagates along the crack plane, the effective thickness of the specimen increases, which reduces the driving force and promotes crack arrest. They reported an improved success rate by employing these modifications but their validation involved only a few specimens.

4. Large-scale crack arrest tests

A number of alternative experimental procedures have been developed to investigate dynamic fracture and crack arrest behavior. Many of these procedures involve large-scale tests to better simulate structural conditions relevant to a particular problem of interest. Others seek to develop alternative methods of measuring the crack arrest fracture toughness for situations where the ASTM standard test method cannot be applied successfully. Several examples of these approaches are presented below.

4.1 *Wide plate tests*

The U.S. Nuclear Regulatory Commission sponsored a series of crack arrest experiments using very large single edge-notch tension, SE(T), specimens which have been called wide-plate tests. The nominal specimens were approximately 10m long by 1m wide with a thickness of 100-150 mm as shown in Figure 20. The objective of this series of tests was to characterize the crack arrest fracture toughness of prototypical nuclear reactor grade pressure vessel steels in the upper region of the ductile-brittle transition region, near the onset of the upper shelf of the Charpy transition curve. These experiments were also designed to provide dynamic fracture data for relatively large amounts of crack extension and serve as validation tests for predictive models of dynamic fracture and crack arrest[30]. The specimen size was selected to minimize the effect of reflected stress waves on the crack propagation and arrest events.

The SE(T) specimen exhibits an increasing driving force with crack length under fixed grip and fixed load conditions. In order to arrest a brittle crack in a rising K-field, such as that in the SE(T) specimen, it is necessary to have the fracture toughness increase across the width of the

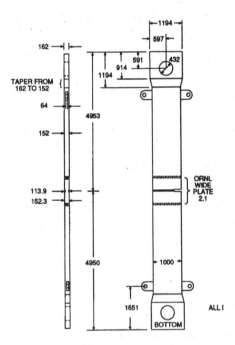

Fig. 20. Wide plate specimen geometry used to determine crack arrest toughness in the upper transition region of reactor pressure vessel steels. (All dimensions are in mm).

specimen so that it exceeds the crack driving force and thus arrests the crack. This was achieved in the wide-plate test series by applying a linear thermal gradient across the test section, cooling the notched edge of the specimen and heating the opposite edge so that the fracture toughness increased across the specimen along with the temperature.

The specimens were instrumented with crack opening displacement gages, strain gages and thermocouples. Strain gages were mounted on the surfaces of the specimen along a line approximately 100mm above and parallel to the crack plane. The strain gages were used to infer the crack tip position and crack velocity during the crack run-arrest event. The tests were analyzed using an elastodynamic, generation-mode, finite element analysis[5,30]. In a generation-mode analysis, the crack tip position as a function of time is used as a known boundary condition in the finite element model. The remote boundary condition was assumed to be either fixed load or fixed displacement.

The results from a typical wide plate test are plotted in Figure 21. This specimen featured seven run-arrest events and crack arrest values were calculated from the dynamic finite element analyses for each arrest

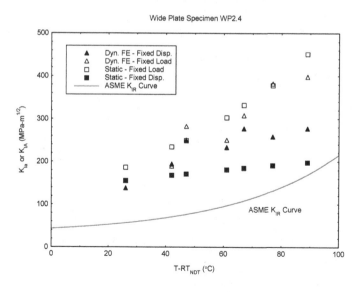

Fig. 21. Crack arrest toughness vs. temperature from static analysis and generation mode dynamic finite element analysis of several run-arrest events in a wide plate test.

event. Crack arrest values were also calculated using static formulas for K based on crack length and both fixed load and fixed displacement conditions. For the first two arrest events occurring at the lowest temperatures, the predictions from the two finite element analyses are in good agreement. At later times, corresponding to the arrival of reflected stress waves from the load points to the crack tip, the finite element results start to diverge with the fixed load case approaching the fixed load static solution. The correct remote boundary condition lies somewhere between the two limiting cases that were analyzed. The results based on a static analysis for either assumed boundary condition do not consistently agree with the dynamic finite element results, indicating that a static analysis is not suitable for this specimen and these test conditions.

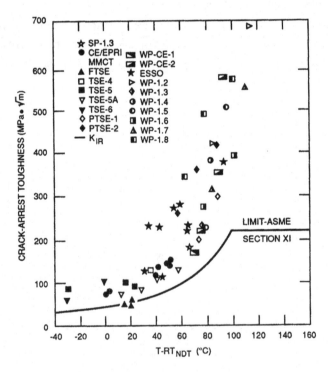

Fig. 22. Summary plot of crack arrest toughness transition curve determined from large-scale wide plate specimen, thermal shock and pressurized thermal shock tests[30].

The wide-plate and thermal shock tests were generally successful in characterizing the crack arrest behavior of a reactor pressure vessel steel in the upper region of the ductile brittle transition as shown in Figure 22.

4.2 Double tension test

The double tension test specimen has been used not only to determine a crack arrest temperature as described previously, but also to determine the crack arrest fracture toughness[31-33]. The test is conducted as an isothermal test and the crack initiates in the embrittled crack starter section by applying a load to the secondary loading tabs. The embrittled crack starter region extends into the main test section and is grooved along one face to help control the crack path and promote a brittle crack initiation. The crack either propagates entirely across the specimen or arrests a short distance into the main test section.

It is possible to determine a crack arrest toughness for tests that exhibit crack arrest. The crack arrest toughness is calculated from a static analysis based on the arrested crack length, the applied stress in the main test section, and a dimensionless function of plate geometry and crack length. The calculated crack arrest toughness will depend on whether constant load or constant displacement boundary conditions are assumed in the static analysis as shown in Table 1, which lists results from double tension crack arrest tests on 1.5% Ni steel reported by Wiesner, *et al.*[33]. The specimens were 880 mm wide and 1500 mm long. The use of a static approach for analyzing the double tension test specimen is based on several arguments. Previous work[2] has demonstrated that dynamic effects are diminishingly small for short crack jumps and low crack velocities. Willoughby and Wood[31] contend that the static approach is valid as long as reflected stress waves from the specimen boundary do not interact with the run-arrest event. A dynamic analysis of the double tension specimen indicated that the dynamic stress intensity factor was less than 15% lower than the static stress intensity factor for short arrested crack lengths[34]. Once reflected stress waves come into play, a dynamic analysis of the specific specimen becomes necessary.

Table 1 - Crack arrest results from double tension specimen tests of 1.5% Ni steel reported by Wiesner, *et al.*[33].

Specimen Identification Code	M3-17	M3-23	M3-24	M3-25
Temperature (°C)	-87	-92	-87	-87
Arrested crack length, a_a (mm)	156	62	280	66
Time at arrest (μs)	260	80	300	66
K_a (constant load) (MPa-m$^{1/2}$)	264	147	288	188
K_a (constant displacement) (MPa-m$^{1/2}$)	209	139	173	176

The time, t_s, for an unloading stress wave to reach a specimen boundary and return to the crack tip is given by $t_s = 2l/c$, where l is the distance from the crack tip to the nearest specimen boundary and c is the velocity of the stress wave (approximately 5000 m/s in steel). The time for the crack to propagate and arrest, t_c, can be estimated from $t_c = \Delta a/v$, where Δa is the crack jump and v is the average crack speed. Typical crack speeds for brittle cracks in steel range from 100-1500 m/s.

For the tests listed in Table 1, the cracks arrest at relatively short distances (arrested crack length/specimen width < 32%) in the main plate. For the shortest arrested cracks in Table 1, the difference between the crack arrest values based on constant load and constant displacement is small, on the order of 6%. The crack propagation times, t_c, listed in the table are less than t_s, which is approximately 300μs. For the other two tests, both exhibiting longer arrested cracks and longer crack propagation times, the difference between the calculations is greater than 25%. Neither of the assumed boundary conditions is correct and dynamic effects may be coming into play.

4.3 Thermal shock and pressurized thermal shock tests

Under certain postulated accident scenarios, such as a loss of coolant accident, nuclear pressure vessels may be subjected to emergency cooling procedures which can result in a thermal shock on the inner surface of the pressure vessel that may also be accompanied by an increase in the internal pressure. Preexisting flaws on the inner surface of the vessel may initiate brittle cracks as a consequence of the high thermal stresses that develop on the inner wall surface and the cooling of the material into the ductile-brittle transition region. The pressure vessels must be able to arrest any crack before it can propagate through the vessel wall thickness. Several large-scale thermal shock experiments (TSE) and pressurized thermal shock experiments (PTSE) have been performed to investigate the crack arrest behavior of cylindrical pressure vessels under loading conditions representative of those predicted for various thermal shock and pressurized thermal shock accident conditions[35-37].

A schematic of the experimental configuration for a typical pressurized thermal shock test is shown in Figure 23. The tests were modeled using finite element models with varying degrees of sophistication to determine the dynamic stress intensity factor and the crack arrest toughness of the material. The analyses assumed either static or dynamic conditions to predict the crack arrest behavior of the vessel. The dynamic analyses were performed using an "application mode" analysis where the dynamic stress intensity vs. crack speed relationship is specified as a material constitutive parameter and the

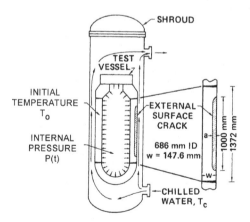

Fig. 23. Example of pressure vessel geometry used in a pressurized thermal shock experiment. (All dimensions are in mm)[36].

crack tip position as a function of time is predicted. Results by Cheverton, *et al.*[38], Brickstad and Nilsson[39,40] and Keeney-Walker and Bass[41] demonstrated that static analyses produced conservative estimates of crack arrest in thermal shock and pressurized thermal shock tests compared with dynamic analyses.

Results from an analysis by Brickstad and Nilsson[40] postulated loss of coolant accident in a pressure vessel, resulting in a pressurized thermal shock, are presented in Figure 24 in which the static and dynamic crack driving force are plotted as a function of crack size. Crack arrest is predicted to occur when the crack driving force falls below the crack arrest curve, K_{Ia}. For the PTS experiment, K_{Ia} is a function of crack size owing to the thermal gradient through the vessel wall and the dependence of K_{Ia} on temperature. In this example, the static analysis, shown by the solid line, predicts a longer, more conservative, estimate of the arrested crack length, $a_a \approx 80$mm, than the dynamic analysis result of $a_a \approx 60$ mm.

Jung and Kanninen[42] analyzed several thermal shock experiments and showed that static analyses could produce slightly non-conservative

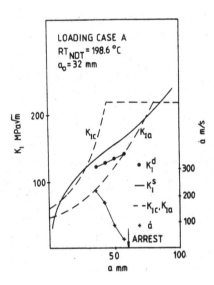

Fig. 24. Comparison of static and dynamic analyses to predict the crack arrest behavior in a pressurized thermal shock test[40].

predictions of crack arrest but suggested that static analyses were nevertheless useful for predicting the dynamic fracture response of pressure vessels under thermal shock conditions.

It is not entirely clear under which conditions a static analysis will be non-conservative. For short crack jumps, static and dynamic analyses appear to produce nearly identical results in both small specimens and in more complex structures[43]. Most of the analyses cited in the literature tend to indicate that a static analysis will lead to a conservative prediction of crack arrest. Static analyses are not generally suitable for predicting the post arrest behavior of a structure when the possibility of reinitiation of the arrested crack must be considered. The post-arrest behavior will be governed by the dynamic response of the structure and the dynamic initiation and propagation toughness of the material [41].

5. Characterization of the crack arrest transition curve

Crack initiation and crack arrest toughness measurements of steels in the ductile-brittle transition region exhibit considerable scatter. In the case of crack initiation toughness, the scatter is a result of the local crack tip stress field, the spatial distribution of cleavage initiation sites in the material surrounding the crack tip and a statistical size effect of the crack front length. A so-called "Master Curve Approach" has been developed by Wallin for characterizing the cleavage fracture initiation toughness of steels[4]. In the master curve approach, the scatter at a fixed temperature in the ductile brittle transition region is described by a three-parameter Weibull distribution where the probability of failure is given by:

$$P_f = 1 - \exp\left\{ -\frac{B}{B_0}\left[\frac{K_I - K_{\min}}{K_0 - K_{\min}} \right] \right\} \qquad (2)$$

where B is the specimen thickness, B_0 is a normalizing thickness, K_I is the fracture toughness corresponding to a failure probability, P_f, $K_{\min}$ is a fitting parameter usually taken as 20 MPa-m$^{1/2}$, and K_0 is the fracture toughness corresponding to a 63.2% probability of failure. The temperature dependence of the mean fracture initiation toughness follows an empirical master curve given by the expression

$$K_I = 30 + 70 \exp\left[0.019\left(T - T_o\right)\right] \tag{3}$$

where K_I is the mean fracture toughness, T is the temperature (deg. C), and T_o is the reference temperature corresponding to a mean fracture toughness of 100 MPa-m$^{1/2}$. The master curve approach has been demonstrated to characterize the cleavage initiation behavior of many ferritic steels and weldments and an ASTM standard, E1921 Test Method for the Determination of Reference Temperature, T_o, for Ferritic Steels in the Transition Range, has been published.

It has been shown that the same master curve used to characterize the initiation toughness can be used to characterize the temperature dependence of the crack arrest toughness of ferritic steels[4,5]. The crack arrest master curve is identical to the crack initiation master curve except that the reference temperature, T_o, is replaced by the crack arrest reference temperature, T_{KIa}. The crack arrest fracture toughness for nine different sets of K_{Ia} data for pressure vessel steel is plotted in Figure 25 along with the crack arrest master curve.

Wallin[45] points out that the scatter and size effects on initiation toughness and the arrest toughness are expected to be different as a result of different fracture mechanisms. Whereas crack initiation is a result of triggering a local initiating particle, crack arrest requires that the local driving force decrease below the crack arrest toughness over a much larger region of the crack front. Therefore, the crack arrest toughness is more a function of the average matrix properties versus local inhomogeneities. Consequently, the crack arrest toughness should exhibit less scatter than the initiation toughness and the statistical size effect seen in initiation should not be present in crack arrest.

At a given temperature, the scatter in crack arrest toughness is assumed to follow a log-normal distribution instead of a three-parameter Weibull distribution. For the data shown in Figure 25, the combined scatter of the data set is $\sigma = 18\%$ compared with $\sigma = 28\%$ for the initiation toughness.

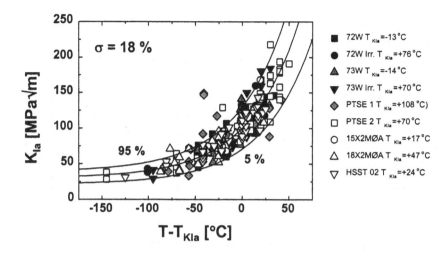

Fig. 25. Crack arrest master curve for nine different pressure vessel steels with a wide range of crack arrest reference temperatures [45].

The relationship between the reference temperature, T_o, corresponding to crack initiation and T_{KIa} for crack arrest is not constant but varies considerably between materials[46]. It is possible to develop relationships between T_o and T_{KIa} for specific materials or classes of materials. Such relationships would facilitate prediction of the crack arrest properties of a material from the crack initiation toughness.

6. Summary

An overview has been presented of the principal investigations that have led up to the current thinking on how best to characterize the crack arrest toughness of a structural material. By its very nature, the crack arrest event must be preceded by a segment of rapid, self-sustaining crack propagation and dynamic fracture and crack arrest behaviors are thus intertwined.

Investigations seeking an improved understanding of the crack arrest phenomenon have always been challenged by the complexity that results from the need to properly account for dynamic effects that necessarily become part of the picture. For this reason, much attention has been given in the past, and will probably continue to be given in the future, to creating test conditions that allow quasistatic analyses to be used without significant error for laboratory-scale experiments. Equal amounts of attention have been given to creating scenarios in which statically-determined values of the crack arrest fracture toughness are conservative estimates of the true dynamic fracture toughness associated with the actual instant of arrest. The compromise that results is nothing other than the age-old one where the desire for full and complete accuracy has to be balanced against the need for simplicity of experimental procedure, accompanying instrumentation, and complexity of associated analyses.

Much progress has been made in the past four decades on better understanding how dynamic fracture and crack arrest concepts can be brought to bear on assuring the safety of critical structures such as nuclear reactor pressure vessels. It can reasonably be expected both that such efforts will continue and that such efforts will extend into other arenas as the costs of replacing an aging infrastructure become increasingly higher in many industrialized nations.

Crack arrest fracture toughness is an attractive parameter for both design and safety assessments. If designers can ensure that the driving force associated with any fast-running crack that develops in a structure will fall below the crack arrest fracture toughness of the material from which the structure is fabricated, and does so before the crack exits the structure, then an arrest event is guaranteed. If conditions can be created in which no re-initiation of the arrested crack takes place, then the integrity of the structure is assured. The resulting avoidance of catastrophic structural failures, with their accompanying unacceptable economic and societal consequences, is a worthwhile goal indeed.

References

1.   Kobayashi, T. and Dally, J.W., "Relation Between Crack Velocity and the Stress Intensity Factor in Birefringent Polymers," *Fast Fracture and Crack Arrest, ASTM STP 627*, G.T. Hahn and M.F. Kanninen, Eds., American Society for Testing and Materials, pp. 257-273, 1977.

2.   Kalthoff, J.F., Beinert, J. and Winkler, S., "Measurements of dynamic stress intensity factors for fast running and arresting cracks in double-cantilever-beam

specimens," *Fast Fracture and Crack Arrest, ASTM STP 627*, G.T. Hahn and M.F. Kanninen, Eds., American Society for Testing and Materials, pp.161-176, 1977.

3. Rosakis, A.J., "Analysis of the Optical Method of Caustics for Dynamic Crack Propagation," *Engineering Fracture Mechanics*, Vol. 13, No. 2, pp. 331-347, 1980.

4. Berger, J.R., Dally, J.W and Sanford, R.J., "Determining the Dynamic Stress Intensity Factor with Strain Gages Using a Crack Tip Locating Algorithm," *Engineering Fracture Mechanics*, Vol. 36, No. 1, pp. 145-156, 1990.

5. Bass, B.R. et al., "Fracture Analysis of Heavy-Section Steel Technology Wide-Plate Crack-Arrest Experiments," *Fracture Mechanics, Nineteenth Symposium, ASTM STP 969*, T.A. Cruse, Ed., American Society for Testing and Materials, Philadelphia, pp. 691-723, 1980.

6. Wells, A.A. and Post, D., "The Dynamic Stress Distribution Surrounding a Running Crack – A Photoelastic Analysis," *S.E.S.A. Proceedings*, Vol. 16, No. 1, pp. 69-92, 1958.

7. Irwin, G.R., Discussion of "The Dynamic Stress Distribution Surrounding a Running Crack – A Photoelastic Analysis," *S.E.S.A. Proceedings*, Vol. 16, No. 1, pp. 93-96, 1958.

8. Kobayashi, T. and Dally, J.W., "Dynamic Photoelastic Determination of the a-dot - K Relation for 4340 Alloy Steel," *Crack Arrest Methodology and Applications, ASTM STP 711*, American Society for Testing and Materials, Philadelphia, pp.189-210, 1980.

9. Kalthoff, J.F., Beinert, J., Winkler, S. and Klemm, W., "Experimental analysis of dynamic effects in different crack arrest test specimens," *Crack Arrest Methodology and Applications, ASTM STP 711*, American Society for Testing and Materials, Philadelphia, pp.109-127, 1980.

10. Rosakis, A.J. and Zehnder, A.T., "On the dynamic fracture of structural metals," *International Journal of Fracture*, 27, pp. 169-186, 1985.

11. Videon, F.F., Barton, F.W., Hall, W.J., "Brittle Fracture Propagation Studies," Ship Structure Series Report SSC-148, Office of Technical Services, Washington, D.C., Aug. 1963.

12. Eftis, J. and Krafft, J.M., "A Comparison of the Initiation With the Rapid Propagation of a Crack in a Mild Steel Plate," Journal of Basic Engineering, Transactions of the ASME, pp. 257-263, March 1965

13. Berger, J.R., Dally, J.W., deWitt, R. and Fields, R.J., "A Strain Gage Analysis of Fracture in Wide Plate Tests of Reactor Grade Steel," *Journal of Pressure VEssel Technology, Transactions of the ASME*, Vol. 115, pp. 398-405, 1993.

14. Robertson, T.S., "Propagation of brittle fracture in steels," *Jour. Of the Iron and Steel Inst.*, 175, pp. 361-374, 1953.

15. Feely, F.J., Jr., Northup, M.S., Kleppe, S.R., and Gensamer, M., "Studies on brittle failure of tankage steel plates," Welding Journal, 34, pp. 596s-607s, 1955.

16. Pellini, W.S. and Puzak, P.P., "Fracture analysis diagram procedures for fracture-safe design of steel structures," *Welding Research Council Bulletin*, No. 88, May 1963.

17. Yoshiki, M. and Kanazawa, T., "On the mechanism of propagation of brittle fracture in mild steel," *J. Soc. Naval Arch. Japan*, 102, p. 39, 1958.

18.   Irwin, G.R. and Wells, A.A., "A continuum mechanics view of crack propagation,"
      Metallurgical Reviews, 91, pp. 223-270, 1965
19.   Crosley, P.B. and Ripling, E.J., "Dynamic fracture toughness testing of A533B
      steel," *Jour. Basic Eng., Trans. ASME D*, 91, p. 525, 1969.
20.   Crosley, P.B. and Ripling, E.J., "Crack arrest toughness of pressure vessel steels,"
      Nuc. Eng. Des., 17, pp. 32-45, 1971.
21.   Hahn, G.T., Hoagland, R.G., Kanninen, M.F., and Rosenfield, A.R., "A
      preliminary study of fast fracture and arrest in the DCB specimen," *Proc. Conf.
      Dynamic Crack Propagation*, Noordhoff, pp. 679-692, 1973.
22.   Kanninen, M.F., Popelar, C., and Gehlen, P.C., "Dynamic Analysis of Crack
      Propagation and arrest in the double-cantilever-beam specimen," *Fast Fracture
      and Crack Arrest, ASTM STP 627*, G.T. Hahn and M.F. Kanninen, Eds., American
      Society for Testing and Materials, pp. 19-38, 1977.
23.   Hoagland, R.G., Rosenfield, A.R., Gehlen, P.C., and Hahn, G.T., "A crack arrest
      measuring procedure for K_{Im}, K_{ID} and K_{Ia} properties," *Fast Fracture and Crack
      Arrest, ASTM STP 627*, G.T. Hahn and M.F. Kanninen, Eds., American Society for
      Testing and Materials, pp. 177-202, 1977.
24.   Hahn, G.T., Hoagland, R.G., Rosenfield, A.R., and Barnes, C.R., "A Cooperative
      Test Program for Evaluating Crack Arrest Test Methods," *Crack Arrest
      Methodology and Applications, ASTM STP 711*, G.T. Hahn and M.F. Kanninen,
      Eds., American Society for Testing and Materials, pp. 248-269, 1980.
25.   Crosley, P.B., et al., "Final Report on Cooperative Test Program on Crack Arrest
      Toughness Measurements," NUREG/CR-3261, U.S. Nuclear Regulatory
      Commission, Washington, D.C. April 1983.
26.   Barker, D.B., Chona, R., Fourney, W.L., and Irwin, G.R., "A Report on the Round-
      Robin Program Conducted to Evaluate the Proposed ASTM Standard Test Method
      for Determining the Plane Strain Crack Arrest Fracture Toughness, K_{Ia}, of Ferritic
      Materials," NUREG/CR-4996, U.S. Nuclear Regulatory Commission, Washington,
      D.C., January 1988.
27.   Pussegoda, L.N., Malik, L., Morrison, J., "Measurement of crack arrest fracture
      toughness of a ship steel plate," *Journal of Testing & Evaluation*, 26(3), p 187-
      197, 1998.
28.   Burch, I.A., Ritter, J.C., Saunders, D.S., Underwood, J.H. "Crack arrest fracture
      toughness testing of naval construction steels," *Journal of Testing & Evaluation*,
      26(3), p269-276, 1998.
29.   Bonenberger, R.J. and Dally, J.W., "On Improvements in Measuring Crack Arrest
      Toughness," *Int. J. Solids Structures*, 32(6/7), pp. 897-909, 1995.
30.   Naus, D.J., et al., ""Crack-Arrest Behavior in SEN Wide Plates of Low-Upper-
      Shelf Base Metal Tested Under Nonisothermal Conditions: WP-2 Series,"
      NUREG/CD-5451, US Nuclear Regulatory Commission, Washington, D.C., 1990.
31.   Willoughby, A.A., and Wood, A.M., "Crack Initiation and Arrest in 9% Ni Steel
      Weldments at Liquefied Natural Gas Temperatures," *Proc. Conf. Transport and
      Storage of LPG and LNG*, Koninklijke Vlaamse Ingenieurs Vereniging,
      Antwerpen (Belgium), pp. 111-118, 1984.

32. Watanabe, I. and Yamagata, S., "Propagation and arrest of brittle cracks in 9% Ni steel structures for LNG service," *Proc. Conf. Transport and Storage of LPG and LNG*, Koninklijke Vlaamse Ingenieursvereniging, Antwerpen (Belgium), pp. 127-136, 1984.

33. Wiesner, C.S., Hayes, B., Smith, S.D. and Willoughby, A.A., "Investigations into the Mechanics of Crack Arrest in Large Plates of 1.5% Ni TMCP Steel," *Fatigue and Fracture of Engineering Materials and Structures*, 17(2), pp. 221-233, 1994.

34. Curr, R.M. and Turner, C.E., "The Numerical Modeling of Crack Propagation and Arrest in Wide Plates of C-Mn Steel," TWI Report 5554/10/87, Appendix A, The Welding Institute, November 1987.

35. Cheverton, R.D., Canonico, D.A., Iskander, S.K., Bolt, S.E., Holz, P.P., Nanstad, R.K., Stelzman, W.J., "Fracture Mechanics Data Deduced From Thermal-Shock and Related Experiments With LWR Pressure Vessel Material," *Journal of Pressure Vessel Technology*, 105, pp. 102-110, May 1983.

36. Bryan, R.H., et al., "Pressurized-Thermal-Shock Test of a 6-inch-thick Pressure Vessels, PTSE-1: Investigation of Warm Prestressing and Upper Shelf Arrest," NUREG/CR-4106, Oak Ridge National Laboratory, Oak Ridge, TN, April, 1985.

37. Bryan, R.H., Merkle, J.G., Nanstad, R.K., and Robinson, G.C., "Pressurized Thermal Shock Experiments with Thick Vessels," *Fracture Mechanics: Nineteenth Symposium ASTM STP 969*, T.A. Cruse, Ed., American Society for Testing and Materials, Philadelphia, pp. 767-783, 1988.

38. Cheverton, R.D., Gehlen, P.C., Hahn, G.T., and Iskander, S.K., "Application of crack arrest theory to a thermal shock experiment," *Crack Arrest Methodology and Applications, ASTM STP 711*, American Society for Testing and Materials, Philadelphia, pp. 392-421, 1980.

39. Brickstad, B., and Nilsson, F., "Dynamic Analysis of Crack Growth and Arrest in a Pressure Vessel Subjected to Thermal and Pressure Loading," *Engineering Fracture Mechanics*, 23(1), pp. 61-70, 1986.

40. Brickstad, B., and Nilsson, F., "A Dynamic Analysis of Crack Propagation and Arrest in a Pressurized Thermal Shock (PTS) Experiments," *Engineering Fracture Mechanics*, 23(1), pp. 99-102, 1986.

41. Keeney-Walker, J. and Bass, B.R., "An evaluation of analysis methodologies for predicting cleavage arrest of a deep crack in an RPV subjected to PTS loading conditions," *Jour. of Pressure Vessel Tech., Trans. ASME*, 116, pp. 128-135, May 1994.

42. Jung, J., and Kanninen, M.F., "An Analysis of Dynamic Crack Propagation and Arrest in a Nuclear Pressure Vessel Under Thermal Shock Conditions," *Journal of Pressure Vessel Technology*, 105, pp. 111-116, May 1983.

43. Xu, W., Wintle, J.B., Wiesner, C.S., and Turner, D.G., "Analysis of crack arrest event in NESC-1 spinning cylinder experiment," *International Journal of Pressure Vessels and Piping*, 79, pp. 777-787, 2002.

44. Wallin, K., "A Simple Charpy-V-K_{Ic} Correlation for Radiation Embrittlement," ASME Pressure Vessel and Piping Conference, PVP-Vol. 170, American Society of Mechanical Engineers, New York, July 1989.

45.   Wallin, K., "Application of the Master Curve Method to Crack Initiation and Crack Arrest," ASME Pressure Vessel and Piping Symposium, PVP-Vol. 393, American Society of Mechanical Engineers, pp. 3-9, August 1999.

46.   Wallin, K., "Descriptive Characteristic of Charpy-V Fracture Arrest Parameter with Respect to Crack Arrest K_{Ia}," ESIS 20, E. van Walle, Ed., Mechanical Engineering Publications, pp. 165-176, 1996.

Chapter 7

Dynamic Fracture in Graded Materials

Arun Shukla[1] and Nitesh Jain[2]

[1] Simon Ostrach Professor and Chair, [2] Research Assistant
[1,2] Dynamic Photomechanics Laboratory,
Department of Mechanical Engineering, University of Rhode Island
Kingston, RI 02881
shuklaa@egr.uri.edu

The dynamic crack propagation in materials with varying properties, i.e., functionally graded materials is presented. First, an elastodynamic solution for a propagating crack inclined to the direction of property variation is introduced. Crack tip stress, strain and displacement fields are obtained through an asymptotic analysis coupled with displacement potential approach. Next, a systematic theoretical analysis is provided to incorporate the effect of transient nature of growing crack-tip on the crack-tip stress, strain and displacement fields. The analysis revealed that crack tip stress fields retain the inverse square root singularity and only the higher order terms in the expansion are influenced by material inhomogeneity. Using these stress, strain and displacement fields, contours of constant maximum shear stress, constant first stress invariant and constant in- plane displacements are generated and the effect of nonhomogeneity and transient nature of crack tip on these contours is discussed.

1. Introduction

The combination of several materials in one component offers, in many cases, significant improvements to its functional performance. Optimally, material properties throughout a component should be tailored to its specific application. This requires combinations of properties that are unattainable with a single homogeneous material. Functionally graded materials (FGMs) offer an advantageous means of combining materials,

273

providing a spatial variation in composition and properties, as an alternative to homogeneous materials and bimaterial interface structures.

Functionally graded materials (FGMs) are a new generation of engineered materials wherein the microstructural details are spatially varied through nonuniform distribution of the reinforcement phase(s). The resulting microstructure produces continuously or discretely changing thermal and mechanical properties at the macroscopic or continuum scale. FGMs may be used as joints between different types of materials, graded surface coatings or simply as graded components. More recently, graded composites have become a focus for tailoring thermal, mechanical and other properties for enhanced performance in structural, high temperature and other specialized applications. Man-made functionally graded materials (FGMs) copy natural biomaterials, such as bamboo (see Fig. 1), bone and shell etc., with a gradient in chemical composition or microstructure from one side to the other side in the material. The concept of functionally graded materials was proposed in 1984 by material scientists in Sendai (Japan). Since then studies to develop high-performance heat-resistant materials using functionally graded technology have continued. Various techniques have been employed to the fabrication of FGMs, including Chemical Vapor Deposition (CVD) / Physical Vapor Deposition (PVD), powder metallurgy, plasma spraying, electro-plating and combustion synthesis.

The principal motivation for FGM development is that a spatial variation in composition and properties at a joint between two different materials has the potential to reduce the stresses at the joint as compared to a bimaterial interface. Additionally, there is an increased interfacial strength at the joint and thus the likelihood of debonding is reduced. This is illustrated in Fig. 2 in the context of a potential application of FGMs, as thermal barrier coatings for high-speed space access vehicle applications. In this situation, a heat resistant ceramic must be strongly bonded to the metal structure to prevent spalling or surface cracking due to thermally induced stresses. A graded ceramic–metal composite interface has been shown to reduce thermal damage and delamination. Even though the initial research on FGMs was largely motivated by the practical applications of the concept on a wide variety of thermal shielding problems, materials with graded physical properties have almost unlimited potential in many other technological applications. Mechanics research on FGMs is needed to provide technical support for material scientists and for design and manufacturing engineers to take full advantage of their certain favorable properties in new product development.

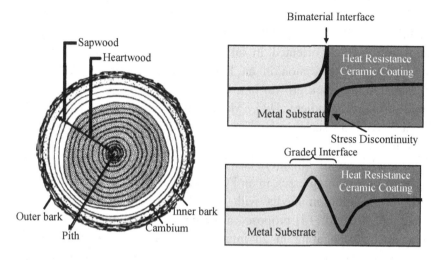

Fig. 1. Gradation in wood.   Fig. 2. FGM advantage.

The absence of sharp interfaces in continually graded FGMs largely reduces material property mismatch, which has been found to improve resistance to interfacial delamination and fatigue crack propagation [1]. However, the microstructure of FGMs is generally heterogeneous, and the dominant type of failure in FGMs is crack initiation and growth from inclusions. A crack in an FGM may exhibit complex behavior. The degree of complexity of this behavior is governed by the function describing the variation in the mechanical properties of the material. Extensive efforts have been made to characterize the crack tip stress field in FGMs under quasi-static loading conditions. Delale and Erdogan [2] studied various fracture mechanics problems of FGMs and presented a number of stress intensity factor solutions. Jin and Noda [3] showed that the inverse square root singularity at a crack tip still prevails in FGMs provided that Young's modulus and Poisson's ratio are continuous and piecewise continuously differentiable, which is an extension of Eischen's result [4] for nonhomogeneous materials with sufficiently smooth material properties. Gu and Asaro [5] and Becker et al. [6] studied crack deflections in FGMs. Jin and Batra [7] and Anlas et al. [8] investigated the K-dominance problem in FGMs. Parameswaran and Shukla developed the asymptotic stress field for stationary cracks along the gradient in FGMs [9]. Recently, Jain et al. developed the crack-tip stress fields for a crack oriented in an arbitrary direction from a linearly varying property variaton in an FGM [10]. Li et al. experimentally investigated the crack growth in a UV irradiated polymeric sheet with a small property gradient [11]. Marur and Tippur demonstrated the possibility of measuring mixed-mode SIFs using strain gages near the crack-tip in FGMs [12].

Recently, lot of attention has been focused on characterizing the behavior of a running crack in an FGM. Atkinson [13] was first to study crack propagation in media with spatially varying elastic properties. Wang and Meguid [14] proposed a theoretical and numerical treatment of a crack propagating in an interfacial layer with spatially varying elastic properties under the anti-plane loading condition. Nakagaki et al. [15] addressed a numerical treatment to dynamic crack propagation in the functionally graded materials and determined the effect of gradation on crack-tip severity as it propagates in FGM. Jain and Shukla developed the explicit expression for the stress fields around propagating crack-tip and studied the effect of nonomogeniety on the same [16]. They considered both opening and mixed mode cracks in their investigation. Considering the firm theoretical basis developed for dynamic fracture in FGMs, it appears, by contrast, that the understanding of the dynamic fracture process of FGMs is still limited due to lack of experimental evidence. Parameswaran and Shukla [17] investigated dynamic fracture in FGMs with discrete property variations using photoelasticity. Rousseau and Tippur studied the crack-tip deformation and other fracture parameters in FGMs subjected to low velocity impact [18].

In most of the studies performed on propagating cracks in FGMs, the crack tip speed was assumed to be constant. Freund and Rosakis [19] investigated transient nature of mode-I plane elastodynamic near-tip fields during crack propagation in homogeneous materials. They found that the square-root-singular term alone in the steady state asymptotic expansion of the crack tip fields does not fully describe the crack tip state during transient crack growth, which is attributed to the dependence of local fields on the past history of time dependent quantities such as crack velocity and stress intensity factor. They demonstrated experimentally that by incorporating transient higher order expansions, an accurate description of crack tip fields under fairly severe transient conditions can be achieved. The transient solution for elastodynamic crack growth in FGMs provides a strong foundation for the interpretation of experimental measurement in dynamic fracture testing of these materials. In fact, the usefulness of transient higher order solutions are certainly not confined to propagating cracks or to the interpretation of experimental results; they can be employed to describe the responses of stationary cracks and other singularities under dynamic loading conditions, and can be used in numerical simulations of such singularities during transient dynamic events. It is therefore that one complete section of this chapter is devoted to provide a theoretical analysis of the dynamic behavior of a transient mixed mode crack in an FGM with exponentially varying elastic properties.

In this chapter, first, expressions for individual stress, strain and in-plane displacement fields around a steadily growing crack–tip in an FGM are developed. Such information is required to analyze the full field experimental data obtained from various experimental techniques like photoelasticity, Moiré interferometry, holographic interferometry and coherent gradient sensing (CGS). These stress and displacement fields developed are then used to generate the contours of constant maximum shear stress (Isochromatics), constant in-plane displacements (Moiré fringes) and constant first stress invariant (Isopachic fringes) and the effect of nonhomogeneity on these contours is discussed. Next, a systematic theoretical analysis is provided to incorporate the effect of transient nature of growing mixed mode crack-tip on the elastodynamic stress, strain and displacement fields. In this context, transient crack growth is understood to include processes in which both the crack tip speed and dynamic stress intensity factor are differentiable functions of time. The effect of transient nature of moving crack-tip on crack-tip fields is also discussed.

2. Elastodynamic stress, strain and displacement fields for steady state crack propagation

2.1 *Theoretical considerations*

The elastic properties at any point in FGM's can be assumed to be same in all directions hence at the continuum level FGMs can be considered as isotropic nonhomogeneous solids. The nonhomogeneity makes analytical solution to elastodynamic problems extremely difficult for FGMs. An asymptotic analysis similar to that employed by Freund [20] for homogeneous materials, can be used to obtain the stress, strain and displacement fields around a propagating crack in an FGM. In the present analysis the crack tip stress, strain and displacement fields are obtained by assuming direction of crack propagation to be inclined to the direction of property variation which is presumed to be the case in most of the practical situations. Localized crack-tip plasticity and three-dimensional effects are neglected in this formulation.

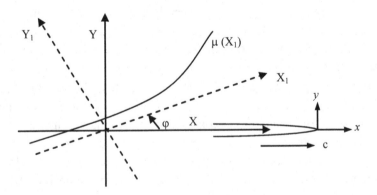

Figure 3. Propagating crack orientation with respect to property variation direction in FGM.

2.2 *Theoretical formulation*

Consider a crack moving with a velocity (c) in an FGM as shown in Fig. 1. The original coordinate system is a spatially fixed Cartesian coordinate system X-Y and a moving coordinate system x-y at the crack-tip is defined such that crack velocity is in X - direction. The shear modulus and mass density are assumed to vary exponentially in $X1$ direction as given in equation (1) and the Poisson's ratio (v) is assumed to be constant. The property gradation direction is at an angle φ to the $Y = 0$:

$$\mu = \mu_0 \exp(\delta X_1), \ \rho = \rho_0 \exp(\delta X_1), \ k = \frac{\lambda}{\mu} \tag{1}$$

where μ_0 and ρ_0 are the shear modulus and mass density at $X = X_1 = 0$ and δ is a constant having dimension (Length)$^{-1}$. Eq. (1) can be written in terms of (X, Y) coordinates by using a simple transformation as shown below.

$$\mu(X,Y) = \mu_o \exp(\alpha X + \beta Y), \ \rho(X,Y) = \rho_o \exp(\alpha X + \beta Y) \tag{2}$$

$$\alpha = \frac{\delta}{\sqrt{1 + \tan^2 \varphi}}, \ \beta = \frac{\delta \ \tan\varphi}{\sqrt{1 + \tan^2 \varphi}} \tag{3}$$

The equations of motion for a plane problem can be written as

$$\sigma_{XX,Y} + \sigma_{XY,Y} = \rho u_{,tt}, \quad \sigma_{XY,X} + \sigma_{YY,Y} = \rho v_{,tt}; \tag{4}$$

where σ_{ij} are the in-plane stress components and u, v are in-plane displacements.

The Hooke's law for plane problem can be written as

$$\begin{aligned}
\sigma_{XX} &= \left((\lambda_0 + 2\mu_0)u_{,X} + \lambda_0 v_{,Y}\right)\exp(\alpha X + \beta Y) \\
\sigma_{YY} &= \left((\lambda_0 + 2\mu_0)v_{,Y} + \lambda_0 v_{,Y}\right)\exp(\alpha X + \beta Y) \\
\sigma_{XY} &= \left(u_{,Y} + v_{,X}\right)\mu_0 \exp(\alpha X + \beta Y)
\end{aligned} \tag{5}$$

Introducing the displacement potentials (Φ and Ψ) as

$$u = \Phi_{,X} + \Psi_{,Y}, \quad v = \Phi_{,Y} - \Psi_{,X} \tag{6}$$

Substituting Eq. (6) in to Eq. (5) and substituting the resulting equation in to Eq. (4), the equation of motion can be expressed in terms of displacement potentials Φ and Ψ as

$$\frac{\partial}{\partial X}\left\{(k+2)\nabla^2\Phi-\frac{\rho_0}{\mu_0}\frac{\partial^2\Phi}{\partial t^2}\right\}+\frac{\partial}{\partial Y}\left\{\nabla^2\Psi-\frac{\rho_0}{\mu_0}\frac{\partial^2\Psi}{\partial t^2}\right\}+\alpha\left\{k\nabla^2\Phi+2\frac{\partial^2\Phi}{\partial X^2}+2\frac{\partial^2\Psi}{\partial X\partial Y}\right\}+$$

$$\beta\left\{2\frac{\partial^2\Phi}{\partial X\partial Y}+\frac{\partial^2\Psi}{\partial Y^2}-\frac{\partial^2\Psi}{\partial X^2}\right\}=0,$$

$$\frac{\partial}{\partial Y}\left\{(k+2)\nabla^2\Phi-\frac{\rho_0}{\mu_0}\frac{\partial^2\Phi}{\partial t^2}\right\}-\frac{\partial}{\partial X}\left\{\nabla^2\Psi-\frac{\rho_0}{\mu_0}\frac{\partial^2\Psi}{\partial t^2}\right\}+\alpha\left\{\frac{\partial^2\Psi}{\partial Y^2}-\frac{\partial^2\Psi}{\partial X^2}+2\frac{\partial^2\Phi}{\partial X\partial Y}\right\}+$$

$$\beta\left\{(k+2)\frac{\partial^2\Phi}{\partial Y^2}-2\frac{\partial^2\Psi}{\partial X\partial Y}+\lambda_0\frac{\partial^2\Phi}{\partial X^2}\right\}=0. \tag{7}$$

Eq. (7) can be rewritten as

$$\frac{\partial}{\partial X}\left((k+2)\nabla^2\Phi-\frac{\rho_0}{\mu_0}\frac{\partial^2\Phi}{\partial t^2}+(k+2)\left\{\alpha\frac{\partial\Phi}{\partial X}+\beta\frac{\partial\Phi}{\partial Y}\right\}+\left\{\alpha\frac{\partial\Psi}{\partial Y}-\beta\frac{\partial\Psi}{\partial X}\right\}\right)+$$

$$\frac{\partial}{\partial Y}\left(\nabla^2\Psi-\frac{\rho_0}{\mu_0}\frac{\partial^2\Psi}{\partial t^2}+k\left\{\alpha\frac{\partial\Phi}{\partial Y}-\beta\frac{\partial\Phi}{\partial X}\right\}+\left\{\alpha\frac{\partial\Psi}{\partial X}+\beta\frac{\partial\Psi}{\partial Y}\right\}\right)=0 \tag{8}$$

$$\frac{\partial}{\partial Y}\left((k+2)\nabla^2\Phi-\frac{\rho_0}{\mu_0}\frac{\partial^2\Phi}{\partial t^2}+(k+2)\left\{\alpha\frac{\partial\Phi}{\partial X}+\beta\frac{\partial\Phi}{\partial Y}\right\}+\left\{\alpha\frac{\partial\Psi}{\partial Y}-\beta\frac{\partial\Psi}{\partial X}\right\}\right)-$$

$$\frac{\partial}{\partial X}\left(\nabla^2\Psi-\frac{\rho_0}{\mu_0}\frac{\partial^2\Psi}{\partial t^2}+k\left\{\alpha\frac{\partial\Phi}{\partial Y}-\beta\frac{\partial\Phi}{\partial X}\right\}+\left\{\alpha\frac{\partial\Psi}{\partial X}+\beta\frac{\partial\Psi}{\partial Y}\right\}\right)=0$$

Eq. (8) can only be satisfied when

$$(k+2)\nabla^2\Phi-\frac{\rho_0}{\mu_0}\frac{\partial^2\Phi}{\partial t^2}+(k+2)\left\{\alpha\frac{\partial\Phi}{\partial X}+\beta\frac{\partial\Phi}{\partial Y}\right\}+\left\{\alpha\frac{\partial\Psi}{\partial Y}-\beta\frac{\partial\Psi}{\partial X}\right\}=0 \tag{9}$$

$$\nabla^2\Psi-\frac{\rho_0}{\mu_0}\frac{\partial^2\Psi}{\partial t^2}+k\left\{\alpha\frac{\partial\Phi}{\partial Y}-\beta\frac{\partial\Phi}{\partial X}\right\}+\left\{\alpha\frac{\partial\Psi}{\partial X}+\beta\frac{\partial\Psi}{\partial Y}\right\}=0$$

Now introducing the moving crack-tip coordinates (x-y) as

$$x=X\text{-}ct,\ y=Y,\ \frac{\partial^2}{\partial X^2}=\frac{\partial^2}{\partial x^2},\ \frac{\partial^2}{\partial t^2}=c^2\frac{\partial^2}{\partial x^2}. \tag{10}$$

Using the transformations given in Eq. (10), Eq. (9) can be written as

$$\alpha_l^2 \frac{\partial^2 \Phi}{\partial x^2} + \frac{\partial^2 \Phi}{\partial y^2} + \left\{ \alpha \frac{\partial \Phi}{\partial x} + \beta \frac{\partial \Phi}{\partial y} \right\} - \frac{1}{k+2} \left\{ \alpha \frac{\partial \Psi}{\partial y} - \beta \frac{\partial \Psi}{\partial x} \right\} = 0$$

$$\alpha_s^2 \frac{\partial^2 \Psi}{\partial x^2} + \frac{\partial^2 \Psi}{\partial y^2} + \left\{ \alpha \frac{\partial \Psi}{\partial x} + \beta \frac{\partial \Psi}{\partial y} \right\} + k \left\{ \alpha \frac{\partial \Phi}{\partial y} - \beta \frac{\partial \Phi}{\partial x} \right\} = 0$$

(11)

where, $\alpha_l = \left(1 - \frac{\rho_0 c^2}{\mu_0 (k+2)} \right)^{1/2}$, $\alpha_s = \left(1 - \frac{\rho_0 c^2}{\mu_0} \right)^{1/2}$

The above equations would reduce to the classical 2-D wave equations of motion by assigning α and β to zero. Due to nonhomogeneity, these equations loose their classical form and remain coupled in two fields Φ and Ψ, through the nonhomogeneity parameters α and β.

2.2.1 *Asymptotic expansion of crack tip fields*

In the asymptotic analysis, a new set of coordinates is introduced as

$$\eta_1 = x/\varepsilon \ , \ \eta_2 = y/\varepsilon \ and \ 0 < \varepsilon < 1 \tag{12}$$

where ε is a small arbitrary positive parameter. The parameter ε is used here so that the region around the crack – tip is expanded to fill the entire field of observation. As ε is chosen to be infinitely small, all the points in the x, y-plane except those near the crack-tip are mapped beyond the range of observation, and the crack line occupies the whole negative η_1 axis. Assuming that the displacement potentials (Φ and Ψ) can be expressed in powers of ε as [20]

$$\Phi(x,y) = \Phi(\varepsilon\eta_1, \varepsilon\eta_2) = \sum_{m=0}^{\infty} \varepsilon^{\frac{m+3}{2}} \phi_m(\eta_1, \eta_2)$$

$$\Psi(x,y) = \Psi(\varepsilon\eta_1, \varepsilon\eta_2) = \sum_{m=0}^{\infty} \varepsilon^{\frac{m+3}{2}} \psi_m(\eta_1, \eta_2)$$

(13)

The first term of series (m=0) corresponds to the expected square root singular contribution proportional to $r^{-1/2}$ in the asymptotic near-tip stress field.

By substituting assumed asymptotic form (13) into the governing equations (11) leads to an infinite series involving differential equations associated with each power of ε as written below,

$$\sum_{m=0}^{\infty}\left\{\varepsilon^{\frac{m+3}{2}}\left(\alpha_l^2\frac{\partial^2\phi_m}{\partial\eta_1^2}+\frac{\partial^2\phi_m}{\partial\eta_2^2}\right)+\varepsilon^{\frac{m+5}{2}}\left(\begin{array}{c}\alpha\left[\frac{\partial\phi_m}{\partial\eta_1}+\frac{1}{k+2}\frac{\partial\psi_m}{\partial\eta_2}\right]\\+\beta\left[\frac{\partial\phi_m}{\partial\eta_2}-\frac{1}{k+2}\frac{\partial\psi_m}{\partial\eta_1}\right]\end{array}\right)+\varepsilon^{\frac{m+7}{2}}[....]\right\}=0$$

$$\sum_{m=0}^{\infty}\left\{\varepsilon^{\frac{m+3}{2}}\left(\alpha_s^2\frac{\partial^2\psi_m}{\partial\eta_1^2}+\frac{\partial^2\psi_m}{\partial\eta_2^2}\right)+\varepsilon^{\frac{m+5}{2}}\left(\begin{array}{c}\alpha\left[\frac{\partial\psi_m}{\partial\eta_1}+k\frac{\partial\phi_m}{\partial\eta_2}\right]\\+\beta\left[\frac{\partial\psi_m}{\partial\eta_2}-k\frac{\partial\phi_m}{\partial\eta_1}\right]\end{array}\right)+\varepsilon^{\frac{m+7}{2}}\{....\}\right\}=0$$

$$(14)$$

For Eq. (14) to be valid, the differential equation corresponding to each power of ε ($\varepsilon^{3/2}$, $\varepsilon^{5/2}$, $\varepsilon^{5/2}$....) should vanish independently. This leads to a system of coupled differential equations in Φ and Ψ [16]. These equations have a general form as

$$\alpha_l^2\frac{\partial^2\phi_m}{\partial\eta_1^2}+\frac{\partial^2\phi_m}{\partial\eta_2^2}+\left\{\alpha\frac{\partial\phi_{m-2}}{\partial\eta_1}+\beta\frac{\partial\phi_{m-2}}{\partial\eta_2}\right\}+\frac{1}{k+2}\left\{\alpha\frac{\partial\psi_{m-2}}{\partial\eta_2}-\beta\frac{\partial\psi_{m-2}}{\partial\eta_1}\right\}=0$$

$$(15)$$

$$\alpha_s^2\frac{\partial^2\psi_m}{\partial\eta_1^2}+\frac{\partial^2\psi_m}{\partial\eta_2^2}+\left\{\alpha\frac{\partial\psi_{m-2}}{\partial\eta_1}+\beta\frac{\partial\psi_{m-2}}{\partial\eta_2}\right\}+k\left\{\alpha\frac{\partial\phi_{m-2}}{\partial\eta_2}-\beta\frac{\partial\phi_{m-2}}{\partial\eta_1}\right\}=0$$

where $\phi_k, \psi_k = \begin{cases}\phi_k, \psi_k & for\ k\geq 0\\ 0 & for\ k<0\end{cases}$

For $m=0$ and 1, we get

$$\alpha_l^2\frac{\partial^2\phi_m}{\partial\eta_1^2}+\frac{\partial^2\phi_m}{\partial\eta_2^2}=0,\ \alpha_s^2\frac{\partial^2\psi_m}{\partial\eta_1^2}+\frac{\partial^2\psi_m}{\partial\eta_2^2}=0$$

$$(16)$$

These partial differential equations are exactly similar to that for a homogeneous material having elastic properties same as that of the elastic properties of the FGM at the crack tip. These equations can be

easily reduced to Laplace's equations in the respective complex domains
$\varsigma_l = \eta_1 + i\alpha_l \eta_2$, $\varsigma_s = \eta_1 + i\alpha_s \eta_2$, $i = \sqrt{-1}$.

Since the crack is propagating at an angle to the direction of property gradation, the stress field near the crack-tip is a combination of both opening and shear modes (mixed mode). For elastic solution the stress field related to opening mode and shear modes can be superposed to obtain the mixed mode solution. The solutions for $m = 0$ and 1 are same as homogenous material [20] and can be written as

$$\phi_m(\rho_l, \theta_l) = A_m \rho_l^{(m+3)/2} \cos\frac{1}{2}(m+3)\theta_l + C_m \rho_l^{(m+3)/2} \sin\frac{1}{2}(m+3)\theta_l \qquad m = 0,\ 1$$

$$\psi_m(\rho_s, \theta_s) = B_m \rho_s^{(m+3)/2} \sin\frac{1}{2}(m+3)\theta_s + D_m \rho_s^{(m+3)/2} \cos\frac{1}{2}(m+3)\theta_s$$

$$(17)$$

where

$$\rho_l = \left[\eta_1^2 + \alpha_l^2\eta_2^2\right]^{1/2}, \quad \tan\theta_l = \frac{\alpha_l\eta_2}{\eta_1}, \quad \rho_s = \left[\eta_1^2 + \alpha_s^2\eta_2^2\right]^{1/2} \text{ and } \tan\theta_s = \frac{\alpha_s\eta_2}{\eta_1}$$

Using the definitions of dynamic stress intensity factors K_{ID} and K_{IID} for opening and shear modes [21] and considering the crack face boundary conditions,

$$A_0 = \frac{4(1+\alpha_s^2)}{3(4\alpha_s\alpha_l - (1+\alpha_s^2)^2)} \frac{K_{ID}}{\mu_c\sqrt{2\pi}}, \quad B_0 = -\frac{2\alpha_l}{1+\alpha_s^2} A_0$$

$$(18)$$

$$C_0 = \frac{8\alpha_s}{3(4\alpha_s\alpha_l - (1+\alpha_s^2)^2)} \frac{K_{IID}}{\mu_c\sqrt{2\pi}}, \quad D_0 = \frac{1+\alpha_s^2}{2\alpha_l} C_0.$$

The solution for higher orders of m can be obtained in a recursive manner [16, 22].

$$
\phi_2 = A_2(t)\rho_l^{5/2}\cos\frac{5}{2}\theta_l - \frac{1}{4\alpha_l^2}\rho_l^{5/2}\left[\begin{array}{l}\cos\frac{1}{2}\theta_l\{\alpha A_0(t)+\beta\alpha_l C_0(t)\}\\[2mm]+\sin\frac{1}{2}\theta_l\{\alpha C_0(t)-\beta\alpha_l A_0(t)\}\end{array}\right]
$$

$$
-\frac{2}{5}\frac{1}{(k+2)(\alpha_l^2-\alpha_s^2)}\rho_s^{5/2}\left[\begin{array}{l}\cos\frac{5}{2}\theta_s\{\alpha\alpha_s B_0(t)-\beta D_0(t)\}\\[2mm]-\sin\frac{5}{2}\theta_s\{\alpha B_0(t)+\beta\alpha_s D_0(t)\}\end{array}\right]
$$

$$
\psi_2 = B_2(t)\rho_s^{5/2}\sin\frac{5}{2}\theta_s - \frac{1}{4\alpha_s^2}\rho_s^{5/2}\left[\begin{array}{l}\cos\frac{1}{2}\theta_s\{\alpha D_0(t)+\beta\alpha_s B_0(t)\}\\[2mm]+\sin\frac{1}{2}\theta_s\{\alpha B_0(t)-\beta D_0(t)\}\end{array}\right]
$$

$$
+\frac{2}{5}\frac{k}{(\alpha_l^2-\alpha_s^2)}\rho_l^{5/2}\left[\begin{array}{l}\cos\frac{5}{2}\theta_l\{\alpha\alpha_l C_0(t)-\beta A_0(t)\}\\[2mm]-\sin\frac{5}{2}\theta_l\{\alpha\alpha_l A_0(t)+\beta C_0(t)\}\end{array}\right]
$$

$$(19)$$

2.2.2 Stress, strain and displacement fields

The stress, strain and displacement fields around the crack tip can now be obtained using displacement potentials (Φ and Ψ) found in the previous sections (Eq. 17-19). These are found to be the functions of the material properties, nonhomogeneity parameters, fracture parameters (A_0, A_1... B_0, B_1..) and coordinates (r, θ). The method for finding the detailed expressions for stress, strain and displacement fields are given in Appendix I.

2.3 Discussions on solutions

The asymptotic representation of crack-tip stress and displacement fields is many times used to extract fracture parameters from experimental data. The fracture parameters such as stress intensity factor and non-singular stress components are related to the coefficients of the asymptotic expansion of the crack-tip stress, strain and displacement fields. These coefficients can be obtained by using experimentally obtained full field data such as the isochromatics [23], isopachics [24],

Moiré fringes [25] or CGS contours [26]. To get an insight into the effects of nonhomogeneity on the contours of constant maximum shear stress (isochromatics), contours of constant first stress invariant (isopachics) and contours of constant in-plane displacement (Moiré fringes) are generated for different values of the non-homogeneity parameter using crack tip stress and displacement fields developed in the previous sections. The contours are generated for an assumed value of the dynamic stress intensity factor (coefficient A_0), whereas the higher order coefficients A_1, A_2, B_1 and B_2 are assumed to be zero. However the nonhomogeneity specific part of the higher order term ($r^{1/2}$), which has A_0 and B_0 as the coefficients is retained. The higher order terms can be given specific values, however, this will not add to the discussion, which is to show the influence of non-homogeneity parameter on stress and displacement fields near the crack tip for fixed values of stress field coefficients.

2.3.1 *Contours of constant maximum shear stress*

Constant maximum shear stress (τ_{max}) at each point around the crack-tip can be determined by substituting the stress field equation, obtained in the previous sections, in Eq. (20).

$$\tau_{max} = \sqrt{\left(\frac{\sigma_{yy} - \sigma_{xx}}{2}\right)^2 + \sigma_{xy}^2} \qquad (20)$$

The synthetically generated isochromatics for exponential property variation are shown in Fig. 4. These contours are generated for opening mode loading assuming the opening mode stress intensity factor (K_{ID}) = 1.0 MPa-m$^{1/2}$ and shear modulus at crack tip (μ_c) = 1 GPa.

Figs. (4a-f) shows the effect of nonhmogeneity and crack velocity on the contours of maximum shear stress (isochromatics). These contours are generated for two different ratios of crack velocity to shear wave velocity (c/c_s) = 0.1 and 0.5. The crack is assumed to be on the negative x-axis and the crack tip is located at (0, 0). The figure shows that the nonhomogeneity does not alter the structure of the contours very close to the crack-tip as compared to homogeneous material. However, slightly removed from the crack-tip, the size and tilt of contours changes with the introduction of nonhomogeneity. Figs. (4 a, c, e) show that for low values of c/c_s the contours are similar to their static counterparts. For $c/c_s = 0.1$, the contours are upright for homogeneous materials. However, for $\alpha > 0$ the contours tilt forward and for $\alpha < 0$ the contours tilt backwards. For the case of $\alpha > 0$, the material ahead of the crack tip is stiffer than material behind the crack tip. Thus, for a given loading the material ahead of crack tip offers more resistance to deformation than the material behind the crack tip. This results into positive stress acting parallel to the crack, which causes the fringes to lean forward. As c/c_s increases to 0.5, the number and size of fringes around the crack tip increases as shown in Figs. (4b, d and f). For higher values of values of c/c_s the aspect ratio of fringes increases and fringes become leaner. It can also be observed that in homogeneous materials ($\alpha = 0$) fringes associated with crack propagating at high velocities tilt backwards as shown in Fig. (4b). Both these features are due to inertia effects that are dominant at these high velocities. Figs (4 d and f) show fringes tilt heavily backward in case of $\alpha < 0$ and only slightly forward when $\alpha > 0$, which might be due to the fact that forward tilt due to a positive nonhomogeniety parameter compensates some of the of the backward tilt due to crack speed.

2.3.2 *Contours of constant stress invariant*

Fig. (5) shows the effect of nonhomogeneity and crack velocity on contours of constant first stress invariant ($\sigma_{xx}+\sigma_{yy}$) for materials with exponentially varying elastic properties. These contours are generated for opening mode loading assuming opening mode stress intensity factor (K_{ID}) = 1.0 MPa-m$^{1/2}$ and shear modulus at crack tip (μ_c) = 1 GPa. The contours are plotted for two different c/c_s ratios (0.1 and 0.5) with three different values of nonhomogeniety parameters (0, -0.2, 0.2). These contours represent the isopachic fringe patterns obtained with holographic interferometery. Figs. (5a, c and e) show that for low values c/c_s the contour shape and size resembles their static counterparts [27].

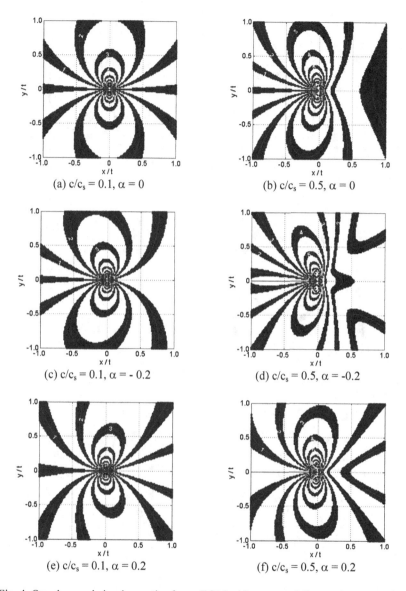

(a) $c/c_s = 0.1$, $\alpha = 0$

(b) $c/c_s = 0.5$, $\alpha = 0$

(c) $c/c_s = 0.1$, $\alpha = -0.2$

(d) $c/c_s = 0.5$, $\alpha = -0.2$

(e) $c/c_s = 0.1$, $\alpha = 0.2$

(f) $c/c_s = 0.5$, $\alpha = 0.2$

Fig. 4. Opening mode isochromatics for an FGM with exponentially varying properties

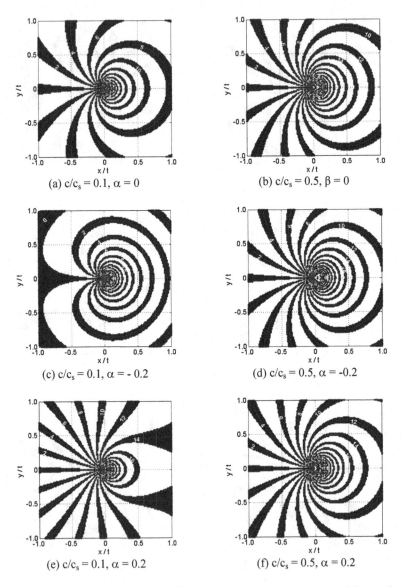

Fig. 5. Contours of constant maximum first stress invariant ($\sigma_{xx}+\sigma_{yy}$) around the crack-tip for opening mode loading in an FGM having exponentially varying properties

It can also be noticed that by introduction of nonhomogeneity the contours shift towards the stiffer side for both the values of c/c_s. One can observe that the size of the contours decreases in the compliant side and increases in the stiffer side, which might be due to the higher resistance offered by the stiffer side. It can also be observed that as crack velocity increases the number and size of the fringes around crack tip increases.

2.3.3 *Contours of constant in-plane displacements*

Another important method to measure fracture parameters from experimental data is Moiré interferometery. Moiré fringes, which typically represent the in-plane displacements u and v, can be used to determine the coefficients in the analytically developed displacement field expressions. The displacement field equations developed in the previous sections are used to generate the contours of constant in-plane displacements (u, v). These contours are generated for an assumed value of $K_I = 1$ MPa-m$^{1/2}$, for exponential varying property gradation. Fig. (6) shows the effect of crack velocity and nonhomogeneity parameter over the contours of in-plane displacement (u) for a propagating crack. It can be observed that the contours tilt away from crack face for $\alpha > 0$ and tilt toward crack face for $\alpha < 0$. This is due to the fact that in case of $\alpha > 0$ material towards crack side is more compliant than the other side which makes it more prone to deform than the other side. Fig. (7) shows the effect of nonhomogeneity parameter over the contours of in-plane displacement (v) for a propagating crack for exponential property gradation. Similar to the case of in-plane displacement (u) the contours tilt away from direction of gradient for $\alpha > 0$ and tilt towards the direction of gradient for $\alpha < 0$ due to the difference in the stiffness of the material ahead and behind the crack tip. It can also be noticed from Figs. (6) and (7) that as the c/c_c ratio increases from 0.1 to 0.5 there is no significant change in the shape of the contours but the contours become more dense which means the value of the in-plane displacements, u and v, at a particular point is more for a crack propagating with a higher velocity.

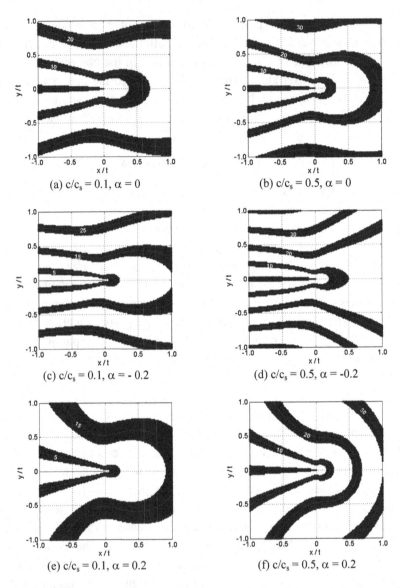

Fig. 6. Contours of constant in-plane displacement (u) around the crack-tip for opening mode loading in an FGM having exponentially varying properties ($K_I = 1.0$ MPa-m$^{1/2}$, Thickness (t) = 0.010 m)

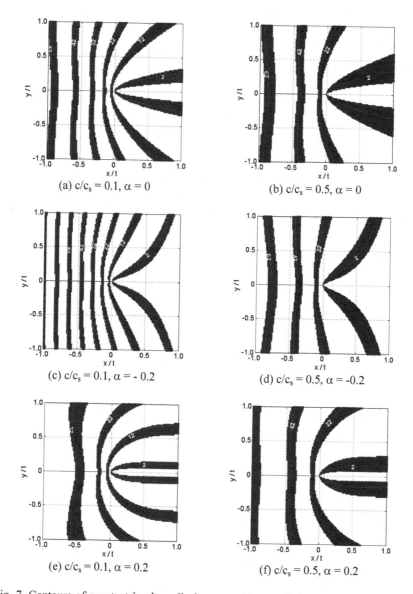

Fig. 7. Contours of constant in-plane displacement (v) around the crack-tip for opening mode loading in an FGM having exponentially varying properties ($K_I = 1.0$ MPa-m$^{1/2}$, Thickness (t) = 0.010 m)

3. Elastodynamic stress, strain and displacement fields for transient crack propagation

3.1 *Theoretical formulation*

Consider a crack moving nonuniformly in an FGM. Suppose that the crack propagates with a nonuniform speed, $c(t)$, and the crack faces satisfy the traction free boundary condition. The shear modulus and mass density are assumed to vary exponentially as given in Eq. (1) and the Poisson's ratio (v) is assumed to be constant.

Following the steady state formulation it can be shown that the equation of motion for the plane elastodynamic problem can be written as Eq. (9)

Now introducing the moving crack-tip coordinates (x-y) as

$$x = X-c(t)t, \quad y = Y$$

$$\frac{\partial^2}{\partial X^2} = \frac{\partial^2}{\partial x^2}; \quad \frac{\partial^2}{\partial t^2} = c^2\frac{\partial^2}{\partial x^2} + \frac{\partial^2}{\partial t^2} - \dot{c}\frac{\partial}{\partial x} - 2c\frac{\partial^2}{\partial x\partial t} \quad \text{where } \dot{c} = \frac{\partial c}{\partial t} \tag{21}$$

Using the transformations given in Eq. (21), the Eq. (9) can be written as

$$\alpha_l^2 \frac{\partial^2 \Phi}{\partial x^2} + \frac{\partial^2 \Phi}{\partial y^2} + \left\{ \alpha \frac{\partial \Phi}{\partial x} + \beta \frac{\partial \Phi}{\partial y} \right\} - \frac{1}{k+2}\left\{ \alpha \frac{\partial \Psi}{\partial y} - \beta \frac{\partial \Psi}{\partial x} \right\}$$

$$+ \frac{\rho_0}{\mu_0(k+2)}\left\{ \dot{c}\frac{\partial \Phi}{\partial x} + 2c\frac{\partial^2 \Phi}{\partial x\partial t} - \frac{\partial^2 \Phi}{\partial t^2} \right\} = 0 \tag{22}$$

$$\alpha_s^2 \frac{\partial^2 \Psi}{\partial x^2} + \frac{\partial^2 \Psi}{\partial y^2} + \left\{ \alpha \frac{\partial \Psi}{\partial x} + \beta \frac{\partial \Psi}{\partial y} \right\} + k\left\{ \alpha \frac{\partial \Phi}{\partial y} - \beta \frac{\partial \Phi}{\partial x} \right\}$$

$$+ \frac{\rho_0}{\mu_0(k+2)}\left\{ \dot{c}\frac{\partial \Psi}{\partial x} + 2c\frac{\partial^2 \Psi}{\partial x\partial t} - \frac{\partial^2 \Psi}{\partial t^2} \right\} = 0$$

The above equations would reduce to the classical 2-D wave equations of motion by assigning α and β to zero. Due to nonhomogeneity, these equations lose their classical form and remain coupled Φ and Ψ, through the nonhomogeneity parameters α and β.

3.1.1 *Asymptotic expansion of crack tip fields*

Using the asymptotic expansion as discussed in previous section the Φ and Ψ can be expressed in powers of ε as given in Eq. (13)

Substituting the assumed asymptotic form (13) into the governing equation (22), we obtain two equations whose left hand side is as infinite power series in ε and whose right hand size vanishes. Since ε is an arbitrary number, the coefficient of each power of ε should be zero. Therefore the governing equation reduces to a system of coupled differential equations in Φ and Ψ. These equations will have a general form as

$$\alpha_l^2 \frac{\partial^2 \phi_m}{\partial \eta_1^{\,2}} + \frac{\partial^2 \phi_m}{\partial \eta_2^{\,2}} + \left\{ \alpha \frac{\partial \phi_{m-2}}{\partial \eta_1} + \beta \frac{\partial \phi_{m-2}}{\partial \eta_2} \right\} + \frac{1}{k+2} \left\{ \alpha \frac{\partial \psi_{m-2}}{\partial \eta_2} - \beta \frac{\partial \psi_{m-2}}{\partial \eta_1} \right\}$$

$$+ \frac{\rho_0 c^{1/2}}{\mu_0 (k+2)} \frac{\partial}{\partial t} \left\{ c^{1/2} \frac{\partial \phi_{m-2}}{\partial \eta_1} \right\} - \frac{\rho_0}{\mu_0 (k+2)} \frac{\partial^2 \phi_{m-4}}{\partial t^2} = 0 \qquad (23)$$

$$\alpha_s^2 \frac{\partial^2 \psi_m}{\partial \eta_1^{\,2}} + \frac{\partial^2 \psi_m}{\partial \eta_2^{\,2}} + \left\{ \alpha \frac{\partial \psi_{m-2}}{\partial \eta_1} + \beta \frac{\partial \psi_{m-2}}{\partial \eta_2} \right\} + k \left\{ \alpha \frac{\partial \phi_{m-2}}{\partial \eta_2} - \beta \frac{\partial \phi_{m-2}}{\partial \eta_1} \right\}$$

$$+ \frac{\rho_0 c^{1/2}}{\mu_0} \frac{\partial}{\partial t} \left\{ c^{1/2} \frac{\partial \psi_{m-2}}{\partial \eta_1} \right\} - \frac{\rho_0}{\mu_0 (k+2)} \frac{\partial^2 \psi_{m-4}}{\partial t^2} = 0$$

$$\text{where} \quad \phi_k, \psi_k = \begin{cases} \phi_k, \psi_k & \text{for } k \geq 0 \\ 0 & \text{for } k < 0 \end{cases}$$

For $m = 0$ and 1 Eqs. (23) are not coupled in Φ and Ψ and reduce to Laplace's equation in coordinates (η_1, $\alpha_l \eta_2$ or $\alpha_s \eta_2$) (similar to that for a homogeneous material having elastic properties equal to the elastic properties of the FGM at the crack tip). Indeed, as will be seen, ϕ_0 and ψ_0 have the same spatial structure in both transient and steady state cases. This is not so, however, for ϕ_m, ψ_m if $m > 1$.

For elastic solution the stress field related to opening mode and shear modes can be superposed to obtain the mixed mode solution. The solutions for $m = 0$ and 1 are same as homogenous material and can be written as

$$\phi_m(\rho_l,\theta_l,t) = A_m(t)\rho_l^{(m+3)/2} \cos\frac{1}{2}(m+3)\theta_l + C_m(t)\rho_l^{(m+3)/2} \sin\frac{1}{2}(m+3)\theta_l$$

$$\psi_m(\rho_s,\theta_s,t) = B_m(t)\rho_s^{(m+3)/2} \sin\frac{1}{2}(m+3)\theta_s + D_m(t)\rho_s^{(m+3)/2} \cos\frac{1}{2}(m+3)\theta_s$$

$$m = 0,\ 1$$

$$(24)$$

where

$$\rho_l = \left[\eta_1^2 + \alpha_l^2\eta_2^2\right]^{1/2}, \quad \tan\theta_l = \frac{\alpha_l\eta_2}{\eta_1}, \quad \rho_s = \left[\eta_1^2 + \alpha_s^2\eta_2^2\right]^{1/2} \text{ and } \tan\theta_s = \frac{\alpha_s\eta_2}{\eta_1}$$

The solution (24) appears to be the same as for steady state crack growth. However, they differ in the fundamental respect that the coordinates (ρ_l,θ_l) now depend upon time. It is the crack speed that determines the degree of distortion of these coordinates, and the crack speed is now a function of time. Also one should note that the coefficients of the series solution given in Eq. (24) are time dependent.

Using the definitions of dynamic stress intensity factors K_{ID} and K_{IID} for opening and shear modes and considering the crack face boundary conditions,

$$A_0(t) = \frac{4(1+\alpha_s^2)}{3(4\alpha_s\alpha_l - (1+\alpha_s^2)^2)}\frac{K_{ID}(t)}{\mu_c\sqrt{2\pi}}, \quad B_0(t) = -\frac{2\alpha_l}{1+\alpha_s^2}A_0(t)$$

$$C_0(t) = \frac{8\alpha_s}{3(4\alpha_s\alpha_l - (1+\alpha_s^2)^2)}\frac{K_{IID}(t)}{\mu_c\sqrt{2\pi}}, \quad B_0(t) = \frac{1+\alpha_s^2}{2\alpha_l}C_0(t)$$

$$(25)$$

The solution for higher orders of m can be obtained in a recursive manner.

$$\phi_2 = A_2(t)\rho_l^{5/2}\cos\frac{5}{2}\theta_l - \frac{1}{4\alpha_l^2}\rho_l^{5/2}\left[\cos\frac{1}{2}\theta_l\{\alpha A_0(t)+\beta C_0(t)\}+\sin\frac{1}{2}\theta_l\{\alpha C_0(t)-\beta A_0(t)\}\right]$$

$$-\frac{2}{5}\frac{\alpha_s}{(k+2)(\alpha_l^2-\alpha_s^2)}\rho_s^{5/2}\left[\cos\frac{5}{2}\theta_s\{\alpha B_0(t)-\beta D_0(t)\}-\sin\frac{5}{2}\theta_s\{\alpha B_0(t)+\beta D_0(t)\}\right]$$

$$+\rho_l^{5/2}\left\{\begin{array}{l}\frac{1}{6}\left[D_l\{A_0(t)\}+\frac{1}{2}B_l^A(t)\right]\cos\frac{\theta_l}{2}-\frac{1}{8}B_l^A(t)\cos\frac{3\theta_l}{2}\\[2mm]+\frac{1}{6}\left[D_l\{C_0(t)\}+\frac{1}{2}B_l^C(t)\right]\sin\frac{\theta_l}{2}+\frac{1}{8}B_l^C(t)\sin\frac{3\theta_l}{2}\end{array}\right.$$

$$(26a)$$

$$\psi_2 = B_2(t)\rho_s^{5/2}\sin\frac{5}{2}\theta_s - \frac{1}{4\alpha_s^2}\rho_s^{5/2}\left[\cos\frac{1}{2}\theta_s\{\alpha B_0(t) + \beta D_0(t)\} + \sin\frac{1}{2}\theta_s\{\alpha B_0(t) - \beta D_0(t)\}\right]$$

$$+\frac{2}{5}\frac{k\alpha_l}{(\alpha_l^2 - \alpha_s^2)}\rho_l^{5/2}\left[\cos\frac{5}{2}\theta_l\{\alpha C_0(t) - \beta A_0(t)\} - \sin\frac{5}{2}\theta_l\{\alpha A_0(t) + \beta C_0(t)\}\right]$$

$$+\rho_s^{5/2}\left\{\begin{array}{l}\frac{1}{6}\left[D_s\{B_0(t)\} + \frac{1}{2}B_s^B(t)\right]\sin\frac{\theta_s}{2} + \frac{1}{8}B_s^B(t)\sin\frac{3\theta_s}{2}\\[2mm]+\frac{1}{6}\left[D_s\{D_0(t)\} + \frac{1}{2}B_s^D(t)\right]\cos\frac{\theta_s}{2} - \frac{1}{8}B_s^D(t)\cos\frac{3\theta_s}{2}\end{array}\right\}$$

(26b)

All the coefficients in the above given solution are defined in the Appendix-I.

3.1.2 *Stress, strain and displacement fields*

The stress, strain displacement fields around the crack tip can now be obtained using displacement potentials (Φ and Ψ) found in the previous sections. The method for finding the detailed expressions for stress, strain displacement fields are given in Appendix-I.

3.2 *Discussion of solutions*

To get an insight into the effects of transient terms on dynamic fracture process, contours of constant maximum shear stress (isochromatics), contours of constant first stress invariant (isopachics) and contours of constant in-plane displacement (Moiré fringes) are generated for opening mode loading conditions. The contours are generated for an assumed value of the dynamic stress intensity factor (coefficient A_0), whereas the higher order coefficients A_1, A_2, B_1 and B_2 are assumed to be zero. However the nonhomogeneity and transient specific parts of the higher order term $r^{1/2}$, which has A_0 and B_0 as the coefficients is retained. The typical values of material properties and material thickness used in generating contours are as follows: Poisson's ratio = 0.3, shear modulus at the crack tip μ_c = 1 GPa, density at the crack tip ρ_c = 2000 kg/m^3 and thickness t = 0.01m. The nonhomogeneity parameter α for plotting these contours was obtained by fitting an exponential curve in the property

variation profile of the in-house
fabricated polymeric FGMs [17]. The
nonhomogeneity parameter α for
FGM fabricated in this study was
0.57.

3.2.1 Contours of constant maximum shear stress

Constant maximum shear stress (τ_{max})
at each point around the crack-tip can
be determined by substituting the
stress field equation, obtained in the
previous section, in Eq. (20). Figure 8
shows the effect of rate of change of
mode-I stress intensity factor ($dK_{ID}(t)$
/dt) on contours of constant maximum
shear stress for opening mode loading
around the crack-tip corresponding to
$\alpha = 0.57$, $K_{ID} = 1.0$ MPa-m$^{1/2}$, $c =$
650 ms^{-1}. The crack is assumed to be
moving with a uniform velocity i.e.
$dc/dt = 0$. As observed by Dally and
Shukla [28] the rate of change of K_{ID}
at crack initiation could be of the
order of 10^5 MPa-m$^{1/2}$-sec^{-1}, the values
of $dK_{ID}(t)$ /dt was varied over six
orders of magnitude for generating the
contours. The crack occupies negative
x-axis and the crack tip is located at
(0, 0). It can be observed from the
figure that as the $dK_{ID}(t)$ /dt increases,
the size and number of the fringes
around the crack tip increases. As can
be seen in figure 8a for $dK_{ID}(t)$ /dt = 0
the fringes have negligible tilt. High

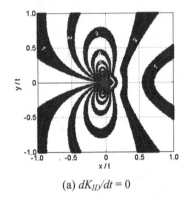

(a) $dK_{ID}/dt = 0$

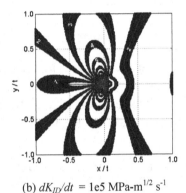

(b) dK_{ID}/dt = 1e5 MPa-m$^{1/2}$ s^{-1}

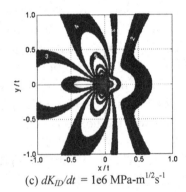

(c) dK_{ID}/dt = 1e6 MPa-m$^{1/2}$s^{-1}

Fig. 8. Effect of rate of change of
stress intensity factor on contours
of constant maximum shear stress
around the crack-tip

velocity has the tendency to cause a backward tilt in fringes but this is compensated by the gradient in Young's modulus, which is increasing in the direction of crack propagation. As the rate of change of dynamic stress intensity factor increases the fringes start tilting backward (Figure 8(b and c)). The significant changes in pattern of these contours around the crack-tip suggest that the transient terms can have a major influence on the crack-tip field.

Figure 9 shows the effect of crack tip acceleration (*dc/dt*) on contours of constant maximum shear stress. The value of *dc/dt* was varied over eight orders of magnitude. Dally and Shukla [28] also showed that the rate of change of velocity at crack initiation could be of the order of 10^7 ms^{-2}. Changes in fringe size and shape up to crack tip accelerations of 10^6 ms^{-2} are negligible (figure 9a). Further increases in acceleration result in a decrease in size of fringes around the crack tip (figures 9 b and c). It can also be observed that the fringes begin to tilt forward as crack tip acceleration increases.

3.2.2 *Contours of constant first stress invariant*

Figure 10 shows the effect of rate of change of mode-I stress intensity factor

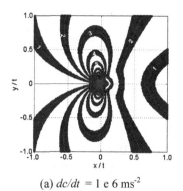

(a) *dc/dt* = 1 e 6 ms^{-2}

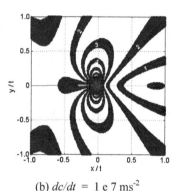

(b) *dc/dt* = 1 e 7 ms^{-2}

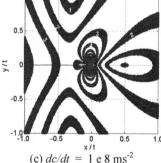

(c) *dc/dt* = 1 e 8 ms^{-2}

Fig. 9. Effect of crack-tip acceleration on contours of constant maximum shear stress around the crack-tip

$(dK_{ID}(t) /dt)$ on contours of constant first stress invariant $(\sigma_{xx}+\sigma_{yy})$ in functionally graded material. These contours represent the isopachic fringe patterns obtained in experimental techniques such as holographic inter-ferometery. These contours correspond to $\alpha = 0.570$, $K_{ID}(t) = 1.0$ MPa-m$^{1/2}$, $c = 650$ ms^{-1} and $dc/dt = 0$. The figure shows that the transient effects do not alter the structure of the contours very close to the crack-tip as compared to homogeneous material [27]. However, slightly removed from the crack-tip, the size and tilt of contours changes significantly by introduction of transient terms in the crack tip stress field. It can be observed from figure 10 that as the rate of change of stress intensity factor increases the size and number of contours around crack tip increases. The aspect ratio of the contours changes and they become elongated as the rate of change of mode-I stress intensity factor $(dK_{ID}(t)/dt)$ increases.

Figure 11 illustrates the effect of crack tip acceleration on contours of constant first stress invariant around the crack tip. The figure shows that as the crack tip acceleration increase beyond the value of 10^6 ms^{-2} the size and number of contours around the crack tip starts decreasing.

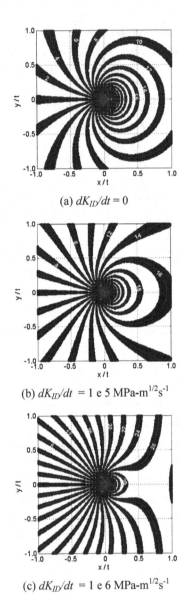

(a) $dK_{ID}/dt = 0$

(b) $dK_{ID}/dt = 1$ e 5 MPa-m$^{1/2}$s^{-1}

(c) $dK_{ID}/dt = 1$ e 6 MPa-m$^{1/2}$s^{-1}

Fig. 10. Effect of rate of change of stress intensity factor on constant maximum first stress invariant $(\sigma_{xx}+\sigma_{yy})$ around the crack-tip

3.2.3 *Contours of constant in-plane displacements*

Figure 12 shows the effect of rate of change of Mode I stress intensity factor on the contours of in-plane displacement (u). The contours of constant in-plane displacements (u and v) represent the Moiré fringes, which are sometimes used to measure fracture parameters. It can be observed from the figure that as rate of change of mode-I stress intensity factor ($dK_{ID}(t)/dt$) increases contours become more dense which indicates an increase in the value of the in-plane displacements, u.

Figure 13 shows the effect of crack-tip accelerations on the contours of in-plane displacement (v). No significant change is noticed in the shape and size of contours until the crack tip acceleration reaches 10^6 ms^{-2}, beyond which number and size of contours around crack tip decreases. It can also be observed that the contours tilt away from crack face as the acceleration increases.

4. Closure

A systematic theoretical analysis is presented for development of crack tip stress, strain and displacement fields for a crack propagating at an angle from the direction of property

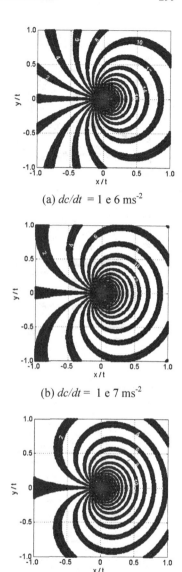

(a) dc/dt = 1 e 6 ms^{-2}

(b) dc/dt = 1 e 7 ms^{-2}

(c) dc/dt = 1 e 8 ms^{-2}

Fig. 11. Effect of crack-tip acceleration on contours of constant maximum first stress invariant $(\sigma_{xx}+\sigma_{yy})$ around the crack-tip

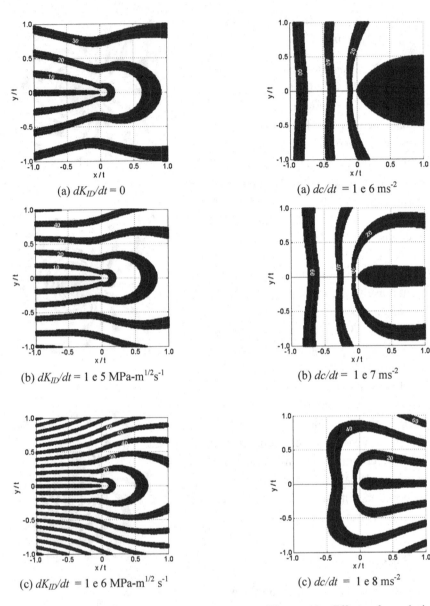

(a) $dK_{ID}/dt = 0$

(a) $dc/dt = 1\ e\ 6\ ms^{-2}$

(b) $dK_{ID}/dt = 1\ e\ 5\ MPa\text{-}m^{1/2}s^{-1}$

(b) $dc/dt = 1\ e\ 7\ ms^{-2}$

(c) $dK_{ID}/dt = 1\ e\ 6\ MPa\text{-}m^{1/2}\ s^{-1}$

(c) $dc/dt = 1\ e\ 8\ ms^{-2}$

Figure 12. Effect of rate of change of mode-I stress intensity factor on constant in-plane displacement (u) around the crack-tip

Figure 13. Effect of crack-tip acceleration on contours of constant in-plane displacement (v) around the crack-tip

gradation in an FGM. The crack tip fields were obtained through an asymptotic analysis coupled with displacement potential formulation. These stress, strain and displacement fields were developed for exponentially varying property gradation. This analysis showed that for a crack propagating in the direction of property variation, the leading term of the expansion still has the classical inverse square root singularity. However, the higher order terms differ considerably from their counterparts in homogeneous materials, which alter the nature of the stresses around the crack tip. These stress and displacements fields are used to generate the synthetic contours of constant maximum shear stress, constant stress invariant and constant in-plane displacement. The contours very close to the crack tip were found to be identical to the homogeneous materials. However, as one moves away from crack tip, contours start deviating from the homogeneous materials showing the affect of nonhomogeneity.

Motivated by the transient nature of the growing crack-tip, an approximation of the near-tip field in the form of an asymptotic expansion has been introduced. The higher order terms in the expansions take into account recent past history of the stress intensity factor and crack motion. The re-examination of the crack-tip fields for elastodynamic crack growth in FGMs under transient conditions has potential to significantly alter the crack-tip fields from the commonly assumed steady state crack propagation. This was shown by discussing the effect of crack tip acceleration and rate of change of stress intensity factor on synthetically generated contours of constant maximum shear stress and constant stress invariant. It was found that transient effects cause significant spatial variation in these contours. Therefore, in studying dynamic fracture of FGMs, it is appropriate to include the higher order transient terms in the crack tip fields for the situations of sudden variation of stress intensity factor or crack tip velocity.

Appendix

$$D_l\{A_0(t)\} = -\frac{3c^{1/2}\rho_0}{\alpha_l^2\mu_0(k+2)}\frac{d}{dt}\left(c^{1/2}A_0(t)\right), \quad B_l^A(t) = \frac{3c^2}{2\alpha_l^4}\left(\frac{\rho_0}{\mu_0(k+2)}\right)^2 A_0(t)\frac{dc}{dt}$$

$$D_l\{C_0(t)\} = -\frac{3c^{1/2}\rho_0}{\alpha_l^2\mu_0(k+2)}\frac{d}{dt}\left(c^{1/2}C_0(t)\right), \quad B_l^C(t) = \frac{3c^2}{2\alpha_l^4}\left(\frac{\rho_0}{\mu_0(k+2)}\right)^2 C_0(t)\frac{dc}{dt}$$

$$D_s\{B_0(t)\} = -\frac{3c^{1/2}\rho_0}{\alpha_s^2\mu_0}\frac{d}{dt}\left\{c^{1/2}B_0(t)\right\}, \quad B_s^B(t) = \frac{3c^2}{2\alpha_s^4}\left(\frac{\rho_0}{\mu_0}\right)^2 B_0(t)\frac{dc}{dt}$$

$$D_s\{D_0(t)\} = -\frac{3c^{1/2}\rho_0}{\alpha_s^2\mu_0}\frac{d}{dt}\left\{c^{1/2}D_0(t)\right\}, \quad B_s^D(t) = \frac{3c^2}{2\alpha_s^4}\left(\frac{\rho_0}{\mu_0}\right)^2 D_0(t)\frac{dc}{dt}$$

On substituting Eqs. (24) and (26) in Eq. (6) the in-plane displacements (u, v) can be obtained as

$$u = \left\{\begin{array}{l}\frac{3}{2}A_0r_l^{1/2}\cos\frac{1}{2}\theta_l + \frac{3}{2}C_0r_l^{1/2}\sin\frac{1}{2}\theta_l + 2A_1r_l\cos\theta_l + 2C_1r_l\sin\theta_l + \frac{5}{2}A_2r_l^{3/2}\cos\frac{3}{2}\theta_l \\[2mm] +\frac{5}{2}C_2r_l^{3/2}\sin\frac{3}{2}\theta_l - r_l^{3/2}\cos\frac{3}{2}\theta_l\left[\frac{3}{2}P_1 + \frac{1}{4}B_l^A\right] - r_l^{3/2}\cos\frac{3}{2}\theta_l\left[P_1 + \frac{5}{2}P_8\alpha_l\right] + 2B_1\alpha_s r_s\cos\theta_s \\[2mm] +r_l^{3/2}\sin\frac{1}{2}\theta_l\left[\frac{3}{2}P_2 + \frac{1}{4}B_l^C\right] - r_l^{3/2}\sin\frac{3}{2}\theta_l\left[P_2 + \frac{5}{2}P_7\alpha_l\right] - r_s^{3/2}\cos\frac{1}{2}\theta_s\left[\frac{3}{2}P_6\alpha_s - \frac{1}{4}B_s^B\right] \\[2mm] +r_s^{3/2}\cos\frac{3}{2}\theta_s\left[\alpha_s P_6 - \frac{5}{2}P_3\right] - r_s^{3/2}\alpha_s\sin\frac{1}{2}\theta_s\left[\frac{3}{2}P_5 - \frac{1}{4}B_s^D\right] - r_s^{3/2}\alpha_s\sin\frac{3}{2}\theta_s\left[P_5 - \frac{5}{2}P_4\right] \\[2mm] +\frac{3}{2}B_0r_s^{1/2}\alpha_s\cos\frac{1}{2}\theta_s + \frac{3}{2}D_0r_s^{1/2}\alpha_s\sin\frac{1}{2}\theta + \frac{5}{2}B_2\alpha_s r_s^{3/2}\cos\frac{3}{2}\theta_s - \frac{5}{2}D_2\alpha_s r_s^{3/2}\sin\frac{3}{2}\theta_{ss} \\[2mm] +2D_1\alpha_s r_s\sin\theta_s - \frac{1}{16}r_l^{3/2}\left[B_l^A\cos\frac{5}{2}\theta_l - B_l^C\sin\frac{5}{2}\theta_l\right] - \frac{1}{16}r_s^{3/2}\left[B_l^B\cos\frac{5}{2}\theta_s + B_l^D\sin\frac{5}{2}\theta_s\right]\end{array}\right\}$$

$$(A\text{-}1a)$$

$$v = \left(\begin{array}{l} -\dfrac{3}{2}A_0\alpha_l r_l^{1/2}\sin\dfrac{1}{2}\theta_l + \dfrac{3}{2}C_0\alpha_l r_l^{1/2}\cos\dfrac{1}{2}\theta_l + 2A_l\alpha_l r_l\sin\theta_l + 2C_l\alpha_l r_l\cos\theta_l - \dfrac{5}{2}A_2\alpha_l r_l^{3/2}\sin\dfrac{3}{2}\theta_l \\[2mm] +\dfrac{5}{2}C_2\alpha_l r_l^{3/2}\cos\dfrac{3}{2}\theta_l - r_l^{3/2}\alpha_l\cos\dfrac{1}{2}\theta_l\left[\dfrac{3}{2}P_2 + \dfrac{1}{4}B_l^C\right] - r_l^{3/2}\cos\dfrac{3}{2}\theta_l\left[\alpha_l P_2 + \dfrac{5}{2}P_7\right] \\[2mm] -\alpha_l r_l^{3/2}\sin\dfrac{1}{2}\theta_l\left[\dfrac{3}{2}P_1 + \dfrac{1}{4}B_l^A\right] - r_l^{3/2}\sin\dfrac{3}{2}\theta_l\left[\alpha_l P_1 - \dfrac{5}{2}P_8\right] + r_s^{3/2}\cos\dfrac{1}{2}\theta_s\left[\dfrac{3}{2}P_6 + \dfrac{1}{4}B_s^D\right] \\[2mm] +r_s^{3/2}\cos\dfrac{3}{2}\theta_s\left[P_5 + \dfrac{5}{2}\alpha_s P_4\right] - r_s^{3/2}\sin\dfrac{1}{2}\theta_s\left[\dfrac{3}{2}P_6 + \dfrac{1}{4}B_s^B\right] + r_s^{3/2}\sin\dfrac{3}{2}\theta_s\left[P_6 + \dfrac{5}{2}P_3\right] \\[2mm] -\dfrac{3}{2}B_0 r_s^{1/2}\sin\dfrac{1}{2}\theta_s - \dfrac{3}{2}D_0 r_s^{1/2}\cos\dfrac{1}{2}\theta_s - 2B_l r_s\sin\theta_s - 2D_l r_s\cos\theta_s - \dfrac{5}{2}B_2 r_s^{3/2}\sin\dfrac{3}{2}\theta_s \\[2mm] -\dfrac{5}{2}D_2 r_s^{3/2}\cos\dfrac{3}{2}\theta_s - \dfrac{1}{16}r_l^{3/2}\left[B_l^A\sin\dfrac{5}{2}\theta_l + B_l^C\cos\dfrac{5}{2}\theta_l\right] - \dfrac{1}{16}r_s^{3/2}\left[B_s^B\sin\dfrac{5}{2}\theta_s - B_l^D\cos\dfrac{5}{2}\theta_s\right] \end{array} \right)$$

$$\text{(A-1b)}$$

where

$$P_1 = -\frac{1}{4\alpha_l^2}\{\alpha A_0(t) + \beta\alpha_l C_0(t)\} + \frac{1}{6}\left[D_l\{A_0(t)\} + \frac{1}{2}B_l^A(t)\right],$$

$$P_2 = \frac{1}{4\alpha_l^2}\{\alpha C_0(t) - \beta\alpha_l A_0(t)\} + \frac{1}{6}\left[D_l\{C_0(t)\} + \frac{1}{2}B_l^C(t)\right],$$

$$P_3 = -\frac{2}{5}\frac{1}{(k+2)(\alpha_l^2-\alpha_s^2)}\{\alpha\alpha_s B_0(t) - \beta D_0(t)\}, \quad P_4 = -\frac{2}{5}\frac{1}{(k+2)(\alpha_l^2-\alpha_s^2)}\{\alpha B_0(t) + \beta\alpha_s D_0(t)\},$$

$$P_5 = -\frac{1}{4\alpha_s^2}\{\alpha D_0(t) + \beta\alpha_s B_0(t)\} + \frac{1}{6}\left[D_s\{B_0(t)\} + \frac{1}{2}B_s^B(t)\right],$$

$$P_6 = \frac{1}{4\alpha_s^2}\{\alpha B_0(t) - \beta D_0(t)\} + \frac{1}{6}\left[D_s\{D_0(t)\} + \frac{1}{2}B_s^D(t)\right],$$

$$P_7 = \frac{2}{5}\frac{k}{(\alpha_l^2-\alpha_s^2)}\{\alpha\alpha_l C_0(t) - \beta A_0(t)\}, \quad P_8 = \frac{2}{5}\frac{k}{(\alpha_l^2-\alpha_s^2)}\{\alpha\alpha_l A_0(t) + \beta C_0(t)\}$$

In-plane strains can be obtained using these displacements in Eq. (A-2) as

$$\varepsilon_{xx} = \frac{\partial u}{\partial x}, \qquad \varepsilon_{yy} = \frac{\partial v}{\partial y}, \qquad \varepsilon_{xy} = \frac{1}{2}\left(\frac{\partial u}{\partial y} + \frac{\partial v}{\partial x}\right). \qquad \text{(A-2)}$$

$$
\varepsilon_{xx} = \left\{
\begin{array}{l}
\dfrac{3}{4}A_0 r_l^{-1/2}\cos\dfrac{1}{2}\theta_l - \dfrac{3}{4}C_0 r_l^{-1/2}\sin\dfrac{1}{2}\theta_l + 2A_1 + \dfrac{15}{4}r_l^{1/2}A_2\cos\dfrac{1}{2}\theta_l + \dfrac{15}{4}r_l^{1/2}C_2\sin\dfrac{1}{2}\theta_l \\[2mm]
- r_l^{1/2}\cos\dfrac{1}{2}\theta_l\left[3P_1 + \dfrac{15}{4}P_8 + \dfrac{1}{4}B_l^A\right] - r_l^{1/2}\cos\dfrac{3}{2}\theta_l\left[\dfrac{3}{4}P_1 + \dfrac{1}{4}B_l^A\right] - \dfrac{3}{4}P_6\alpha_s r_s^{1/2}\cos\dfrac{3}{2}\theta_s \\[2mm]
+ r_l^{1/2}\sin\dfrac{3}{2}\theta_l\left[\dfrac{3}{4}P_2 + \dfrac{1}{4}B_l^C\right] + \alpha_s r_s^{1/2}\sin\dfrac{1}{2}\theta_s\left[\dfrac{15}{4}P_4 - \dfrac{1}{4}B_s^D\right] - \dfrac{15}{4}\alpha_s r_s^{1/2}D_2\sin\dfrac{1}{2}\theta_s \\[2mm]
- \dfrac{3}{4}P_5\alpha_s r_s^{1/2}\sin\dfrac{3}{2}\theta_s + \dfrac{3}{4}B_0\alpha_s r_s^{-1/2}\cos\dfrac{1}{2}\theta_s + \dfrac{3}{4}D_0\alpha_s r_s^{-1/2}\sin\dfrac{1}{2}\theta_s + \dfrac{15}{4}\alpha_s r_s^{1/2}B_2\cos\dfrac{1}{2}\theta_s \\[2mm]
- r_l^{1/2}\sin\dfrac{1}{2}\theta_l\left[3P_2 + \dfrac{15}{4}P_7 + \dfrac{1}{4}B_l^C\right] + 2\alpha_s B_1 + \dfrac{1}{32}r_l^{1/2}\left[B_l^A\cos\dfrac{7}{2}\theta_l - B_l^C\sin\dfrac{7}{2}\theta_l\right] \\[2mm]
+ \dfrac{1}{32}\alpha_s r_s^{1/2}\left[B_s^B\cos\dfrac{7}{2}\theta_s + B_s^D\sin\dfrac{7}{2}\theta_l\right] - \alpha_s r_s^{1/2}\cos\dfrac{1}{2}\theta_s\left[\dfrac{15}{4}P_3 - \dfrac{1}{4}B_s^B\right]
\end{array}
\right\}
$$

$$(\text{A-3a})$$

$$
\varepsilon_{yy} = \left\{
\begin{array}{l}
-\dfrac{3}{4}A_0\alpha_l^2 r_l^{-1/2}\cos\dfrac{1}{2}\theta_l + \dfrac{3}{4}C_0\alpha_l^2 r_l^{-1/2}\sin\dfrac{1}{2}\theta_l - 2\alpha_l^2 A_1 - \dfrac{15}{4}r_l^{1/2}\alpha_l^2 A_2\cos\dfrac{1}{2}\theta_l \\[2mm]
- r_l^{1/2}\alpha_l\cos\dfrac{1}{2}\theta_l\left[3\alpha_l P_1 - \dfrac{15}{4}P_8 - \dfrac{1}{4}\alpha_l B_l^A\right] + r_l^{1/2}\alpha_l^2\cos\dfrac{3}{2}\theta_l\left[\dfrac{3}{4}P_1 - \dfrac{1}{4}B_l^A\right]_s \\[2mm]
- r_l^{1/2}\alpha_l\sin\dfrac{1}{2}\theta_l\left[3\alpha_l P_2 - \dfrac{15}{4}P_7 - \dfrac{1}{4}\alpha_l B_l^C\right] + \alpha_s r_s^{1/2}\cos\dfrac{1}{2}\theta_s\left[\dfrac{15}{4}P_3\alpha_s - \dfrac{1}{4}B_s^B\right] - 2\alpha_s B_1 \\[2mm]
- \alpha_l^2 r_l^{1/2}\sin\dfrac{3}{2}\theta_l\left[\dfrac{3}{4}P_2 - \dfrac{1}{4}B_l^C\right] - \alpha_s r_s^{1/2}\sin\dfrac{1}{2}\theta_s\left[\dfrac{15}{4}P_4\alpha_s - \dfrac{1}{4}B_s^D\right] + \dfrac{3}{4}P_6\alpha_s r_s^{1/2}\cos\dfrac{3}{2}\theta_s \\[2mm]
- \dfrac{3}{4}B_0\alpha_s r_s^{-1/2}\cos\dfrac{1}{2}\theta_s - \dfrac{3}{4}D_0\alpha_s r_s^{-1/2}\sin\dfrac{1}{2}\theta_s - \dfrac{1}{32}\alpha_l^2 r_l^{1/2}\left[B_l^A\cos\dfrac{7}{2}\theta_l - B_l^C\sin\dfrac{7}{2}\theta_l\right] \\[2mm]
+ \dfrac{15}{4}\alpha_s r_s^{1/2}D_2\sin\dfrac{1}{2}\theta_s - \dfrac{1}{32}r_l^{1/2}\left[B_s^B\cos\dfrac{7}{2}\theta_s + B_s^D\sin\dfrac{7}{2}\theta_l\right] - \dfrac{15}{4}\alpha_s r_s^{1/2}B_2\cos\dfrac{1}{2}\theta_s \\[2mm]
- \dfrac{15}{4}r_l^{1/2}\alpha_l^2 C_2\sin\dfrac{1}{2}\theta_l - \dfrac{3}{4}P_5\alpha_s r_s^{1/2}\sin\dfrac{3}{2}\theta
\end{array}
\right\}
$$

$$(\text{A-3b})$$

$$\varepsilon_{xy} = \left\{ \begin{aligned} &\frac{3}{4}A_0\alpha_l r_l^{-1/2}\sin\frac{1}{2}\theta_l + \frac{3}{4}C_0\alpha_l r_l^{-1/2}\cos\frac{1}{2}\theta_l - \frac{3}{8}D_0(1+\alpha_s^2)r_s^{-1/2}\cos\frac{1}{2}\theta_s \\ &-r_l^{1/2}\cos\frac{1}{2}\theta_l\left[\frac{15}{8}P_7(1+\alpha_l^2) - \frac{1}{4}\alpha_l B_l^C\right] - \frac{3}{4}P_2\alpha_l r_l^{1/2}\cos\frac{3}{2}\theta_l + \frac{3}{8}B_0(1+\alpha_s^2)r_s^{-1/2}\sin\frac{1}{2}\theta_s \\ &-\frac{3}{4}P_1\alpha_l r_l^{1/2}\sin\frac{3}{2}\theta_l + r_s^{1/2}\cos\frac{1}{2}\theta_s\left[\frac{15}{4}P_4\alpha_s + \frac{3}{2}P_5(1-\alpha_s^2) + \frac{1}{8}(1+\alpha_s^2)B_s^D\right] \\ &+\frac{(1+\alpha_s^2)}{8}r_s^{1/2}\cos\frac{3}{2}\theta_s[3P_5 + B_s^D] - \frac{(1+\alpha_s^2)}{8}r_s^{1/2}\sin\frac{3}{2}\theta_s[3P_6 + B_s^B] \\ &-\frac{15}{4}A_2\alpha_l r_l^{1/2}\cos\frac{1}{2}\theta_l + \frac{15}{4}C_2\alpha_l r_l^{1/2}\cos\frac{1}{2}\theta_l + r_l^{1/2}\sin\frac{1}{2}\theta_l\left[\frac{15}{8}P_8(1+\alpha_l^2) - \frac{1}{4}\alpha_l B_l^A\right] \\ &+\frac{1}{32}\alpha_l r_l^{1/2}\left[B_l^A\sin\frac{7}{2}\theta_l + B_l^C\cos\frac{7}{2}\theta_l\right] - \frac{1}{64}r_s^{1/2}(1+\alpha_s^2)\left[B_s^B\sin\frac{7}{2}\theta_s + B_s^D\cos\frac{7}{2}\theta_s\right] \\ &+2\alpha_l C_1 - \frac{15}{8}B_2(1+\alpha_s^2)r_s^{1/2}\cos\frac{1}{2}\theta_s r_s^{1/2}\sin\frac{1}{2}\theta_s\left[\frac{15}{4}P_3\alpha_s + \frac{3}{2}P_6(1-\alpha_s^2) + \frac{1}{8}(1+\alpha_s^2)B_s^B\right] \\ &-\frac{15}{8}D_2(1+\alpha_s^2)r_s^{1/2}\sin\frac{1}{2}\theta_s - D_1(1+\alpha_s^2) \end{aligned} \right\}$$

$$(A\text{-}3c)$$

By substituting Eq. (6) in to Eq. (5), in-plane stress component can be written in terms of displacement potentials Φ and Ψ as,

$$\frac{\sigma_{xx}}{\mu_c} = \left[\left(\frac{p+1}{p-1}\right)\frac{\partial^2\Phi}{\partial^2 X} + \left(\frac{3-p}{p-1}\right)\frac{\partial^2\Phi}{\partial^2 Y} + 2\frac{\partial^2\Psi}{\partial X\partial Y}\right]\exp(\alpha x)$$

$$\frac{\sigma_{yy}}{\mu_c} = \left[\left(\frac{3-p}{p-1}\right)\frac{\partial^2\Phi}{\partial^2 X} + \left(\frac{p+1}{p-1}\right)\frac{\partial^2\Phi}{\partial^2 Y} - 2\frac{\partial^2\Psi}{\partial X\partial Y}\right]\exp(\alpha x)$$

$$\frac{\sigma_{xy}}{\mu_c} = \left[-\frac{\partial^2\Psi}{\partial^2 X} + \frac{\partial^2\Psi}{\partial^2 Y} + 2\frac{\partial^2\Phi}{\partial X\partial Y}\right]\exp(\alpha x)$$

$$(A\text{-}4)$$

where $p = 3 - 4\nu$ for plane strain and $p = \dfrac{3-\nu}{1+\nu}$ for plane stress

Substituting for Φ and Ψ in Eq. (A-4) gives

$$
\frac{\sigma_{xx}}{\mu'} = \left\{
\begin{array}{l}
R_1 \left\{
\begin{array}{l}
\frac{3}{4}A_0 r_l^{-1/2}\cos(\frac{\theta_l}{2}) - \frac{3}{4}C_0 r_l^{-1/2}\sin(\frac{\theta_l}{2}) + 2A_1 + \frac{15}{4}r_l^{1/2}\cos(\frac{\theta_l}{2})A_2 \\
+ \frac{15}{4}r_l^{1/2}\sin(\frac{\theta_l}{2})C_2 + \frac{15}{4}P_4 r_s^{1/2}\sin(\frac{\theta_s}{2}) - r_l^{1/2}\cos(\frac{3\theta_l}{2})\left[\frac{3}{4}P_1 + \frac{1}{4}B_l^C\right] \\
- r_l^{1/2}\cos(\frac{\theta_l}{2})\left[3P_1 + \frac{1}{4}B_l^A\right] - r_l^{1/2}\sin(\frac{\theta_l}{2})\left[3P_2 + \frac{1}{4}B_l^C\right] - \frac{15}{4}P_3 r_s^{1/2}\cos(\frac{\theta_s}{2}) \\
r_l^{1/2}\sin(\frac{3\theta_l}{2})\left[\frac{3}{4}P_2 + \frac{1}{4}B_l^A\right] + \frac{1}{32}r_l^{1/2}\left[B_l^A\sin(\frac{7\theta_l}{2}) - B_l^C\cos(\frac{7\theta_l}{2})\right]
\end{array}
\right\} + 4\alpha_s B_1 \\[3em]
R_2 \left\{ \alpha_l^2 \left[
\begin{array}{l}
\frac{3}{4}C_0 r_l^{-1/2}\sin(\frac{\theta_l}{2}) - \frac{3}{4}A_0 r_l^{-1/2}\cos(\frac{\theta_l}{2}) - 2A_1 - \frac{15}{4}r_l^{1/2}\cos(\frac{\theta_l}{2})A_2 \\
- \frac{15}{4}r_l^{1/2}\sin(\frac{\theta_l}{2})C_2 + r_l^{1/2}\cos(\frac{3\theta_l}{2})\left[\frac{3}{4}P_1 - \frac{1}{4}B_l^A\right] - r_l^{1/2}\cos(\frac{\theta_l}{2})\left[3P_1 - \frac{1}{4}B_l^A\right] \\
- \frac{15}{4}P_4 \frac{\alpha_s^2}{\alpha_l^2}r_s^{1/2}\sin(\frac{\theta_s}{2}) - r_l^{1/2}\sin(\frac{3\theta_l}{2})\left[\frac{3}{4}P_2 - \frac{1}{4}B_l^C\right] - \frac{15}{4}P_3 \frac{\alpha_s^2}{\alpha_l^2}r_s^{1/2}\cos(\frac{\theta_s}{2}) \\
- r_l^{1/2}\sin(\frac{\theta_l}{2})\left[3P_2 - \frac{1}{4}B_l^C\right] - \frac{1}{32}r_l^{1/2}\left[B_l^A\cos(\frac{7\theta_l}{2}) - B_l^C\sin(\frac{7\theta_l}{2})\right]
\end{array}
\right]\right\} + \\[3em]
- \frac{3}{2}B_0\alpha_s r_s^{-1/2}\cos(\frac{\theta_s}{2}) + \frac{3}{2}D_0\alpha_s r_s^{-1/2}\sin(\frac{\theta_s}{2}) + \frac{15}{2}B_2\alpha_s r_s^{1/2}\cos(\frac{\theta_s}{2}) - \frac{15}{2}P_7\alpha_l r_l^{1/2}\sin(\frac{\theta_l}{2}) \\
- \frac{15}{2}D_2\alpha_s r_s^{1/2}\sin(\frac{3\theta_s}{2}) - \frac{3}{2}P_5\alpha_s r_s^{1/2}\sin(\frac{3\theta_s}{2}) - \frac{3}{2}P_6\alpha_s r_s^{1/2}\cos(\frac{3\theta_s}{2}) - \frac{15}{2}P_8\alpha_l r_l^{1/2}\cos(\frac{\theta_l}{2})
\end{array}
\right)
$$

(A-5a)

$$\frac{\sigma_{yy}}{\mu'} = \left\{ \begin{aligned} & R_1 \left\{ \alpha_l^2 \begin{bmatrix} -\frac{3}{4}A_0 r_l^{-1/2}\cos(\frac{\theta_l}{2}) + \frac{3}{4}C_0 r_l^{-1/2}\sin(\frac{\theta_l}{2}) - 2A_1 - \frac{15}{4}r_l^{1/2}\cos(\frac{\theta_l}{2})A_2 \\ -\frac{15}{4}r_l^{1/2}\sin(\frac{\theta_l}{2})C_2 - r_l^{1/2}\sin(\frac{3\theta_l}{2})\left[\frac{3}{4}P_2 - \frac{1}{4}B_l^C\right] + r_l^{1/2}\cos(\frac{3\theta_l}{2})\left[\frac{3}{4}P_1 - \frac{1}{4}B_l^A\right] \\ -r_l^{1/2}\sin(\frac{\theta_l}{2})\left[3P_2 - \frac{1}{4}B_l^C\right] + \frac{15}{4}P_3\frac{\alpha_s^2}{\alpha_l^2}r_s^{1/2}\cos(\frac{\theta_s}{2}) - \frac{15}{4}P_4\frac{\alpha_s^2}{\alpha_l^2}r_s^{1/2}\sin(\frac{\theta_s}{2}) \\ -\frac{1}{32}r_l^{1/2}\left[B_l^A\cos(\frac{7\theta_l}{2}) - B_l^C\sin(\frac{7\theta_l}{2})\right] - r_l^{1/2}\cos(\frac{\theta_l}{2})\left[3P_1 - \frac{1}{4}B_l^A\right] \end{bmatrix} \right\} \\ & R_2 \left\{ \begin{bmatrix} -\frac{3}{4}C_0 r_l^{-1/2}\sin(\frac{\theta_l}{2}) + \frac{3}{4}A_0 r_l^{-1/2}\cos(\frac{\theta_l}{2}) + 2A_1 + \frac{15}{4}r_l^{1/2}\cos(\frac{\theta_l}{2})A_2 \\ +\frac{15}{4}r_l^{1/2}\sin(\frac{\theta_l}{2})C_2 - r_l^{1/2}\cos(\frac{3\theta_l}{2})\left[\frac{3}{4}P_1 + \frac{1}{4}B_l^A\right] - r_l^{1/2}\cos(\frac{\theta_l}{2})\left[3P_1 + \frac{1}{4}B_l^A\right] \\ +r_l^{1/2}\sin(\frac{\theta_l}{2})\left[3P_2 - \frac{1}{4}B_l^C\right] - r_l^{1/2}\sin(\frac{3\theta_l}{2})\left[\frac{3}{4}P_2 - \frac{1}{4}B_l^C\right] - \frac{15}{4}P_3\frac{\alpha_s^2}{\alpha_l^2}r_s^{1/2}\cos(\frac{\theta_s}{2}) \\ +\frac{15}{4}P_4\frac{\alpha_s^2}{\alpha_l^2}r_s^{1/2}\sin(\frac{\theta_s}{2}) + \frac{1}{32}r_l^{1/2}\left[B_l^A\cos(\frac{7\theta_l}{2}) - B_l^C\sin(\frac{7\theta_l}{2})\right] \end{bmatrix} \right\} \\ & +\frac{3}{2}B_0\alpha_s r_s^{-1/2}\cos(\frac{\theta_s}{2}) + \frac{3}{2}D_0\alpha_s r_s^{-1/2}\sin(\frac{\theta_s}{2}) + 4\alpha_s B_1 - \alpha_s r_s^{1/2}\cos(\frac{\theta_s}{2})\left[\frac{15}{2}P_8 - \frac{1}{2}B_s^B\right] \\ & -\alpha_s r_s^{1/2}\sin(\frac{\theta_s}{2})\left[\frac{15}{2}P_7 - \frac{1}{2}B_s^D\right] + \frac{15}{2}B_2\alpha_s r_s^{1/2}\cos(\frac{\theta_s}{2}) - \frac{15}{2}D_2\alpha_s r_s^{1/2}\sin(\frac{\theta_s}{2}) \\ & +\frac{1}{16}\alpha_s r_s^{1/2}\left[B_s^B\cos(\frac{7\theta_s}{2}) + B_s^D\sin(\frac{7\theta_s}{2})\right] - \frac{3}{2}P_5\alpha_s r_s^{1/2}\sin(\frac{3\theta_s}{2}) - \frac{3}{2}P_6\alpha_s r_s^{1/2}\cos(\frac{3\theta_s}{2}) \end{aligned} \right.$$

$$(A\text{-}5b)$$

$$\frac{\sigma_{xy}}{\mu'} = \left\{ \begin{array}{l} \dfrac{3}{2}A_0\alpha_l r_l^{-1/2}\sin(\dfrac{\theta_l}{2}) + \dfrac{3}{2}C_0\alpha_l r_l^{-1/2}\cos(\dfrac{\theta_l}{2}) + \dfrac{3}{4}(1+\alpha_s^2)B_0 r_s^{-1/2}\sin(\dfrac{\theta_s}{2}) \\[2mm] -\dfrac{3}{4}(1+\alpha_s^2)D_0 r_s^{-1/2}\sin(\dfrac{\theta_s}{2}) + 4\alpha_l C_1 + 2(1+\alpha_s^2)D_1 - \dfrac{15}{2}A_2\alpha_l r_l^{1/2}\sin(\dfrac{\theta_l}{2}) \\[2mm] +\dfrac{15}{2}C_2\alpha_l r_l^{1/2}\cos(\dfrac{\theta_l}{2}) - \dfrac{15}{4}B_2 r_s^{1/2}\sin(\dfrac{\theta_s}{2}) - \dfrac{15}{4}D_2(1+\alpha_s^2)r_s^{1/2}\sin(\dfrac{\theta_s}{2}) \\[2mm] -\dfrac{15}{4}P_7 r_l^{1/2}(1+\alpha_l^2)\cos(\dfrac{\theta_l}{2}) - \dfrac{3}{2}P_2 r_l^{1/2}\alpha_l\cos(\dfrac{3\theta_l}{2}) + \dfrac{15}{4}P_8 r_l^{1/2}(1+\alpha_l^2)\sin(\dfrac{\theta_l}{2}) \\[2mm] -\dfrac{3}{2}P_1 r_l^{1/2}\alpha_l\sin(\dfrac{3\theta_l}{2}) + r_s^{1/2}\cos(\dfrac{\theta_s}{2})\left(3P_5[1-\alpha_s^2] + \dfrac{15}{2}P_4\alpha_s + \dfrac{(1+\alpha_s^2)}{4}B_s^D\right) \\[2mm] +\dfrac{1}{16}\alpha_l r_l^{1/2}\left[B_l^A\sin(\dfrac{7\theta_l}{2}) + B_l^C\cos(\dfrac{7\theta_l}{2})\right] - r_s^{1/2}\dfrac{(1+\alpha_s^2)}{4}\sin(\dfrac{3\theta_s}{2})[3P_6 - B_s^B] \\[2mm] -\dfrac{1}{2}\alpha_l r_l^{1/2}\left[B_l^A\sin(\dfrac{\theta_l}{2}) - B_l^C\cos(\dfrac{\theta_l}{2})\right] + r_s^{1/2}\dfrac{(1+\alpha_s^2)}{4}\cos(\dfrac{3\theta_s}{2})[3P_5 + B_s^D] \\[2mm] +r_s^{1/2}\sin(\dfrac{\theta_s}{2})\left(3P_6[1-\alpha_s^2] + \dfrac{15}{2}P_3\alpha_s + \dfrac{(1+\alpha_s^2)}{4}B_s^B\right) \\[2mm] +\dfrac{1}{32}(1+\alpha_s^2)r_s^{1/2}\left[B_s^B\sin(\dfrac{7\theta_s}{2}) - B_s^D\cos(\dfrac{7\theta_s}{2})\right] \end{array} \right.$$

(A-5c)

where

$$R_1 = \frac{p+1}{p-1}, \quad R_1 = \frac{3-p}{p-1} \text{ and } \mu' = \mu_c \exp(\alpha x + \beta y)$$

Acknowledgements

Arun Shukla would like to acknowledge the support of the National Science Foundation, Air Force Office Of Scientific Research and the Office of Naval Research over the years for his research on dynamic failure.

References

1   F. Erdogan, F., *Comp. Engg.*, **5**, 7, 753-770 (1995).

2   F. Delale. and F. Erdogan, *J. Appl. Mech.* **50**, 67-80 (1983).

3   Z. H. Jin, and N. Noda, *J. Appl. Mech.*, **61**, 738-739, (1994).

4   J. W. Eischen, *Int. J. Fract e*, **34:3**, 3-22, (1987).

5   P. Gu, and R. J. Asaro, *Int. J. Sol. Struct.*, **32**, 1017-1046, (1997).

6　T. L. Becker, R. M. Cannon, and R. O. Ritchie, *International Journal of Solids and Structures*, **38**, 5545–5563, (2001).

7　Z. H. Jin and R. C. Batra, *J. Mech. Phy. Solid.*, **44:8**, 1221-1235, (1996).

8　G. Anlas, J. Lambros, and M. H. Santare, *Int. J. Fract.*, **115**, 193–204, (2002).

9　V. Parameswaran and A. Shukla, *Trans. of ASME*, 69, 240-243, (2002).

10　N. Jain, C. E. Rousseau, and A. Shukla, A., *Theor. Appl. Frac. Mech.*, **42 (2)**, 155-170, (2004).

11　H. Li, J. Lambros, B. Cheeseman and M. H. Santare, *Int. J. Sol. Struct.*, **37**:3715–32, (1999).

12　P. R. Marur, and H. V. Tippur, *Int. J. Sol. Struct.*, **37**, 5353–5370, (2000).

13　C. Atkinson and R. D. List, *Int. J. Eng. Sc.*, **16**, 717-730, (1978).

14　X. D. Wang and S. A. Meguid, *Int. J. Fract.*, **69**, 87–99, (1995).

15　M. Nakagaki, H. Sasaki, and S. Hagihara, *PVP-Dynamic Fracture, Failure and Deformations, ASME*, **300**, 1-6, (1995).

16　N. Jain and A. Shukla, *Act. Mech.*, **171 (1)**, 75-103, (2004).

17　V. Parameswaran and A. Shukla, *J. Mat. Sc.*, **33**, 3303-3311, (1998).

18　C. E. Rousseau and H. V. Trippur, *Act. Mater.*, 48, 4021-4033, (2000).

19　L. B. Freund and A. J. Rosakis, *J. Mech. Phy. Sol.*, **40(3)**, 699-719, (1992).

20　L .B. Freund, Cambridge University Press, Cambridge, (1990).

21　A. Shukla and R. Chona., Fracture Mechanics: Eighteenth Symposium, *ASTM STP 945*, D. T. Read and R. P. Reed, editors, 86-99 (1987).

22　V. Parameswaran and A. Shukla, *Mech. Mat.* **31**, 579-596.

23　R. J. Sanford and J. W. Dally, *Eng. Fract. Mech.* **11**, 621-633 (1979).

24　T. D. Dudeerar, H. J. Gorman, *Exp. Mech*, **11**, 41-247 (1989).

25　D. Post, W. A. Bacarat, *Exp. Mech*, **21 : 3**,100–104 (1981).

26　A. J. Rosakis, *Opt. Las. Engg.*, 19 3-41 (1993).

27　V. B. Chalivendra, A. Shukla and V. Parmeswaran, J. Elast. **69**, 99-199 (2004).

28　J. W. Dally and A. Shukla, *Mech. Res. Comm.*, **54(1)**, 72-78, (1987).

Chapter 8

Dynamic Fracture Initiation Toughness at Elevated Temperatures With Application to the New Generation of Titanium Aluminides Alloys

Mostafa Shazly[*], Vikas Prakash[*] and Susan Draper[+]

[*]*Department of Mechanical and Aerospace Engineering*
Case Western Reserve University
Cleveland, OH 44106-7222
[+]*NASA Glenn Research Center*
Cleveland, OH 44135

Recently, a new generation of titanium aluminide alloy, named Gamma-Met PX, has been developed with better rolling and post-rolling characteristics. Previous work on this alloy has shown the material to have higher strengths at room and elevated temperatures when compared with other gamma titanium aluminides. In particular, this new alloy has shown increased ductility at elevated temperatures under both quasi-static and high strain rate uniaxial compressive loading. However, its high strain rate tensile ductility at room and elevated temperatures is limited to ~ 1%.

In the present chapter, results of a study to investigate the effects of loading rate and test temperature on the dynamic fracture initiation toughness in Gamma-Met PX are presented. Modified split Hopkinson pressure bar was used along with high-speed photography to determine the crack initiation time. Three-point bend dynamic fracture experiments were conducted at impact speeds of ~1 m/s and tests temperatures of up-to 1200°C. The results show that the dynamic fracture initiation toughness decreases with increasing test temperatures beyond 600°C. Furthermore, the effect of long time high temperature air exposure on the fracture toughness was investigated. The dynamic fracture initiation toughness was found to decrease with increasing exposure time. The reasons behind this drop are analyzed and discussed.

1. Introduction

The design of structures against catastrophic failure requires the knowledge of material fracture properties over a wide range of loading rates and service temperatures. These properties include crack initiation, propagation and arrest. For aerospace applications, events such as soft foreign object damage (bird strike), hard foreign object damage (debris), and engine fan-blade containment under catastrophic failure require a detailed knowledge of the fracture behavior of all materials considered in such events. However, these events involve both high loading rates combined with elevated temperatures.

Over the last two decades, the gamma titanium aluminides have been evaluated as candidate aerospace materials to replace the heavy nickel-based superalloys. However, poor ductility and low fracture toughness have remained the primary road blocks for the full utilization of these alloys. Over the last decade, a large number of experimental studies have been performed to evaluate the fracture toughness of gamma titanium aluminides; the results of these studies have provided valuable information to relate the fracture toughness to the chemical composition[1-3], microstructure and microstructural features[2-7].

However, most of the aforementioned previous work has focused on determining the quasi-static fracture toughness of gamma titanium aluminides, and the dynamic fracture toughness of these alloys has been studied by relatively few researchers. Enoki and Kishi[8] investigated the effect of loading rate on the fracture toughness in fully lamellar and duplex Ti-48Al (at.%) microstructures using the Charpy impact machine. They also investigated the relationship between dynamic fracture toughness, grain size, and the volume fraction of the γ phase. The fracture toughness was found to decrease with increasing colony size, and increase with increasing volume fraction of the γ phase. Sun *et al.*[9] studied the fracture toughness of fatigue pre-cracked three point bend specimens of Ti-45Al-1.6Mn (at.%) at impact velocities up to 10 m/s in a servo hydraulic test machine. They showed that the fracture toughness of the material was about 25 MPa.m$^{1/2}$, and was independent of the impact speed.

At high strain rates and elevated temperatures, Matsugi *et al.*[10] studied the Charpy impact properties at elevated temperatures for two different types of spark sintered Ti-53Al (at.%) microstructures in the notch region. The test temperature ranged from room to 1100°C, and the impact-speed employed in these tests was about 2.5 m/s. It was found that the Charpy impact energy increased monotonously with an increase in test temperature, However, the maximum load measured by the instrumented Charpy machine showed an increase at test temperatures up to 600°C for first microstructure and 520°C for second microstructure, and was almost the same at each temperature for the two microstructures. However, beyond these test temperatures the maximum load shows a continuous drop up to 1127°C. Although this behavior could be explained in terms of a change in fracture mode from trans-granular to mixed inter-/trans-granular to predominantly intergranular with microvoids at the higher temperature, no attempt was made to correlate the uniaxial behavior of the material to the measured Charpy energy. Similarly, Fukumasu *et al.*[11] studied the dynamic fracture toughness of Ti-45Al-1.6Mn (at.%) alloy at elevated temperatures and at loading velocities of 0.1 m/s and 1 m/s using a servo hydraulic machine. The fracture toughness at 0.1 m/s was observed to be always slightly higher than that measured at 1 m/s. Moreover, the fracture toughness increased with increasing test temperatures for all impact velocities up to 600°C, followed by drop in fracture toughness. This behavior was also consistent with uniaxial test data for this material, which showed a yield anomaly at 600°C.

Although gamma titanium aluminides is targeted for an operating temperature of 900°C, embrittlement by oxidation has limited the maximum temperature to 700°C. Several authors had attempted to study the effect of chemical composition, microstructure, and environment on the oxidation behavior of gamma titanium aluminide systems. Most of these studies have focused on oxidation kinetics and scale structure while little has focused on oxidation effects on mechanical properties. Studies on the scale structure had shown that the outermost scale layer is TiO_2 and the subsequent layers are a mixture of TiO_2 and Al_2O_3 [12-14]. The effects of various alloying elements have been studied with Nb appeared to be the most influential element as it provides an effective barrier to the oxidation rate[15]. Other alloying elements such as Si, Cr and V were also found to improve oxidation resistance[16, 17].

With regards to oxide scale effects on the mechanical properties, Cheng[18] and Huang et al.[19] studied the effect of thermal exposure at 700°C for up to 3000 hours on the stability of various gamma microstructures. Their studies showed that although there was evidence of micro-structural instability, it did not lead to degradation in mechanical properties of these alloys. However, this was contrary to the findings of Loretto et al.[20], who observed that the microstructural changes were accompanied by a significant drop in mechanical properties as α_2 phase changes to the γ phase. Similarly, Wright and Rosenberger[21] reported degradation in fatigue properties after high temperature air exposure of certain gamma alloys tested at elevated temperatures, with the largest effect being in the temperature range 500-550°C. A detailed study by Pather et al.[22] showed that including the surface scale lead to a significant drop in tensile strength. Moreover, this drop was dependent on the alloy composition and the materials restored their strength after machining the surface layer.

Recently, due to the ingoing efforts, a new generation of gamma titanium aluminides, named Gamma-Met PX, has been developed by GKSS of Germany in cooperation with Plansee AG of Austria. Gamma-Met PX is available in both sheet and bulk forms. It has better rolling characteristics and improved post-rolling mechanical properties. Moreover, it does not need the costly fabrication processes typically used for intermetallic components. Shaping and forming can be carried out at relatively low temperatures to produce parts which are more uniform than those obtained with other methods. Components can be economically fabricated in normal production settings with the same equipment as that used for conventional titanium alloys[23]. The lack of information about this material, and its potential use for air frames as well as aero-engine components makes it important to understand its fracture behavior not only due to impact loads but also coupled with elevated temperatures. Recent study by Shazly et al.[24] has shown that Gamma-Met PX has high strength at room and elevated temperatures when compared with other gamma titanium aluminides. It also showed increased ductility at elevated temperatures under quasi-static uniaxial loading and under high strain rate compressive loading. However, the high strain rate tensile ductility at room and elevated temperatures was limited to a maximum of 1%. In a recent study, Draper *et al.*[25] observed that the ductility of Gamma-Met PX falls off after exposure at 700 °C and 800°C for 200 hrs in air and dry oxygen. However, after removal of the surface layer, the material's ductility was restored. Different embrittlement mechanisms were investigated. While no microstructural instability was found, hydrogen embrittlement was found not to be responsible for the loss of ductility. The most probable source for the loss of ductility was attributed to subsurface oxidation; however, microhardness indentations close to the surface did not reveal any information.

In the work presented in this paper, a series of experiments was conducted to determine the dynamic fracture initiation toughness of Gamma-Met PX at room and elevated temperatures. Moreover, the role of oxide scale, formed by long time high temperature air exposure, on the dynamic fracture initiation toughness was investigated. Three point bend sample were used to obtain conduct the dynamic fracture tests using the modified split Hopkinson pressure bar at both room and elevated temperatures. For the study involving the effect of high temperature air exposure on fracture initiation toughness, the samples were exposed at 700°C for 100 hours, 200 hours, and 300 hours, in air prior to testing at room temperature.

2. The Modified Split Hopkinson Pressure Bar (MSHPB)

For years the Charpy test has been the method of choice for classifying materials and to determine their ductile to brittle transition temperatures; this has primarily been because the Charpy test is relatively inexpensive and easy to perform. The instrumentation of Charpy test, by attaching a strain gage on the tub and/or on the specimen, has permitted the determination of the load-time history upon impact. The time-load history can be used along with the classical quasi-static facture mechanics analysis to obtain the relevant fracture mechanics parameters. However, due to the effects of inertia, the load obtained from the strain gages oscillates about the actual load. Moreover, the time at which the maximum load is reached does not coincide with the crack initiation time[26]. Therefore, the determination of the dynamic fracture initiation toughness based on the maximum load and simple quasi-static formula for stress intensity factor may lead to incorrect results. As a better alternative, the Modified Split Hopkinson Pressure Bar (MSHPB) technique, shown in Fig. 1, which makes use of the one dimensional wave propagation theory in elastic bars, can be used to determine the load-displacement history at the load point on the specimen during the impact loading process. The advantages of MSHPB over Charpy methods include (a) its relatively simple set-up, (b) accurate determination of both the load and the load-point displacement using one-dimensional wave analysis, and (c) minimization of material inertia effects.

The first implementation of the MSHPB began in the mid 1980's with the work of Ruiz and Mines[27]. They used two strain gage stations on the incident bar to measure the incident and reflected stress waves and used them to infer the load and displacement at the specimen. The same technique was employed by Bacon *et al.*[28] for the determination of the dynamic fracture toughness of PMMA. In the late 1980's, Yokoyama and Kishida[29] used the same technique but supported the specimen in two transmitted bars. Four strain gage stations were employed, one on the incident bar, two on the transmitted bars, and one on the specimen near the crack to detect the onset of crack propagation. In their work, based on the arrival of the transmitted waves, they concluded that the specimen lost contact with the support for very short period of time. Weisbrod and Rittel[30] and Rittel *et al.*[31] used the MSHPB in a one point bend impact configuration to determine the fracture toughness of relatively small sized specimens.

The MSHPB technique has been successfully used by several researchers to determine the fracture initiation toughness in different systems of materials. For example, Irfan *et al.*[32] in the dynamic fracture of discontinuously reinforced aluminum composites, Popelar *et al.*[33] in the dynamic fracture toughness of 4340 steel, Martins and Prakash[34] in the dynamic fracture of linear medium density polyethylene, and more recently Evora and Shukla[35] in the characterization of plyester/TiO_2 nanocomposites.

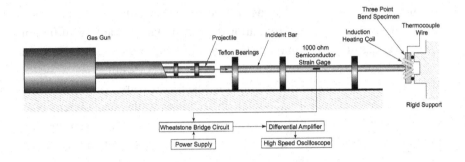

Fig. 1. Schematic of the Modified Split Hopkinson Pressure Bar utilized in the present study.

2.1 *MSHPB theory*

Upon the impact of the striker bar to incident bar, a compressive stress-pulse propagates down the incident bar towards the three-point-bend specimen. The amplitude of this stress-pulse is controlled by the impact velocity while the duration is controlled by the length of the striker bar. The impact force is transmitted to the specimen at the incident bar specimen interface, also referred to as the load-point of the specimen. The wave propagation in the incident and the striker bars is shown schematically in Fig. 2. In all experiments the impact-velocity must be controlled such that the striker and incident bars remain essentially elastic during the entire duration of the experiment.

The stress and particle velocity relationship that governs the propagation of elastic longitudinal stress waves in the striker and the incident bars are given by a system of 1st order hyperbolic partial differential equations, i.e.

$$F \pm (\rho c)_L A V = const \text{ along } \frac{dx}{dt} = \mp c .$$ (1)

where, F is the axial force; ρ is the mass density; c is the longitudinal wave-speed in the bar; $(\rho c)_L A$ is the longitudinal impedance of the striker and the incident bars; and V is the particle velocity.

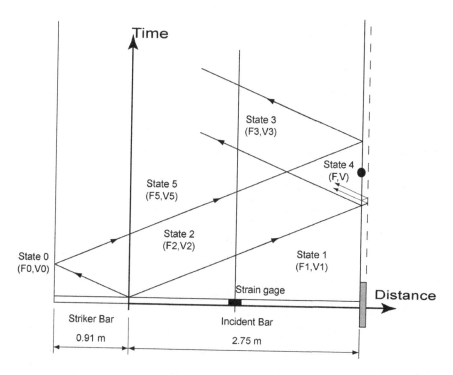

Fig. 2. Time distance diagram for the wave propagation in the MSHPB bars.

Using the method of the characteristics, the force and particle velocity $\left(F, V\right)$ at the loading point (State (4)) can be related to the force and particle velocity in States (2), (3) and (5) by

$$F(t) = F_2 + F_3 - F_5, \text{ and } V(t) = \frac{\left(F_3 - F_5 - F_2\right)}{(\rho c)_L A}.$$ (2)

Note that in the present case, State 2 represents the incident wave, State 3 represents the reflected wave and State 5 is a zero state, i.e. F_5=0, and V_5=0. It is interesting to note that both force and load point displacement can be expressed solely in terms of the forces F_2, F_3 and F_5, and the longitudinal mechanical impedance of the incident bar. The force and displacement at the loading point are expressed in terms of the measured incident and reflected waves as

$$F(t) = AE\left[\varepsilon_{in}(t) + \varepsilon_{ref}(t)\right], \tag{3}$$

and

$$u(t) = c\int_0^t \left[\varepsilon_{in}(\tau) - \varepsilon_{ref}(\tau)\right]d\tau . \tag{4}$$

2.2 Calculation of dynamic fracture initiation toughness

Nakamura *et al.*[36] have quantified the effects of material inertia in laboratory three-point-bend fracture specimens. Their work suggested several avenues to minimize such effects. In particular, they have shown that the behavior of a dynamically loaded three-point bend specimen can be characterized by a short time response dominated by discrete waves, and a long time response that is essentially quasi-static. At intermediate times, the global inertia effects are significant but the local oscillations at the crack tip are small due to kinetic energy being absorbed by the crack tip plasticity. Furthermore, to distinguish short time response from the long time behavior, they defined a transition time, t_τ, as the time at which the kinetic energy and deformation energy in the specimen are approximately equal. From this analysis they concluded that the effects of material inertia dominate prior to the transition time, but the deformation energy dominates at times significantly greater than t_τ. Thus, for $t \gg t_\tau$ inertia effects can be neglected and quasi-static models can be applied in the interpretation of the experimental results.

From the analysis of Nakamura *et al.*[36], the transition time t_τ, is given by

$$t_\tau = D\Lambda\frac{W}{c_o} \tag{5}$$

where, W is the width of the specimen; Λ is a geometric factor; c_o is the wave speed; and D is a non-dimensional constant denoted by

$$D = t\left.\frac{\dot{\Delta}}{\Delta}\right|_{t_\tau} \tag{6}$$

In Eq. 6, Δ is the instantaneous load point displacement, and $\dot{\Delta}$ is the rate of load point displacement, both being evaluated at $t = t_\tau$.

The quasi-static analysis conditions can be met by minimizing t_τ. This can be achieved by (a) decreasing the applied displacement rate, i.e. by decreasing the impact velocity and/or employing a relatively low impedance incident bar instead of the conventional maraging steel bar employed in split Hopkinson bar apparatus, and (b) by decreasing the width of the specimen. Following this, the fracture toughness initiation can be calculated using the quasi-static formula[37]

$$K_I = \frac{P}{B\sqrt{W}} f(a/W), \tag{7}$$

and

$$f(a/W) = \frac{3\frac{S}{W}\sqrt{\frac{a}{W}}}{2\left(1+2\frac{a}{W}\right)\left(1-\frac{a}{W}\right)^{3/2}}\left[1.99 - \frac{a}{W}\left(1-\frac{a}{W}\right)\left\{2.15 - 3.93\left(\frac{a}{W}\right) + 2.7\left(\frac{a}{W}\right)^2\right\}\right] \tag{8}$$

In Eq. 7, P is the maximum load and is obtained from the experimental load versus displacement plot, B is the specimen thickness, W is the specimen width, S is the span, and a is the crack length.

3. Experimental work

3.1. Material

Gamma-Met PX has a chemical composition of Ti-45Al-5Nb-B-C (at.%) with Nb content higher than other classical gamma titanium aluminides, which have usually a 2% (at.) of Nb. The material used in the present study was supplied by Plansee AG in the form of sheets having a thickness of 5.5 mm. Microstructural investigation (Fig. 3) shows that the material consists of a duplex microstructure with some elongated grains in the rolling direction. The duplex microstructure has been shown to provide these alloys with the best tensile elongation combined with a reasonable strength. The toughness is ranged between 10 and 16 MPa.m$^{1/2}$ with low resistance to crack propagation[4].

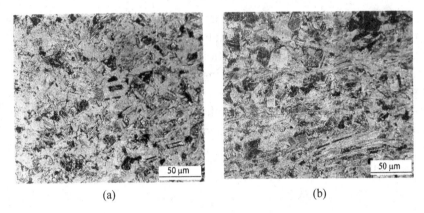

(a)   (b)

Fig. 3. Microstructure of the as received Gamma-Met PX: (a) in rolling direction and (b) normal to rolling direction.

3.2 *Specimen design*

The specimen used in the present study, shown in Fig. 4, was designed according to ASTM E 812-91[38]. Due to the brittle nature of this material no fatigue pre-cracking was performed, however, the notch was extended 1 mm beyond the V-notch using an EDM wire of diameter 0.006" giving a notch radius of 75 μm. A detailed study was performed by Pu *et al.*[5] to determine the effect of notches and microstructure on the fracture toughness of Ti-46.5Al-2.5V-1Cr (at.%) alloy. The fracture toughness was found to be independent of the notch radius up to a certain critical value after which the fracture toughness increases linearly with the square root the notch radius. This critical notch radius for each microstructure was comparable the grain size of the corresponding microstructure. For the current study the average colony size is about 25 μm which is three times less than the notch radius. However, Gamma-Met PX has a yield stress of about 1000 MPa, which is twice as that for the material used by Pu *et al.*[5]. This and the fact that when Pu *et al.*[5] used a notch radius of one and half of the average colony size they reported an increase in fracture toughness by 8.8% from the precracked specimen, assure that the notch radius will have a minimal effect of the measured fracture toughness of the material.

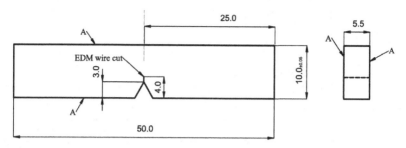

All Dimensions in mm
"A" surfaces should be perpendicular and parallel as appicable within 0.01 mm TIR

Fig. 4. Three-point bend specimen used in the present study

3.3 *Experimental setup*

The schematic of the Modified Hopkinson Pressure Bar (MSHPB) apparatus is shown in Fig. 1. In order to minimize the effects of impedance mismatch between the incident bar and the specimen, the incident bar is made from Aluminum 7075-T6 alloy. Also, in order to generate a single well defined loading pulse, the striker bar is also made from the same material as the incident bar. The diameter of the incident and the striker bars is 19.05 mm and the radius at the end of the incident bar that is in contact with the specimen, i.e. at the load point, is machined to be approximately 50mm.

The incident and the reflected strain in the incident bar are measured by a pair of strain gages attached to the surface of the incident bar at a gage station. A pair of diametrically opposite semiconductor strain gages (BLH SPB3-18-100-U1) is used to minimize bending effects. A 15V Tektronix PS-280 power supply is used to power the Wheatstone bridge circuit. The signal from the Wheatstone bridge circuit is amplified by a differential amplifier having a cut off frequency of 1 MHz (Tektronix 5A22N) and recorded on a digital Oscilloscope (Tektronix TDS-420) having a minimum frequency bandwidth of 100 MHz. The measured strain profiles are used in conjunction with one-dimensional elastic wave theory to obtain the load versus displacement history at the incident bar specimen interface. The load versus displacement history is used along with principles of linear elastic fracture mechanics (LEFM) to obtain the dynamic fracture initiation toughness.

For the high temperature experiments, the specimens were heated using induction heating technique by using a water cooled Hüttinger TIG 10/100 RF generator. This generator uses 3-phase 230-V power line and delivers a maximum power of 10 kW at 100 kHz. The power is coupled to the specimen by using 0.125" copper tubing coils. In the present study, a coil of 1.0" inside diameter and four round turns around the specimen was used. The coil was shaped such that it allows the incident bar to pass though it and contact the specimen. The temperature of the specimen was monitored by thermocouple wires attached to the specimen surface near the notch. A 0.015"chromel-alumel wire is spot welded to the specimen to monitor the specimen temperature prior to the test. This allows testing to be carried out over the range from room to 1200°C.

Prior to testing, the bar is moved away from the specimen, and the heating process is started. The temperature is increased slowly until it reaches 50 to 100°C above the desired test temperature. At this moment, the heating system is shut down and the bar is moved manually and brought in contact with the specimen. When the specimen is close to the desired test temperature (usually 5°C above the test temperature) the projectile is fired and the test is performed.

4. Results and discussions

4.1 *Assessment of inertia effects*

In the present study, in order to determine the influence of inertia, a series of experiments were conducted at room temperature and impact velocity ranging from 1.0 m/s to 3.6 m/s. The high speed camera (DRS Hadland ULTRA 17) with a framing rate of 150,000/sec was used to determine the crack initiation time and correlate it to the strain gage measurements. Figure 5 shows an example of the strain gage signals measured in one of these experiments at low impact speed (1 m/sec). Using Eqs. 4 to 7, the force-time diagrams is constructed and shown in Fig. 6. The corresponding high speed camera frames are shown in Fig. 7

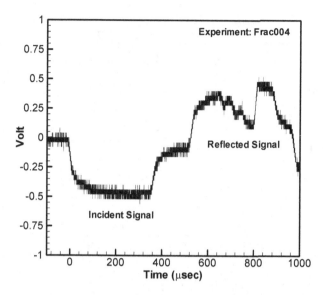

Fig. 5. Strain gage output signal in MSHPB experiment no. Frac004.

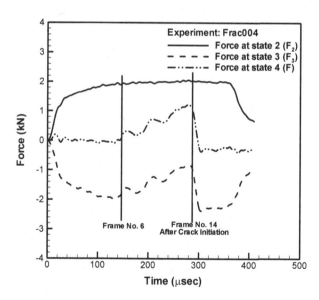

Fig. 6. Forces at states 2, 3 and 4 inferred from strain gage measurements for experiment Frac004.

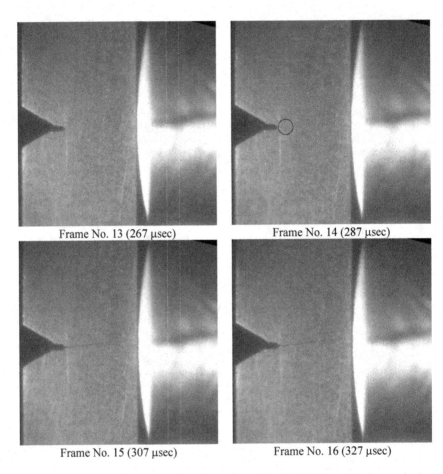

Frame No. 13 (267 μsec)   Frame No. 14 (287 μsec)

Frame No. 15 (307 μsec)   Frame No. 16 (327 μsec)

Fig. 7. Selected frames from a video of experiment Frac004. Frames are also marked on Fig. 6.

which shows clearly the coincidence of the load drop as shown in Fig. 6 to the crack initiation as observed from the high speed camera photographs. Figures 8 and 9 show the force time diagram for experiments performed at impact speed of 3.6 m/sec and the corresponding high speed camera frames in Figs. 10 and 11. The results of this series of experiments are summarized in Fig. 12. From Fig. 12, it can be clearly observed that for impact speeds less than 2.5 m/sec, the

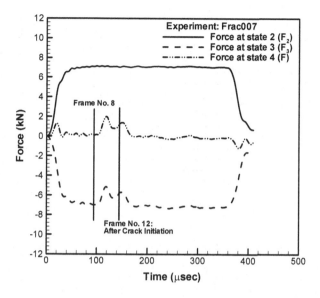

Fig. 8. Forces at states 2, 3 and 4 inferred from strain gage measurements for experiment Frac007.

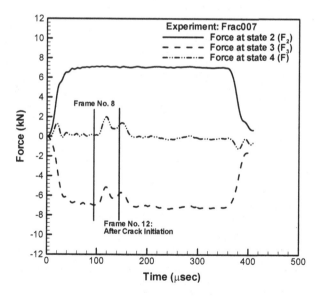

Fig. 9. Forces at states 2, 3 and 4 inferred from strain gage measurements for experiment Frac008.

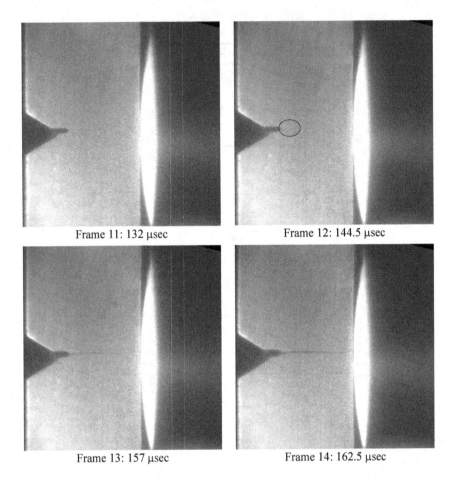

Fig. 10. Selected frames from a video for experiment Frac007. Frames are also marked on Fig. 8.

crack initiation time coincides with the maximum load while for higher impact speeds inertia effects dominate and the crack initiation time no longer coincides with the peak load. The agreement of crack initiation time with peak force in low impact velocity experiments is similar with recent work conducted by Evora and Shukla[35] where they employed the photo-elastic technique along with the high speed camera to determine the crack initiation time, while for high impact velocities the

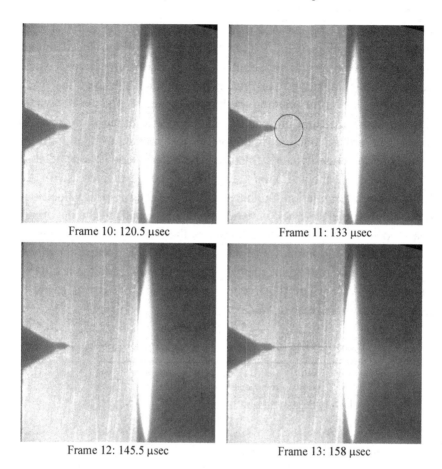

Frame 10: 120.5 µsec   Frame 11: 133 µsec

Frame 12: 145.5 µsec   Frame 13: 158 µsec

Fig. 11. Selected frames from a video for experiment Frac008. Frames are also marked on Fig. 9.

disagreement of crack initiation time with peak force is similar to the work conducted by Kalthoff *et al.*[26] using a drop-weight machine and Rittel *et al.*[31] using the MSHPB. This disagreement can be attributed to either a complex stress wave pattern that is developed in the specimen and/or inertia effects under the high impact velocity loading. However, the results we obtained are consistent with the finite element analysis by Nakamura *et al.*[36] where for the current specimen design and slow loading rate ($\Lambda=1$ and $c_o=6622$ m/sec) the transition time is $t_\tau \sim 15\mu s$. Times corresponding to the maximum load for the high speed impact experiments (~3.6 m/sec) are about 15 and 21 μsec, respectively, which are close to the transition times given by Eqs. 8 and 9 (shown in Figs. 8 and 9) and the corresponding high speed camera frames shown in Figs. 10 and 11. In the present study, in order to avoid the effects of inertia and assure fracture initiation at the maximum load, all the elevated temperature dynamic fracture experiments were conducted at an impact speed of about 1.0 m/s. Figure 12 shows an increase in the dynamic fracture initiation toughness with increasing impact velocity up to the point where the crack initiation time corresponds to the maximum measured load. However, this contradicts the reported results for other gamma titanium aluminides by Sun et al.[9] where the fracture toughness initiation was found to be independent on the load rate and the results by Fukumasu et al.[11] where the fracture initiation toughness was found to decrease with an increase in loading rate.

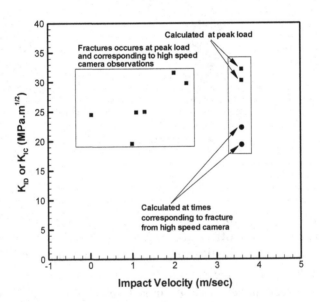

Fig. 12. Effect of impact speed on the crack initiation time.

4.2 Elevated temperature dynamic fracture toughness of Gamma-Met PX

To study the effect of test temperature on the dynamic fracture initiation toughness of Gamma-Met PX, the elevated temperature MSHPB was utilized. The temperatures were increased in steps of 200°C from room temperature all the way up to 800°C after which the temperature was increased in 100°C steps. Figure 13 shows the effect of test temperature on the dynamic fracture toughness at crack initiation. The room temperature dynamic fracture initiation toughness is about 22.5 MPa.m$^{1/2}$ and drops to about 13 MPa.m$^{1/2}$ at 1200°C. Moreover, the dynamic fracture initiation toughness is independent of the test temperature up to about 650°C. Although, for most metallic materials the fracture toughness is expected to increase with temperature, the observed fracture behavior is consistent with results obtained Gamma-Met PX under uniaxial dynamic tensile loading where the material shows a drop in the flow stress while maintaining the same room temperature ductility[24]. Under dynamic bend loading, where the material fails in a tensile mode, the specimen is expected to fail at the same crack-tip opening

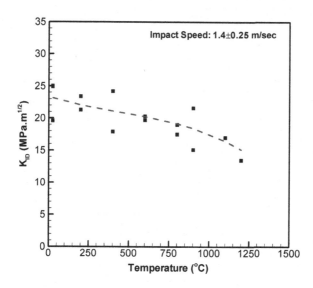

Fig. 13. Effect of test temperature on the dynamic fracture toughness initiation.

displacement regardless of the test temperature. In addition, since with increasing test temperatures, the material's flow stress-drops, it is expected that the material will require less force to initiate the crack. These observations are also consistent with the results of Matsugi *et al.*[10] and Fukumasu *et al.*[11].

Figure 14 shows the fracture surface of sample fractured at relatively low impact velocity and test temperatures ranging from room to 1200°C. The room temperature fracture is cleavage like fracture, as shown in Fig. 14(a). With an increase in test temperature the fracture mode changes from brittle cleavage to ductile like fracture with void growth. At 600°C, Fig. 14(b), the fracture surface shows a mixture of cleavage and ductile modes of fracture, while at higher temperatures the fracture is essentially ductile.

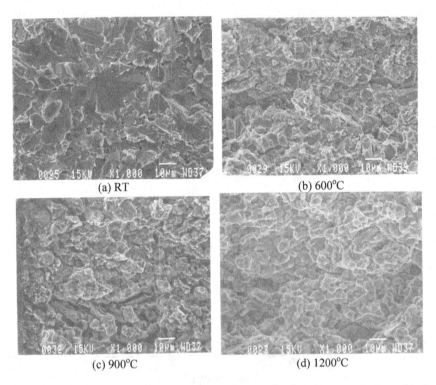

(a) RT (b) 600°C

(c) 900°C (d) 1200°C

Fig. 14. Fracture surface morphology for samples tested at low impact velocity and test temperatures ranging from room to 1200°C

4.3 *Effect of long-time high temperature exposure on dynamic fracture toughness of Gamma-Met PX*

In this part of the study, the high temperature exposed specimens were tested at room temperature using the same low impact velocity that was employed in testing the as-received specimens at room and elevated temperatures. Figure 15 shows the effect of exposure time on the room temperature dynamic fracture initiation toughness. The toughness drops from approximately 22 MPa.m$^{1/2}$ at no exposure to about 12-13 MPa.m$^{1/2}$ at a long-time exposure of 300 hours at 700°C.

To understand the reasons behind the reduction in the dynamic fracture toughness, the oxide morphology on the exposed samples was investigated. In order to examine the exposed specimens under the SEM, the samples were cut using ISOMET 4000 precision saw and with an abrasive wheel #4207. The cutting speed was 2500 rpm and the cutting force was set to 330 grams. These samples were then set in Buehler Ultra-Mount ™ room temperature fast-setting epoxy. They were then polished using Buehler Metaserv 2000 polishing machine. The SEM micrographs for these samples are shown in Fig. 16. It is clearly shown

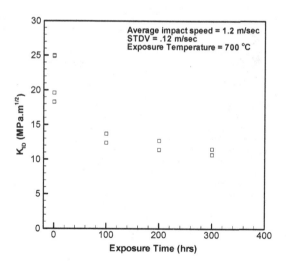

Fig. 15. Effect of high temperature air exposure times on the dynamic fracture initiation toughness.

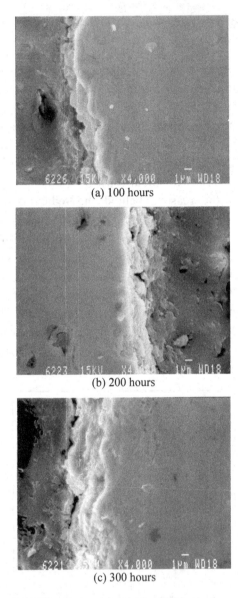

(a) 100 hours

(b) 200 hours

(c) 300 hours

Fig. 16. SEM micrographs of oxide scale thickness formed after exposure at 700°C for different times.

that the oxide scale increases from 1.43±0.64 μm after 100 hours of exposure to 2.63±0.66 μm after 200 hours and reaches 4.15±1.13 μm after 300 hours of exposure. However, the structure of the oxide scale, analyzed using a microprobe, showed that it eventually consists of a layer of TiO_2 and Al_2O_3, Ti/Al-N layer at the specimen-scale interface, and an Nb enriched phase at the grain boundaries, as shown in Fig. 17. Finally, nanoindentations on the subscale were performed on samples prepared according the procedure listed above. A Hysitron Triboscope head, mounted on Digital Instruments Dimension 3100™ atomic force microscope, was used. The head transducer is capable of providing a maximum force of 11000 μN and a maximum displacement of 4 μm. The nanoindenter tip used was a Berkovich tip with a radius of curvature of 100 nm to 200 nm. Three different force levels, i.e. 3000, 5000 and 7000 μN, were used to obtain the hardness at different locations. A typical trapezoidal loading cycle of 5 seconds loading, 5 seconds dwell at maximum load and 5 seconds for unloading was used in the indentation process. Figures 18 to 20 show the hardness profile as function of distance measure from the sample surface at different exposure times. Although no distinction is found for samples exposed for 100 hrs and those exposed for 200 or 300 hrs, the hardness profile in all cases extended up to about 30-40 μm. The simple fracture mechanics calculation[37] shows that the plastic zone (process zone) size under plane strain conditions is about 21 μm for a fracture toughness of 20 MPa.m$^{1/2}$ and yield stress of 1000 MPa, which is well inside the region of influence of the hardness profile. Although the maximum scale thickness was 4.15±1.13 μm, the extension of the region of "hardness-influence" is attributed to the diffusion of alloying elements in this region. The most influential element is most likely Nb, as shown in Fig. 17, and reported by Schmitz-Niederau and Schütze[14], where under elevated temperature air exposure it had a higher concentration at the subscale than at the bulk material. This could be the reason for the drop in fracture toughness of the exposed samples. Also the extension of the profile to between 30-40 μm is somewhat consistent with what Draper *et al.*[25] reported for the restoration of ductility upon the removal of 25-50 μm of the specimen surface layer. However, the drop in toughness as exposure time increases could be attributed to the change in critical flaw size as the oxide scale increases.

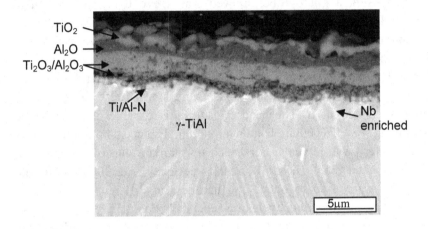

Fig. 17. Near surface microstructure of an as-extruded sample exposed to 800 °C for 200 hours in air

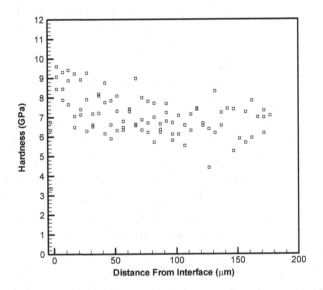

Fig. 18. Subscale nanohardness measured from the surface for a sample exposed at 700°C for 100 hours.

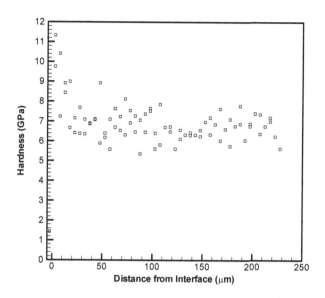

Fig. 19. Subscale nanohardness measured from the surface for a sample exposed at 700°C for 200 hours.

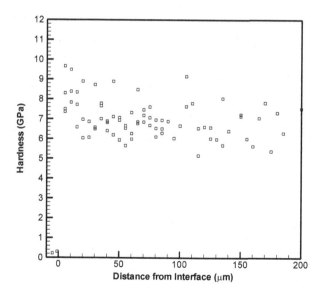

Fig. 20. Subscale nanohardness measured from the surface for a sample exposed at 700°C for 300 hours.

5. Summary

A new technique has been presented to determine the dynamic fracture toughness initiation at elevated temperature. In the relatively low impact velocity (less than 2.5 m/sec) experiments the fracture initiation corresponds to the maximum load measured in the samples. At higher impact velocities, however, the initiation occurs after the peak-load as inertia effects become more dominant. The use of a notched specimen has a minimal effect on the fracture toughness. The dynamic fracture toughness initiation was found to decrease with increasing test temperature and this can be correlated to uniaxial dynamic tensile properties of Gamma-Met PX.

Long-time high-temperature air exposure were found to decrease the fracture toughness in Gamma-Met PX. This drop could be attributed to subscale embrittlement by Nb diffusion and an increase of the defect size as the scale thickness increases with exposure time.

Acknowledgments

The authors would like to acknowledge financial support from NASA's advanced aero-propulsion research program through grant No. NAG 3-2677. The authors would also like to thank Andreas Venskutonis of Plansee, Austria, for providing the Gamma-Met PX material used in the present study. The authors would also like to acknowledge the financial support from the National Science Foundation through grants CMS-9908189 and CMS-0079458 in setting up the elevated temperature test facility at CWRU.

References

1. S-C. Huang and E.L. Hall, Metall. *Mater. Trans.*, **A22**, 427(1991).
2. B. Dogan, D. Schöneich, K.-H. Schwalbe and R. Wagner, *Intermetallics*, 4, 61 (1996).
3. C.T. Liu, J.H. Schneibel, P.J. Maziasz, J.L. Wright and D.S. Easton, *Intermetallics*, 4, 429 (1996).
4. Y-W Kim, *Mater. Sci. Eng.* **A192/193**, 519 (1995).
5. Z.J. Pu, K.H. Wu, J. Shi and D. Zou, *Mater. Sci. Eng.* **A192/193**, 347 (1995).

6. M.A. Morris and M. Leboeuf, *Mater. Sci. Eng.* **A224**, 1 (1997).
7. N.J. Rogers, P.D. Crofts, I.P. Jones and P. Bowen, *Mater. Sci. Eng.* **A192/193**, 379 (1995).
8. M. Enoki and T. Kishi, *Mater. Sci. Eng.* **A192/193**, 420 (1995).
9. Z.M. Sun, T. Kobayashi, H. Fukumasu, I. Yamamoto and K. Shibue, K., *Metall. Mater. Trans.* **A29**, 263 (1998).
10. K. Matsugi, T. Hatayama and O. Yanagisawa, *Intermetallics*, 7, 1049 (1999).
11. H. Fukumasu, T. Kobayashi, H. Toda and K. Shibue, *Metall. Mater. Trans.* **A31**, 3053 (2000).
12. T. Shimizu, T. Iikubo and S. Isobe, *Mater. Sci. Eng.* **A153**, 602 (1992).
13. A. Takasaki, K. Ojima, Y. Taneda, T. Hoshiya and A. Mitsuhasi, *J. Mater. Sci.*, 28, 1067 (1993).
14. M. Schmitz-Niederau and M. Schütze, *Oxid. Met.*, **52**(314), 225 (1999).
15. V.A.C. Haanappel, H. Clemens and M.F. Stroosnijder, *Intermetallics*, 10, 293 (2002).
16. K. Maki, M. Shioda, M. Sayashi, T. Shimizu and S. Isobe, *Mater. Sci. Eng.* **A153**, 591 (1992).
17. R.A. Perkins, K.T. Chiang and G.H. Meier, *Scr. Metall.*, 21, 1505 (1987).
18. T.T. Cheng, *Intermetallics*, 7, 995 (1999).
19. Z.W. Huang, W. Voice and P. Bowen, *Intermetallics*, 8, 417 (2000).
20. M.H. Loretto, A.B. Godfrey, D. Hu, P.A. Blenkinsop, I.P. Jones and T.T. Cheng, *Intermetallics*, 6, 663 (1998).
21. P.K. Wright and A.H. Rosenberger, *Mater. Sci. Eng.* **A329/331**, 538 (2002).
22. R. Pather, W.A. Mitten, P. Holdway, H.S. Ubhi, A. Wisbey and J.W. Brooks, *Intermetallics*, 11, 1015 (2003).
23. A. Venskutonis and K. Rißbacher, *Prep. Future*, **10** (2), 2 (2000).
24. M. Shazly, V. Prakash and S. Draper, *Int. J. Solids Struct.*, **41**(22-23), 6485 (2004).
25. S. Draper, S., B.A. Lerch, I.E. Locci, M. Shazly and V. Prakash, *Intermetallics*, **13** (9), 1014 (2005)
26. J.F. Kalthoff, S. Winkler and J. Beinert, *Int. J. Fract.*, 13, 528 (1977).
27. C. Ruiz and R.A.W. Mines, *Int. J. Fract.*, 29, 101 (1985).
28. C. Bacon, J. Färm and J.L. Lataillade, *Exp. Mech.*, **34**(3), 217 (1994).
29. T. Yokoyama and K. Kishida, *Exp. Mech.*, **29**(2), 188 (1989).
30. G. Weisbrod and D. Rittel, *Int. J. Fract.*, 104, 89 (2000).
31. D. Rittel, A. Pineau, J. Clisson and L. Rota, *Exp. Mech.*, **42**(3), 247 (2002).
32. M. A. Irfan, N-S. Liou and V. Prakash, *Material Research Society Symposium Proceedings*, 434, 219 (1996).
33. C. H. Popelar, C.E. Anderson Jr. and A. Nagy, A., *Exp. Mech.*, **40**(4), 401 (2000).
34. C. Martins and V. Prakash, *Proceedings of the TMS Fall Meeting*, 105 (2002).
35. V. Evora and A. Shukla, *Society of Experimental Mechanics Conference*, Charlotte, Noth Carolina, June 2-6.
36. T. Nakamura, C.F. Shih and L.B. Freund, *Eng. Fract. Mech.*, **25**(3), 323 (1986).

37. T.L. Anderson, *Fracture Mechanics*: Fundamentals and Applications, CRC Press (1995).
38. ASTM Standards, Standard Test Method for Crack Strength of Slow-Bend Precracked Charpy Specimens of High Strength Metallic Materials, Section Three: Metals Test Methods and Analytical Procedures, 03.01, 677 (2002).

Chapter 9

Dynamic Fracture of Nanocomposite Materials

Arun Shukla[1], Victor Évora[2], and Nitesh Jain[3]

[1]*Simon Ostrach Professor and Chair,* [2]*Mechanical Engineer,*
[3]*Sr. Research Engineer*
[1] *Dynamic Photomechanics Laboratory,*
Department of Mechanical Engineering, University of Rhode Island
Kingston, RI 02881
shuklaa@egr.uri.edu
[2] *Naval Undersea Warfare Center Division,*
Newport, RI 02841
[3] *Corporate Research,*
The Goodyear Tire & Rubber Company,
Akron, OH 44305

The fabrication of nanocomposites using various techniques is presented. In particular, the coupling of ultrasonics with an *in-situ* polymerization technique to produce nanocomposite with excellent particle dispersion, as verified by transmission electron microscopy (TEM), is discussed in detail. Dynamic fracture toughness testing is carried out on three-point bend polyester/TiO_2 nanocomposite specimens using a modified split-Hopkinson pressure bar, and results are compared to those of the matrix material. An increase in dynamic fracture toughness relative to quasi-static fracture toughness is observed. Scanning electron microscopy (SEM) of fracture surfaces is carried out to identify toughening mechanisms. A relationship between dynamic stress intensity factor, K_I, and crack tip velocity, $\dot{a}$, is established. Dynamic photoelasticity coupled with high speed photography has been used to obtain crack tip velocities and dynamic stress fields around the propagating cracks. Single-edge notch tension and modified compact tension specimens were used to obtain a broad range of crack velocities. Fractographic analysis was carried out to understand fracture processes. Results showed that crack arrest toughness in the nanocomposites was 60% greater than in the matrix material. Crack propagation velocities prior to incipient branching in the nanocomposites were found to be 50% greater than those in polyester.

1. Introduction

Advancements in material performance depend on the ability to synthesize new materials that exhibit enhancements in their structural, thermal, electrical, and optical properties, to name a few. Nanocomposite materials are ideal materials to meet this challenge, as it has been shown that they have the potential to deliver the aforementioned enhancements oftentimes with minimal increase in weight, as well as with minimum effect on other desired properties; a luxury not always realized with conventional composites or metals. The building blocks of nanocomposites are in the nanometer length scale, typically defined as 1-100nm. This is the range where phenomena associated with atomic and molecular interactions strongly influence material properties.

Nanocomposites were first conceived by a group of researchers led by Koichi Niihara, at the Toyota Central R & D Laboratories in Japan just over 15 years ago. Observing that common reinforcement materials for polymers such as glass fibers and clays were not homogeneously dispersed at the macroscopic level, researchers hypothesized that a nylon/clay hybrid using a layered silicate clay might have improved results. Silicate stacks were separated into individual 1nm thick platelets and dispersed throughout the resin. This dramatically increased the polymer stiffness and strength without sacrificing its toughness. This early pioneering research generated several commercial products for Toyota as well as other companies, and simultaneously sparked excitement in the scientific community, thus stimulating a great deal of interest in expanding the technology to other areas such as electronics, medicine and health, space exploration, environment, energy, etc. The expansion to these areas was facilitated by the invention in the 1980s of instruments such as scanning tunneling, atomic force, and nearfield microscopes which provided the "eyes" and "fingers" required for nanostructure measurement and manipulation. As stated by Horst

Stormer, Nobel Laureate, "Nanotechnology has given us the tools...to play with the ultimate toy box of nature – atoms and molecules...The possibilities to create new things appear limitless."[1].

Polymer resins have been, and continue to be, the matrix of choice for many high-performance components in the aerospace, automobile, and electronics industry owing to their mechanical, electrical, and chemical properties. However, some highly cross-linked thermosetting polymeric materials such as unsaturated polyester for instance, are extremely brittle owing to their covalently bonded network structure, and thus are poor inhibitors of crack initiation and propagation. Nevertheless, researchers have been able to improve their toughness with the addition of soft, as well as rigid fillers [2-4]. These fillers typically take the form of spheres, platelets, or tubes (see Fig. 1).

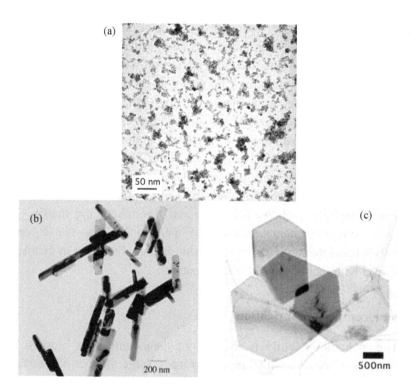

Fig. 1. Nanoparticle morphologies: (a) spherical; (b) tubular; (c) platelet.

Fillers with tubular morphologies, in particular, carbon nanotubes (CNT), have drawn a great deal of attention over the last decade. Polymer/CNT composites have shown high mechanical strength and good electrical conductivity at low filler concentrations [5,6]. The enhanced electrical conductivity stems from the high aspect ratios of CNTs [7,8]. It should be pointed out that, irrespective of the filler morphology used, the degree of enhancement of a particular property in a nanocomposite is highly dependent on the matrix/filler type system used, the extent of filler adhesion to the matrix, and the level of dispersion of the filler throughout the matrix.

Most of the mechanical characterization of nanocomposites done to date has dealt primarily with measurements of elastic modulus, strength, and quasi-static fracture toughness, in other words, they've dealt primarily in the quasi-static regime [9,10]. However, an understanding of the principles governing dynamic fracture is very important as it helps to explain the response of materials to dynamic stimuli. Dynamic fracture studies of conventional monolithic and composites materials often tend to focus attention on the characterization of crack initiation and/or propagation, with less attention paid to run-arrest behavior. However, an understanding of the principles governing the latter is of great importance, as it lends insight as to how the catastrophic failure of materials can be stopped, or at the very least, subdued.

In this chapter, the fabrication of nanocomposites using various techniques is presented and discussed. Particular attention is paid to the in-situ polymerization technique which results in nanocomposites with a uniform distribution of the filler material throughout the matrix. The dynamic fracture characterization of nanocomposites via a crack velocity/stress intensity factor relationship is then discussed in detail and compared with that of the monolithic matrix.

2. Nanocomposite fabrication

The nanometer building blocks of nanocomposites possess an enormous surface area which translates into a large number of interfaces between the intermixed phases. It is believed that the special properties of nanocomposites arise from the interactions of its phases at the interfaces. Consequently, in order to tap into this "nano effect" it is essential that nanocomposite materials possess a well-dispersed filler throughout the matrix. This, however, is not an easy task, as nanocomposites tend to form agglomerates, or clusters, containing numerous nanoparticles, that tend to negate the sought after "nano effect"[11].

2.1 *Fabrication techniques*

Several techniques have been employed over the years to synthesize nanocomposite materials. Sol-gel is a traditional method in which a precursor of the filler, in the form of a solution, is mixed with a solvent to form a solution. The monomer of the desired polymer is added to the solution to undergo polymerization to form a gel. The gel is subsequently dried to evaporate the solvent to form the nanocomposite [12]. This method has been found to produce nanocomposites with a high level of porosity, which may not be desired for commercial applications [13]. Melt intercalation is an approach that is typically used for the synthesis of polymer/layered silicate nanocomposites. In this process, the polymer and clay are mixed and heated for several hours above the polymer-melt temperature, thus allowing the polymer chains to diffuse into the spaces between the parallel layers of clay platelets [14]. Other researchers have used hot pressing of powders and pressureless sintering as fabrication methods [15-17].

In recent years the use of acoustic energy, or more specifically, ultrasonics, has become a popular method to help disperse nano-fillers within a matrix [9,10,18,19]. Some researchers have actually fabricated nanocomposites by coupling ultrasonics with in-situ polymerization, a technique in which the resin and filler are cast in a mold, and subsequently polymerized [9,10,18]. A detailed description of such a technique is presented in the next section.

2.2 Coupling of ultrasonics with in-situ polymerization to fabricate nanocomposites

The nanocomposite fabrication procedure reported here was developed by Évora et al. [9]. Few nanocomposites are commercially available and are very expensive. Furthermore, they are typically available from the manufacturer in small pallets the size of a grain of rice that must be melted and extruded into desired shapes suitable for mechanical testing. On the other hand, general purpose resins are readily obtained from manufacturers that are castable and easily cured using relatively simple procedures. The sample resin described in this section is a general purpose unsaturated polyester manufactured by Ashland Chemical Company (MR17090), which is highly cross-linked, transparent, and very brittle. The filler in this case was spherical TiO_2 (titania) nanoparticles with 36 nm average diameter, obtained from Nanophase Technologies Corporation. The particle size distribution was such that 95% of the nanoparticles had diameters smaller than 60nm. Mechanical and optical properties of the polyester resin and polyester/1% vol. TiO_2 nanocomposites are shown in Table 1. This table also shows properties of birefringent coatings which will be discussed later.

Table 1 Mechanical and optical properties of polyester and polyester/1% vol. TiO_2 nanocomposites, and PS-1A birefringent coatings.

Property	Polyester	Nanocomposite	Coating
Young's Modulus, E (GPa)	3.62	3.78	2.48
Initiation Fracture toughness, K_{IC} (MPa-m$^{1/2}$)	0.54	0.85	N/A
Shear Modulus, G (GPa)	1.53	1.58	0.91
Fringe Value, f_σ (N/m)	28,000	N/A	7,000

Titania nanoparticles were added to 230 mg of polyester resin at varying volume percentages, and mechanically mixed in a glass beaker for approximately 5 min. The mixture was then placed in a vacuum chamber at 28 Torr for 5 min to remove trapped air bubbles generated during the mechanical mixing process. After the deaeration process, the mixture was poured into a stainless steel beaker surrounded by an ice enclosure, and a 20 mm diameter acoustic probe from a Vibra-Cell ultrasonic processor resonant at 20 kHz was used to disperse the nanoparticles throughout the matrix. Ultrasonic energy was employed for 70 min in the pulsing mode (10 sec on, 10 sec off) for an effective sonication time of 35 min. Although pulsing was primarily used to inhibit heat build-up in the specimen, it additionally enhances material processing by allowing the material to settle back under the probe after each ultrasonic burst. The ice enclosure was simultaneously used to inhibit heat build up. After the sonication process, the catalyst methyl ethyl ketone peroxide (MEKP) and the accelerator cobalt octate were added separately to the mixture and mixed thoroughly at 0.85 and 0.03% by weight of polyester, respectively, to initiate and accelerate the polymerization process. The mixture was then very briefly deaerated at 28 Torr to remove trapped air bubbles generated during the sonication process. The brevity of the second deaeration process was necessary as polymerization commences almost immediately after the catalyst and the accelerator are added. The final mixture was poured into a rectangular mold lined with thin (0.18 mm) mylar sheets to obtain smooth flat surfaces. In order to prevent particle settlement at the bottom, the mold was rotated at 2 rpm for at least 6 h until the mixture was rigid enough so that particles could no longer migrate. The resin mixture was allowed to cure at room temperature for 72 hours. After this time, the specimen was taken out from the mold and post-cured in an air-circulating oven for 4 hours at 52°C followed by 5 hours at 63°C. After allowing the oven to come down to room temperature, the specimens were taken out and machined to desired shapes. The above procedure was perfected after numerous iterations of different combinations of sonication times, intervals, and resin quantities. The fabrication evolution is shown schematically in Fig. 2.

Transmission electron microscopy (TEM) analysis of the cast specimens was conducted to verify the level of nanoparticle dispersion in the matrix. TEM micrographs of specimens containing 1, 2, and 3 vol.% TiO_2 are shown in Fig. 3.

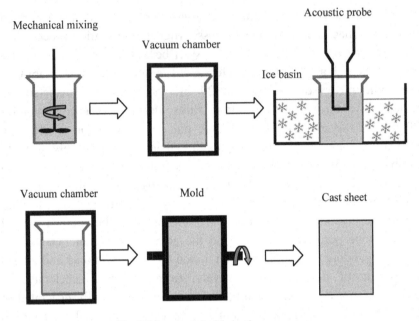

Fig. 2. Nanoparticle fabrication process.

As can be observed, the fabrication procedure employed produced nanocomposites with excellent particle dispersion, particularly in specimens containing 1, 2, and 3 vol.% TiO_2. Ultrasonics and in-situ polymerization can therefore be easily combined to obtain adequate filler dispersion within a matrix, so long as manufacturer's curing recommendations are followed, and initial trials are conducted to determine optimum sonication times and intensities for desired filler dispersion levels.

3. Dynamic fracture

Dynamic fracture studies involve the characterization of crack initiation, propagation, and run-arrest behavior in materials. In linear elastic brittle solids under small scale yielding, nonlinear deformations and crack front irregularities are usually negligible, so that the stress field surrounding a moving crack tip can be represented by a single parameter, the stress intensity factor, K. This fact has lead to the suggestion of a dependence of crack velocity, à, on K. Although some debate has surrounded this dependence [20], researchers have nonetheless characterized monolithic materials via the à-K relationship [21,22]. In this section, the dynamic fracture behavior of polyester/TiO$_2$ nanocomposites is presented via an à-K characterization, and via dynamic fracture toughness testing.

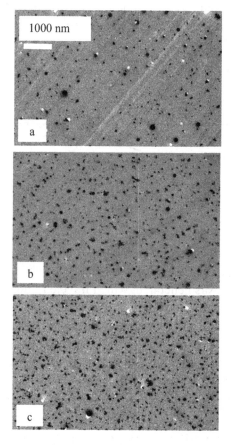

Fig. 3 TEM showing nanoparticle dispersion. Volume fraction: (a) 1%; (b) 2%; (c) 3%.

3.1 *Dynamic fracture toughness characterization of nanocomposites*

3.1.1 *Testing setup*

Dynamic fracture toughness testing was carried out by loading three-point bend specimens to failure in an instrumented modified split-Hopkinson pressure bar (SHPB) apparatus (see Fig. 4). Polyester/TiO$_2$ nanocomposite specimens were fabricated using the procedure described in the preceding section. A description of the testing methodology is now presented. In-depth theoretical details of the test methodology may be found elsewhere [23].

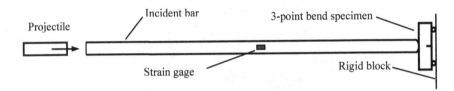

Fig. 4 Schematic of modified SHPB.

A cylindrical projectile was propelled down the barrel of a gas gun by means of compressed air. Upon impact with the cylindrical bar, a compressive pulse was generated, which traveled down the length of the bar to the bar/specimen interface. Upon reaching the specimen, the impact force of the pulse was transmitted to the specimen, which ultimately loaded the crack tip. Incident and reflected strain-pulse histories were obtained from two diametrically opposed strain gages located at the mid-point of the bar. In an effort to minimize the effects of impedance mismatch between the polyester-based nanocomposites and the bar, the latter was made from polymeric LEXAN®. The projectile and bar were both made from the same material in order for a well-defined single pulse to be generated. The force history at the bar/specimen interface was obtained using one-dimensional elastic wave propagation theory according to the following relation [24]:

$$F(t) = [\varepsilon_i(t) + \varepsilon_r(t)]EA \tag{1}$$

where F is the force, ε_i and ε_r the incident and transmitted strain pulses, E the dynamic Young's modulus, and A is the cross-sectional area of the bar. Since the time to fracture was allowed to be sufficiently long, inertia effects could be neglected, and the following mode I stress intensity factor relation [25] could be used to calculate the dynamic fracture toughness:

$$K_I(t) = \frac{F(t)}{B\sqrt{W}} f(a/W) \tag{2}$$

where K_I is the stress intensity factor, B the specimen thickness, W the specimen width, a the initial crack length, and $f(a/W)$ is a geometric factor. The dynamic fracture toughness (K_{ID}) then corresponded to the value of stress intensity factor at the time of crack initiation, i.e.,

$$K_{ID} = K_I(t_{initial}) \tag{3}$$

Photoelastic analysis using high-speed photography was initially used on transparent polyester specimens to ascertain that the crack initiation time thus obtained coincided with the peak force obtained using the modified SHPB technique. This analysis also served to validate the value of K_{ID} calculated using Eq. (3).

3.1.2 *Testing results*

Fig. 5 shows the variation in dynamic fracture toughness obtained as a function of volume percentage of TiO_2 nanoparticles. Also included in the figure for comparison is the variation obtained by quasi-static loading of specimens using an INSTRON machine. The error bars represent a 95% confidence interval resulting from the average of five tests conducted at each volume fraction.

It should also be pointed out that in all the experiments using the SHPB the projectile velocity was such that the loading rate was approximately $6.9 MPa\text{-}m^{1/2}$ $(ms)^{-1}$. The dynamic profile obtained is similar to that obtained using quasi-static loading, with the exception of an upward shift in the former. Moreover, in the dynamic case, there is no precipitous drop in fracture toughness with the 4% vol. specimens. Specimens containing 1% vol. (3.5% wt.) of TiO_2 particles yielded the greatest increase in fracture toughness vis-à-vis the virgin polyester: 57% and 75% respectively, for quasi-static and dynamic testing. Postmortem scanning electron microscopy (SEM) analysis of the region near crack initiation in specimens tested quasi-statically were conducted and are shown in Fig. 6 for the case of virgin polyester and a nanocomposite specimen containing 1% vol. TiO_2. It was speculated that the increase

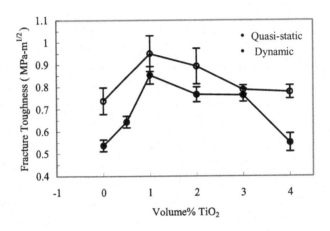

Fig. 5. Quasi-static and dynamic fracture toughness as a function of %vol. TiO_2 nanoparticles.

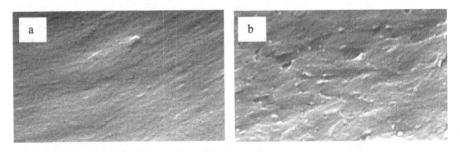

Fig. 6. SEM photographs of fracture surfaces resulting from quasi-static fracture toughness testing. Volume fraction of TiO2 nanoparticles: (a) 0%; (b) 1%.

in fracture toughness observed in Fig. 5 was related to the apparent presence of toughening mechanisms, such as crack trapping and crack pinning, believed to be observed in the SEM micrograph of the nanocomposites.

3.2 Crack velocity – Stress intensity factor characterization of Nanocomposites

Studies involving à -K_I characterization to study dynamic crack propagation have been conducted in monolithic materials [26-29]. To date, however, to the best of the author's knowledge, there has only been one such study done in nanocomposites [2]. Moreover, dynamic fracture studies often tend to center attention on the characterization of crack initiation and/or propagation, with somewhat less attention paid to run-arrest behavior. An understanding of the principles governing crack arrest is nonetheless of great importance, as it lends insight into how the catastrophic failure of materials can be stopped, or at the very least subdued, once crack propagation has occurred. Details surrounding the experimental procedure and analysis of such a characterization conducted in polyester/TiO$_2$ nanocomposites are now presented.

3.2.1 *Specimen geometry*

In order to capture the large velocity gradient (crack arrest and crack branching) inherent in an à -K characterization of a material, more than one specimen geometry is typically needed. Modified compact tension (MCT) and single-edge notch tension (SENT) are geometries commonly used for this purpose; photographs of both geometries are shown in Fig. 7.

The MCT specimen geometry is known as a decreasing-K geometry, whereas the SENT specimen geometry is known as an increasing-K geometry. Thus, MCT specimens were used to capture velocities approaching crack arrest values, and SENT specimens higher propagating velocities.

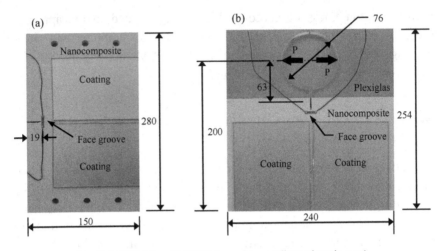

Fig. 7. SENT (a) and MCT (b) specimens (dimensions in mm).

The MCT specimen was composed of sheets from two different materials, Plexiglas and polyester/TiO$_2$ nanocomposite. The reason for the two-part specimen had to do with the limitation on the amount of nanocomposite material that could be fabricated using the technique previously discussed, while still maintaining a good dispersion of particles within the matrix. The large circular hole was made on the separate sheet of Plexiglas, and the two pieces were bonded using extra fast setting epoxy adhesive. The crack was subsequently machined. Plexiglas was chosen as a suitable mating part because of its compatibility with polyester in terms of density and modulus. Starter notches were made on all specimens with a band saw. Sandpaper was then used to round the blunt notch and to eliminate microcracks induced during the cutting operation. The normalized crack length a/W, where W is the specimen width, used for the SENT and MCT specimens were 0.125 and 0.31, respectively. In order to ensure failure in tension and to guide the propagating crack in a straight line, shallow face grooves were made on both sides of the specimens. Crack detection gages (CD-02-10A, Micro Measurements, Inc.) were bonded near the starter notch to detect the initiation of fracture.

3.2.2 *Birefringent coatings*

Birefringent coatings are typically employed to conduct photoelastic studies on opaque specimens, such as the polyester/TiO$_2$ nanocomposite discussed herein. Birefringent coatings allow for displacements at the specimen/coating interface to be transmitted without amplification or attenuation. Unlike strain gages, birefringent coatings allow whole field response. In this study, a split birefringent coating technique is employed, in which the coating is placed on both sides of the specimen, and a small distance (2mm) away from the anticipated crack path (Fig. 7). Birefringent coatings consisted of 3mm thick polycarbonate sheets with vacuum deposited aluminum on the back surface. The sheets were cut to desired dimensions while using liberal amounts of cooling fluid to ensure minimum development of residual stresses. Extra fast setting epoxy adhesive was used to bond the sheets to the specimens. Coating properties are shown in Table 1. Methods to account for the influence of the coatings, as well a dynamic validation technique employed are discussed later.

3.2.3 *Specimen loading*

An INSTRON 5585 apparatus and a crack-line-loading frame were used to conduct experiments on SENT and MCT specimens, respectively. SENT specimens were statically loaded to a predetermined force value corresponding to initiation stress intensity factor, K_Q. Upon reaching the prescribed load values, crack propagation was initiated by tapping a sharp razor blade on the specimen notch. Once crack propagation was initiated, crack detection gages mounted near the starter notch triggered an electronic circuit that caused a high-speed camera to commence taking a sequence of photographs of the isochromatics. MCT specimens were loaded by expanding two semicircular split D's to create the loading configuration depicted in Fig. 7. The D's were forced apart by forcing a wedge into the split opening with the aid of a hydraulic cylinder. The wedge force was monitored with an in-line load cell calibrated to 0.23mV/N. Crack propagation was initiated using the same procedure used with SENT specimens.

3.2.4 *Photoelastic analysis*

3.2.4.1 Background

The method of photoelasticity has been used for many years to conduct dynamic fracture studies [2,21,22,30-32]. It is based on the stress-optic law that relates the optical properties of the material to the stress field components [33]. The relationship is given as:

$$\tau_{max} = \frac{\sigma_1 - \sigma_2}{2} = \sqrt{\left(\frac{\sigma_{yy} - \sigma_{xx}}{2}\right)^2 + \sigma_{xy}^2} = \frac{Nf_\sigma}{2h} \tag{4}$$

where τ_{max} is the maximum in-plane shear stress, σ_1 and σ_2 are in-plane principal stresses, N is the fringe order, f_σ is material fringe value, and h is the length of the optical path through the material. In the case where photoelastic coatings are employed, h corresponds to twice the coating thickness. The stress intensity factor is obtained by combining the stress optic law with dynamic stress field equations [32]. If the assumption is made that the strains in the coating are equal to those in the specimen due to adequate bonding, and that strain gradients through the coating thickness are negligible, the stress intensity factor in the specimen can be related to the stress intensity factor in the coating by [33]

$$K_{Id}^S = F_{CR} \frac{E^S}{E^C} \frac{1+\upsilon^C}{1+\upsilon^S} K_{Id}^C \tag{5}$$

where E is the Young's modulus, υ is Poisson's ratio, and the subscripts c and s refer to the coating and specimen, respectively. F_{CR} is a reinforcement correction factor that accounts for the fact that the coating carries a portion of the load, causing the strain on the specimen to be reduced by a certain amount. F_{CR} is given as

$$F_{CR} = 1 + \frac{h^C}{h^S} \frac{E^C}{E^S} \frac{1+\upsilon^S}{1+\upsilon^C} \tag{6}$$

where h^S and h^C are the thickness of the specimen and coating, respectively.

3.2.4.2 Testing setup

The schematic of the photoelastic configuration used to test MCT and SENT nanocomposite specimens is shown in Fig. 8. Xenon flash lamps were used as light sources to illuminate the specimen in the reflection mode. Upon crack initiation, crack detection gages mounted near the starter notch triggered the circuit and caused the camera to commence taking a sequence of 16 photographs. Camera interframe times ranged from 8-15μsec depending on the specimen type and geometry used. Interframe times were thus chosen to ensure that the dynamic fracture event of interest was captured within the 16 frames.

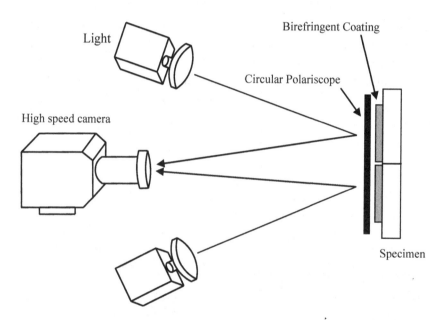

Fig. 8. Photoelastic setup for dynamic fracture study of nanocomposites.

3.2.5 *Results and discussion*

3.2.5.1 Validation of use of birefringent coatings

A birefringent coating validation experiment was initially conducted using an innovative setup which permitted, in a single experiment, the simultaneous capture of isochromatics on the coating mounted on a transparent specimen, and on the specimen itself (Fig. 9).

The method involved the mounting of a coating on just one half of an SENT polyester specimen. Polarizers were placed on either side of the specimen so as to simultaneously form a light-field circular polariscope on the transparent end, while allowing the other end containing the coating to end up with a single polarizer. Flash lamps were then positioned to illuminate the transparent end of the specimen from one side, and the end containing the coating from the other side. Tensile load was applied with the INSTRON machine until a load of 9600N was reached. A sharp razor was then tapped in the notch to initiate fracture. Figure 10 shows isochromatic fringes generated in the specimen and in the corresponding coating as the crack propagated. It can be seen that the fringes in the coating were of the same nature (Mode-I fringes) as those developed in the clear specimen. Furthermore, Fig. 11 shows good agreement (within 5%) between the stress intensity factors obtained from analysis of the isochromatics developed in the specimen and in the coating.

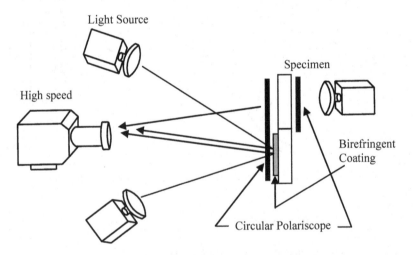

Fig. 9. Photoelastic setup for validation of use of birefringent coatings.

crack propagation

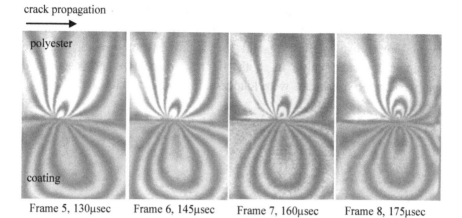

Frame 5, 130μsec   Frame 6, 145μsec   Frame 7, 160μsec   Frame 8, 175μsec

Fig. 10. Coating validation experiment showing simultaneous isochromatic fringes obtained on a polyester specimen and on the coating during crack propagation.

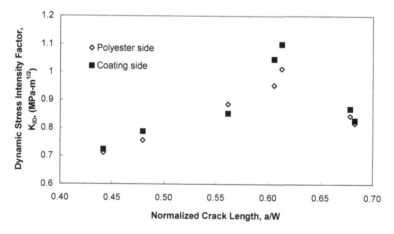

Fig. 11. Good agreement of results from birefringent coating dynamic validation experiment conducted on a polyester specimen.

3.2.5.2 Crack propagation and run-arrest profiles

Experiments were conducted on SENT and MCT specimens in order to obtain crack propagation and run-arrest profiles. Crack tip velocities were obtained by the curve-fitting and subsequent differentiation of experimental data from crack position versus time profiles. Typical crack position versus time profiles pertaining to run-arrest and propagation events are shown in Figs. 12 and 13 for polyester and nanocomposite specimens, respectively. In both figures, respective polyester and nanocomposite specimens were loaded to the same load levels.

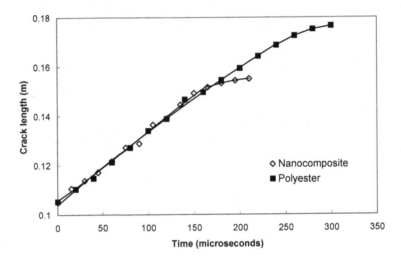

Fig. 12. Typical crack length versus time profile obtained using the MCT specimen.

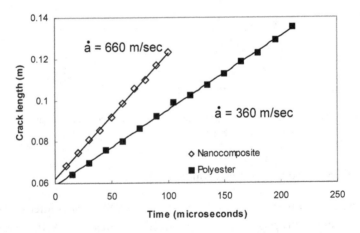

Fig. 13. Typical crack length versus time profile obtained using the SENT specimen.

For the run-arrest case, the crack propagated in the nanocomposite with an average speed of 260m/sec up to a crack length of 0.14m, decelerating rapidly to 67m/sec in the following 0.01m. In the polyester, the average crack velocity was 278m/sec up to 0.16m, decelerating to 72m/sec in the next 0.02m. Figure 13 shows typical propagation profiles obtained using SENT specimens. In these particular cases, cracks propagated with nearly constant velocities of 360m/sec and 660m/sec in polyester and nanocomposites, respectively. Photographs of typical fracture surfaces obtained using SENT polyester and nanocomposite specimens are shown in Fig. 14.

Fracture surfaces in SENT polyester specimens were typically characterized by three distinct patterns: mirror, mist, and hackle. The emergence and severity of the hackle pattern was dependent on the level of stress intensity factor. The fracture surfaces of nanocomposites, on the other hand, were characterized by two patterns: mirror and rough (hills and valleys).

Figures 15 and 16 show typical isochromatic fringes obtained using polyester and nanocomposite SENT specimens.

(a)

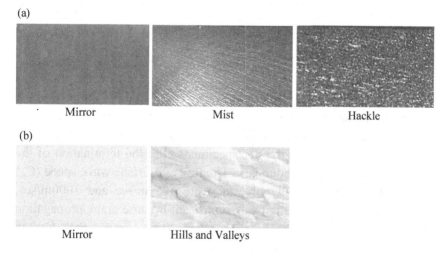

(b)

Fig. 14. Typical fracture surface patterns obtained using the SENT specimen. (a) polyester; (b) nanocomposite.

crack propagation

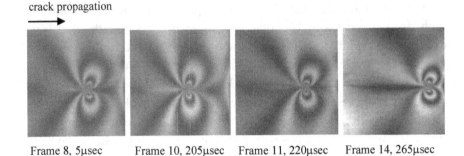

Frame 8, 5µsec   Frame 10, 205µsec   Frame 11, 220µsec   Frame 14, 265µsec

Fig. 15. Typical isochromatic fringes obtained using SENT polyester specimens.

crack propagation

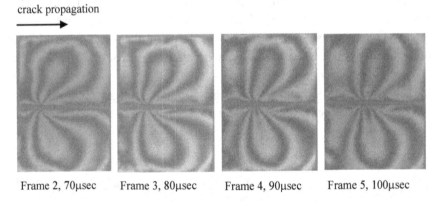

Frame 2, 70µsec   Frame 3, 80µsec   Frame 4, 90µsec   Frame 5, 100µsec

Fig. 16. Typical isochromatic fringes obtained using SENT nanocomposite specimens.

Isochromatics from each of the frames were analyzed using the procedure described earlier. Raleigh surface waves were also observed in the nanocomposite and polyester specimens after the termination of the main crack at the frontal boundary. The Raleigh surface wave speed (C_R) in the nanocomposite and polyester were 1050m/sec and 1000m/sec, respectively. This observation was significant because crack propagation velocities discussed above approached 0.62 times the C_R in nanocomposite, whereas in isotropic homogeneous materials crack speeds are generally limited to 0.4 times C_R in the absence of superimposed stresses. The high velocity ratio obtained indicates that the nanocomposite specimens can sustain higher velocities before crack branching occurs.

3.2.5.3 à -K Relationship

Results obtained above were compiled so that a constitutive relationship between crack velocity and dynamic stress intensity factor could be developed for polyester and polyester/TiO$_2$ nanocomposites. Such relationship is shown in Fig. 17. The à - K$_I$ profile consists of three distinct velocity regions: low, intermediate, and high. In the low-velocity region, the velocity is very sensitive to changes in K$_I$, with the slightest increase in K$_I$ resulting in large velocity gradients. This region is commonly referred to as the "stem" of the à - K$_I$ profile. In the intermediate-velocity region the velocity becomes less sensitive to changes in K$_I$, whereas in the high-velocity region large increases in K$_I$ are required to produce small increases in velocity. The low-velocity region ranges approximately from 70m/sec to 270m/sec and 70m/sec to

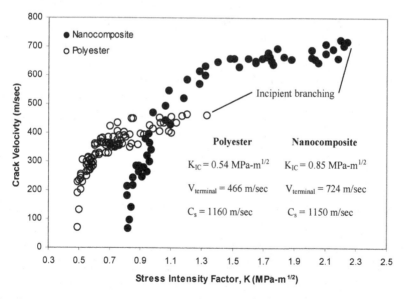

Fig. 17. Crack velocity vs. stress intensity factor relationship for polyester/TiO$_2$ nanocomposites and polyester.

300m/sec for polyester and nanocomposite, respectively, while the intermediate region ranges from 270m/sec to 430m/sec and 300 to 630m/sec for polyester and nanocomposite, respectively. The high-velocity region runs all the way to incipient branching at 460m/sec and 730m/sec for polyester and nanocomposite, respectively. It should be pointed out that the à - K_I profile obtained in this study is similar to those obtained by other researchers [21,21,28] on Homalite 100 and 4340 alloy steel, in that the three aforementioned velocity regions were observed.

Crack initiation fracture toughness, K_{IC}, was higher than the corresponding arrest toughness, K_{Im}, by 8% and 4% in polyester and nanocomposites, respectively. Crack arrest toughness observed in polyester and nanocomposites was 0.5MPa-m$^{1/2}$ and 0.82MPa-m$^{1/2}$, respectively, obtained from the curve fitting of experimental data points. The increase in K_{Im} was significant as it indicated premature crack arrest in nanocomposites vis-à-vis the virgin material once catastrophic failure had occurred. Ideally, one hopes to obtain a value of K_{Im} based on an observed zero crack velocity. However, in this case, the lowest velocities observed were around 70m/sec. Comparison of polyester and nanocomposite propagation profiles in the high-velocity region revealed that propagation velocities were approximately 50% higher in the nanocomposites. An initial attempt to branching occurred at K_I values of 1.3MPa-m$^{1/2}$ and 2.2MPa-m$^{1/2}$, respectively, for polyester and nanocomposites; a 70% increase. These incipient branching values represent 2.4 and 2.6 times the corresponding values of K_{IC} in polyester and nanocomposites, respectively.

4. Closure

A discussion has been presented on the fabrication and dynamic fracture testing aspects of nanocomposite materials; particular attention has been paid to a polyester/TiO$_2$ system. Various conventional nanocomposite fabrication techniques were briefly discussed, and an in-situ polymerization fabrication technique using direct ultrasonification was discussed in detail. TEM observations of specimens fabricated using this latter technique revealed that specimens containing 1, 2, and 3% volume fractions of TiO_2 possessed excellent dispersion of nanoparticles throughout the polyester matrix. It was also evident that in specimens containing 4% vol. TiO_2, more and larger nanoparticle agglomerations were present, although the vast majority of these were still in the nanometer size range. Ultrasonics and in-situ polymerization were thus shown to be easily combined to obtain adequate filler dispersion within a matrix.

Dynamic fracture testing aspects of using a modified SHPB on polyester/TiO_2 nanocomposite were discussed. Test results showed that dynamic fracture toughness was higher than quasi-static fracture toughness for all volume fractions tested (0, 1, 2, and 3%), with specimen containing 1vol.% TiO_2 nanoparticles possessing the highest value ($0.95MPa\text{-}m^{1/2}$). Fracture surface features such as thumbnail markings and out-of-plane flaking, as observed by SEM, indicated that toughening mechanisms such as crack pinning and crack trapping are inherent to the polyester/TiO_2 nanocomposite system; crack pinning being the dominant type.

Experimental aspects surrounding the dynamic fracture constitutive characterization of polyester/TiO_2 nanocomposites were discussed in detail. Birefringent coatings were used to obtain isochromatics in the opaque nanocomposite specimens. Dynamic photoelasticity combined with high-speed photography was used to study crack propagation and run-arrest events. Dynamic fracture constitutive characterization was subsequently performed and results were compared with those obtained using virgin polyester. A relationship was established between crack tip velocities and dynamic stress intensity factors. Crack arrest toughness increased by 64% in the nanocomposite relative to the virgin polyester, indicating that the nanocomposite is a better suppressor of crack propagation once catastrophic failure has occurred. The inability to obtain a K_{Im} value based on a zero velocity was due to the abrupt deceleration of the crack ($2x10^5g$). At such high rates, interframe times were not sufficiently low to capture a zero velocity arrest. Crack propagation velocities in nanocomposites were found to be 50% greater than those in the virgin polyester. The high ratio (62%) of propagation velocity in nanocomposites to Raleigh wave velocity indicated that the nanocomposites could sustain higher velocities before the onset of branching. In fact, initial attempt of branching values in the nanocomposites were found to be 1.7 times those in polyester. Incipient branching values were found to be 2.4 and 2.6 times the corresponding values of K_{IC} in polyester and nanocomposites, respectively.

Acknowledgements

Victor Évora would like to acknowledge the financial support of the Naval Undersea Warfare Center ILIR Program. The authors would like to acknowledge the support of the Office of Naval Research under grant number N000140410268, and that of the National Science Foundation under grant number INT-0422767.

References

1   M. C. Roco, R. S. Williams, and P. Alivisatos (Eds.), WIGN Workshop Report: Nanotechnology Research Directions, Kluwer Academic Publishers, Boston, p. XI (2002).

2   V.M.F. Évora, N. Jain, and A. Shukla, *Exp. Mech.*, **45**, 2, 153-159 (2005).

3   M. Hussain, A. Nakahira, and S. Nishijima, *Mater. Lett.*, **27**, 21-25 (1996).

4   A.C. Moloney, H.H. Kausch, and H.R. Stieger, *J. Mater. Sc.*, **19**, 1125-1130 (1984).

5   E. Dujardin, T.W. Ebbesen, A. Krishnan, P.N. Yanils, and M.M.J. Treacy, *Phys. Rev. B*, **58**, 14013-14019 (1998).

6   M.F. Yu, B.S. Files, S. Arepalli, and R.S. Ruoffet, *Phys. Rev Let.*, **84**, 5552-5555 (2000).

7   S. Frank, P. Poncharal, Z.L. Wang, and W.A. de Heer, *Science*, **280**, 1744-1746 (1998).

8   S.A. Curran, P.M. Ajayan, W.J. Blau, D.L. Carroll, J.N. Coleman, A.B. Dalton, A.P. Davey, A. Drury, B. McCarthy, S. Maier, and A. Strevens, *Adv. Mater.*, **10**, 1091-1093 (1988).

9   V.M.F. Évora and A. Shukla, *Mater. Sc. and Eng.*, **A361**, 358-366 (2003).

10   R.P. Singh, M. Zhang, and D. Chan, *J. Mater. Sc.*, **37**, 781-788 (2002).

11   R. Dagani, *Chem. Eng. News*, **77**, 23, 25-37 (1999).

12   D. F. Shriver, Inorganic Chemistry, Freeman, NY (1994).

13   S. Maeda and S. P. Armes, *Synthetic Metals*, **73**, 2, 151-155 (1995).

14   E. Giannelis, *Appl. Organomettalic Chem.*, **12**, 675-680 (1998).

15   J. Zhao, L.C. Stearns, M.P. Harmer, H.M. Chan, and J.H. Miller, *J. Amer. Ceram. Soc.*, **76**, 503-510 (1993).

16   C. Anya, *J. Mater. Sc.*, **34**, 5557-5567 (1999).

17   Sung-Tag Oh, M. Sando, and K, Niihara, *J. Amer. Ceram. Soc.*, **81**, 11, 3013-3028 (1998).

18   C.B. Ng, B. J. Ash, L.S. Schadler, and R.W. Siegel, *Adv. Comp. Lett.*, **10**, 3, 101-111 (2001).

19   T.E. Chang, L.R. Jensen, A. Kisliuk, R.B. Pipes, R. Pyrz, and A.P. Sokolov, *Pol.*, **46**, 439-444 (2005).

20   J.W. Dally, W. L. Fourney, and G. R. Irwin, *Intl. J. of Frac.*, **27**, 159-168 (1985).

21   G. R. Irwin, J. W. Dally, T. Kobayashi, W. L. Fourney, M. J. Etheridge, and H. P. Rossmanith, *Exp. Mech.*, **19**, 121-128 (1979).

22   T. Kobayashi and J. W. Dally, *ASTM STP 711*, 182-210 (1980).

23   C.F. Martins and V. Prakash, in: T. S. Srivatsan, D. R. Lesuer, E. M. Taleff (Eds.) *The Minerals, Metals and Materials Society*, 105-122 (2002).

24   H. Kolsky, Stress Waves in Solids, Dover Publications, New York, p.41 (1963).

25   T. L. Anderson, Fracture Mechanics: Fundamentals and Applications, CRC Press, Boca Raton, FL, p. 63 (1995).

26   G. R. Irwin, *U.S. NRC Report NUREG-75-107*, University of Maryland (1975).

27   G. R. Irwin, *U.S. NRC Report NUREG-0072*, University of Maryland (1976).

28   A. Shukla and H. Nigam, *Eng. Fract. Mech*, **25**, 91-102 (1986).

29   J. F. Kalthoff, Workshop on Dynamic Fracture, California Institute of Technology, 110-118 (1983).

30   A. A. Wells and D. Post, *Proc. of SESA*, **16**, 69-93 (1958).

31   W. B. Bradley and A. S. Kobayashi, *Exp. Mech.*, **10**, 106-113 (1970).

32   A. Shukla, R. K. Agarwal, and H. Nigam, *Eng. Fract. Mech.*, **31**, 501-515 (1988).

33   J. W. Dally and W. F. Riley, Experimental Stress Analysis, McGraww-Hill, New York (1991).

Printed in the United States
By Bookmasters